T0293486

Understanding Statistics
and Statistical Myths

How to Become a Profound Learner

Understanding Statistics
and Statistical Myths

How to Become a Profound Learner

Kicab Castañeda-Méndez

CRC Press
Taylor & Francis Group
Boca Raton London New York

CRC Press is an imprint of the
Taylor & Francis Group, an **informa** business

A PRODUCTIVITY PRESS BOOK

CRC Press
Taylor & Francis Group
6000 Broken Sound Parkway NW, Suite 300
Boca Raton, FL 33487-2742

© 2016 by Taylor & Francis Group, LLC
CRC Press is an imprint of Taylor & Francis Group, an Informa business

No claim to original U.S. Government works

Printed on acid-free paper
Version Date: 20150515

International Standard Book Number-13: 978-1-4987-2745-7 (Hardback)

Visit the Taylor & Francis Web site at
http://www.taylorandfrancis.com

and the CRC Press Web site at
http://www.crcpress.com

Contents

Preface

Experience by itself teaches nothing... Without theory, experience has no meaning. Without theory, one has no questions to ask. Hence, without theory, there is no learning.

W. Edwards Deming
The New Economics for Industry, Government, Education—2nd Edition

Every myth listed here has been stated in several course materials my clients or employers used to train thousands of people or by several experts who by their own account have trained thousands of people. Some myths have occurred in peer-reviewed articles or books. Via Internet searches, you can find the myths stated as if true in books, articles, and forums.

Why are these statements myths? Each is an unconditional statement that, taken literally and at face value, is false. All are false under some conditions while a few are not true under any condition. In this book, we explore the conditions that render false the universality of the statements to understand why.

It may seem strange that these myths persist. But, as Deming was wont of saying, "How could they know?"

By being profound learners, they could know.

This book is about three types of learning behaviors: rote learning, inflexible learning, and profound learning.

With respect to photography, I am a rote learner although it is a hobby in which I engage heavily once or twice a year on vacations. While I have produced some very nice photos, I do not spend any time learning about photography to be better.

But in the area of problem solving, I consider myself a profound learner. I was fortunate enough that my father, a philosophy professor, gave my siblings and me logic puzzles when we were quite young. The earliest I remember was when I was in second grade. I enjoyed the challenge and attention and reward of trying to solve them, trying to solve them faster than my siblings, and trying to understand the puzzles.

The problem-solving skills I developed starting in childhood and throughout my life opened a career for me in process improvement and teaching. When Six Sigma came around, it was a natural application for me because of my advanced degrees in math and statistics. Also, I had been applying process improvement using the tools even before Six Sigma was invented and the term coined at Motorola. My first project after getting my statistics degree was at Whirlpool. My manager give me a pile of documents containing 10 years of history of the problem. Four months later, I completed installing the solution. My 80+ page report was indistinguishable from Six Sigma projects I came to know 15 years later.

Since then, I have worked with numerous people trained or certified as Six Sigma Belts, Six Sigma consultants, and Six Sigma instructors, and seen and used numerous Six Sigma materials from leading organizations (businesses who applied Six Sigma and consultancies who provided expertise). To my surprise, I began to notice here and there a mistaken understanding or two, particularly about statistics, which is the backbone of Six Sigma.

As I noticed their persistence, I mentally stored these misunderstandings as I took on the task of countering, addressing, or correcting them. It wasn't easy. That was a second surprise. I was puzzled by people who readily and publicly admitted they were not statisticians but were so sure of their statistical beliefs. Even a simple concept of whether a number is continuous or discrete would become a long debate—or worse, no discussion at all.

After a few of these debates (I really should have learned to dialogue), I recalled one of my favorite childhood authors, Edgar Allen Poe, and his story "The Murders in the Rue Morgue." The murderer is not human. Yet when detective Dupin questions witnesses, he encounters an intriguing paradox.

"That was the evidence itself," said Dupin, "but it was not the peculiarity of the evidence. You have observed nothing distinctive. Yet there was something to be observed. The witnesses, as you remark, agreed about the gruff voice; they were here unanimous. But in regard to the shrill voice, the peculiarity is—not that they disagreed—but that, while an Italian, an Englishman, a Spaniard, a Hollander, and a Frenchman attempted to describe it, each one spoke of it as that of a foreigner. Each is sure that it was not the voice of one of his own countrymen. Each likens it—not to the voice of an individual of any nation with whose language he is conversant—but the converse. The Frenchman supposes it the voice of a Spaniard, and 'might have distinguished some words had he been acquainted with the Spanish.' The Dutchman maintains it to have been that of a Frenchman; but we find it stated that 'not understanding French this witness was examined through an interpreter.' The Englishman thinks it the voice of a German, and 'does not understand German.' The Spaniard 'is sure' that it was that of an Englishman, but 'judges by the intonation' altogether, 'as he has no knowledge of the English.' The Italian believes it the voice of a Russian, but 'has never conversed with a native of Russia.' A second Frenchman differs, moreover, with the first, and is positive that the voice was that of an Italian; but, not being cognizant of that tongue, is, like the Spaniard, 'convinced by the intonation.'"*

The murderer, an orangutan, spoke neither French nor Spanish nor Dutch nor Italian nor Russian.

After recording several of these statistical misunderstandings, I decided to call them myths. I began wondering why they persisted and why people were so fiercely defending them. Why would someone who readily admits he or she missing an essential element to be an expert on the topic (a la Dupin's witnesses not familiar with a language they thought they heard) be so sure of their beliefs and so unwilling to discuss or learn more than what they were taught, and apparently taught without explanation? The irony was managers, consultants, and trainers who were change agents but unwilling to change themselves.

It was then that I began noticing the connection between these myths and other topics. So, I developed a theory. From that theory came an application.

This book is about that theory of rote, inflexible, and profound learning applied to statistics. Rote and inflexible learning lead to statistical myths. Profound learning leads to understanding why the myths are myths. By understanding why, one is in a position to apply more useful and powerful statistical analyses and create analyses that are more generally applicable.

---

* http://poestories.com/read/murders.

Prologue

We do not so much know what we see—rather, we see what we know.

Goethe
Common Sense, Science, and Skepticism (A. Musgrave)

What do Six Sigma, the new mammography guidelines, politics, adult learning, art training, Deming's last book, and the God gene have in common?

Rote learning.

ROTE LEARNING

Rote learning is essential for survival. All babies, whether human or not, learn first by rote. A distressed (we presume this via her behavior) mother baboon shrieks and grabs her baby as it reaches for a poisonous berry or wanders in the direction of a snake. The baby baboon doesn't understand why but learns not to touch that type of berry and to avoid snakes. Few infant baboons would survive if understanding were required before learning such lessons. If baboons cannot understand the concepts of danger and poison, only rote learning is possible.

The same occurs with humans. A distressed parent grabs an infant before it reaches for a hot stove and, ironically, may even say "If you ever touch that stove, I'll...." The parent's distress is clearly conveyed to the infant and the infant learns a lesson without understanding. Before a certain age, a child is incapable of understanding, yet learns the lesson.

As with baboons, human infant mortality rates would be significantly higher without rote learning, perhaps forcing human evolution into routinely having multiple births to compensate. Humans, however, are capable of understanding the lessons of danger and poison and substantially more; perhaps not when they learn these survival lessons, but certainly later at school, at play, at work.

Even when humans reach the age of understanding, most of what we learn is by rote. It is not just that no one has the time to research everything to understand, but some knowledge is simply beyond most people's capacity. I reached my mathematical capability with real analysis in graduate school—there was nothing real about it (I thought). Other times, people may not have the interest to understand even if curious. That's my position on the nutritional amounts and calories found on packaged foods. Nevertheless and more importantly, it is not essential for everyday living to understand most things let alone everything.

While young children ask why incessantly, they do not have the capacity to understand all answers to that question. Nor do parents and teachers always have the knowledge to explain.

For example, as a child, I learned that Columbus discovered America in 1492. I learned this by rote. While an elementary school child, I did not have the ability to ask critical questions or to understand what I was taught. I did not ask, for example, what does *discovered* mean in the statement *Columbus discovered America*? If someone already lived in America and they were not "lost" (what does *lost* mean?), how can America be discovered by someone else? How do we know that no one "discovered" America earlier? How can "America" be discovered before it existed? Does discovering "America" require only that one part be known—like the seven blind men "discovering" what an elephant is by just the trunk or ear or tail?

Rote learning is not limited to children.

Adult Learning

Malcolm Knowles developed his adult learning theory[*][†] based on certain characteristics of adults including being goal oriented, practical, and self-directed; relevancy; autonomy; need for respect; and need to connect to years of life's experiences. Associated with these characteristics are motivators that include socialization, personal advancement, not be bored, altruistic application, and pleasing authority.

Having taught math and statistics and coached sports to both adults and children, I believe that except for the greater years of experience adults have, there is often little difference between how adults learn and children

---

[*] Malcolm S. Knowles et al., *The Adult Learner: The Definitive Classic in Adult Education and Human Resource Development*, 7th ed. (Milton Park, UK: Taylor & Francis, 2011).

[†] Malcolm S. Knowles, *Andragogy in Action: Applying Modern Principles of Adult Learning* (San Francisco: Jossey-Bass, 1984).

learn. When a child says "why do I need to learn algebra [or history or chemistry, etc.], I'm not going to be using it," the child is addressing the issues of relevancy, boredom, personal advancement, goal orientation, autonomy, and self-directedness. And, today, especially, when kids learn at an early age the phrase "don't disrespect me," the issues of need for respect and socialization are also a part of learning. Surely, the numerous movies about fictional and actual teachers who successfully made their subject relevant and pertinent to their students should be strong evidence that children can and do learn the way adults learn.

Adult learning theory also includes the need to learn for learning's sake. Here, also, adult learning theory, in my opinion, has identified a difference that doesn't exist. Few adults in my classes asked why. Few children after a certain age also ask why. Compared to young children, far fewer adults and older children ask why.

However, there are learning differences between adults and children.

But asking why is how one learns more profoundly. Even if they do not ask why, young children may be more likely than adults to "research" by trial-and-error. For example, children will use trial-and-error and help from friends to learn how to use new technologies. In my experience, adults unfamiliar with a technology will not do that on their own—even when their job depends on it. The youngest of children are not afraid of making mistakes—of learning from mistakes.

Unfortunately, young children's curiosity occurs when they have less capacity to understand; as they mature, their capacity to understand and develop theories grows as their curiosity decreases. I believe that this desire to know is drained from kids by a teaching paradigm of there is only one correct answer. This "only one correct answer" disease plagues students of Six Sigma and eventually experts of Six Sigma.

The irony is that in Six Sigma courses, participants are taught a tool called 5 Whys. I and every instructor I have seen teach this tool always explained the tool by referring to children—2-year-olds typically—and how they incessantly ask why. Parents knowingly nod their heads.

DEMING'S PROFOUND KNOWLEDGE

What is the point of this? First, it is to make clear that not only is rote learning not an insult, it is a necessity. Hence, rote learners are not stupid or less capable since we are all rote learners for most topics and disciplines. In areas where we specialize through hobbies, profession, and other ways, we transcend to varying degrees the rote learning mode.

Second, we all transition through stages of learning from rote to some amount of profound learning. This is true whether a person is an infant or an adult.

In *The New Economics*, Deming detailed a management system called profound knowledge.* It consists of four components: appreciation of a system, knowledge of variation, theory of knowledge, and psychology. Here and in the chapters that follow, the focus is on the third component, theory of knowledge or epistemology. More specifically, the book addresses the following points Deming makes

- "Without theory, experience has no meaning. Hence, without theory, one has no questions to ask. Hence without theory, there is no learning."
- "To put another way, information, no matter how complete and speedy, is not knowledge. ... Knowledge comes from theory."†

To gain knowledge, Deming ascribed to a process attributable to Walter Shewhart, called the Plan–Do–Check–Act (PDCA) cycle. This application of PDCA can be shown as the following sequence:

Theory → questions → answers → understanding → profound knowledge

A profound learner, as defined here, is the person who habitually initiates and cycles through this application of PDCA. On the other hand, the rote learner is substantially more passive: skipping the theory, not asking the questions, and accepting the answers received. As a result, rote learners are prone to thinking they understand because they know the "truth." The profound learner isn't just someone who understands but someone who seeks to understand.

Perhaps no one has profound knowledge, but we can all acquire profound learning habits.

Rote learning is learning without asking questions and, therefore, without understanding. We all do it. My reasons for being a rote learner vary, depending on the topic.

---

* W. Edwards Deming, *The New Economics for Industry, Government, Education*. 2nd ed. (Cambridge, MA: The MIT Press, 2000), Chapter 4.
† Deming, *The New Economics*, 103 and 106.

- I have an interest in the topic, but don't believe I am capable of understanding.
- I have no interest at all in the topic, so I accept what others say.
- I am interested and ignorant about the topic but do not have or do not want to spend the time to research it. I accept what I hear or read—perhaps, limiting my sources to those who are or appear to know more than I.
- I am interested in a different issue and I think this issue is related. For example, I may or may not believe that global warming exists and therefore I accept or reject the claim as I see it fitting with my other beliefs (e.g., we should or should not drill in Alaska). In either case, I do not seek to understand the claim. I merely accept or reject.

A person can apply profound learning on one topic and rote learning on another. For example, I read that three species per hour are becoming extinct. I did not research the answers to many questions that would get at a theory. Nor did I start with a theory and ask questions to check or test my theory. A profound learner would ask questions—and seek the answers—to understand. Some initial questions might include the following:

- What is a species?
- Do scientists agree on the definition of species?
- Is the definition sufficiently specific and measurable to make an estimate?
- What is the procedure for estimating?
- If a species is extinct, how would one know?
- If a species became extinct, how would one know when it occurred?
- Over what period is this average calculated?
- Did some cataclysmic event occur so the number of extinct species was averaged over a longer period?

Because I do not ask these questions or seek these answers, I am a rote learner on this issue—and many other issues. Rote learning permeates everything we do as adults. It makes us susceptible and gullible.

Politics

Political campaigning *is* about rote learning—reducing complex issues to simple slogans. Obama reduced his campaign to the slogan "Change we can believe in." The vagueness of the slogan allowed millions of people to interpret the message in whatever way they wanted. Surely not everyone understood it the same way. Most likely, individuals and groups had opposing or contradictory interpretations of the slogan. Most did not have an understanding because we did not ask Obama the candidate what he meant. Would we have gotten a specific answer?

The GOP reduced a vast, complex proposal on health care reform to one- or two-word phrases, for example, socialism or death panels or simply Obamacare. As with Obama's slogan, these labels allow multiple interpretations. Since most people don't have the time to read a few-page document let alone one of thousands of pages like the Health Care Reform Act care reform act and few people would have the ability to understand it, we are susceptible to rote learning. We should not (although we do) expect congressional members to read and understand these huge proposals. We may or may not expect (or want) them to set aside politics to understand it.

Interestingly, in Kentucky, Obamacare was relabeled kynect to avoid the connotations associated with the term Obamacare. Some rote learners had harsh words for Obamacare but good words for kynect—even though they were the same thing.

As a result, we rely on single sentences or, worse, on two- or three-word slogans and labels to explain entire policies, entire proposals, and entire subjects. By not understanding what we are voting for, except as a slogan, we may push our congressional leaders to pass laws that actually harm us financially and physically. If they are supposed to be the experts on these laws but are rote experts, then they may unwittingly cause the harm they wish to avoid.

While I am not suggesting we have philosopher kings, as Plato recommended, we would do better to have more profound learning legislators—and voters.

Art as a Process

In a recent *Harvard Business Review* article,* the authors proposed that not every process improvement needs to follow a procedure or methodology. They made the contrast of process-oriented problem solving and

---

* Joseph M. Hall and M. Eric Johnson, When should a process be art, not science? *Harvard Business Review* March (2009): 58–65.

problem solving as art. I am not an artist although I do have some artistic skills from painting and drawing when I was young. Yet, those skills were enough for me to question whether the authors' analogy is correct. I know there are artists, just as there are R&D scientists and physicians, who do not believe they follow processes. Perhaps they have this belief for the same reason the authors do—process- or methodology-based problem solving is not creative or stifles creativity.

A few months after reading that article, my mother (who is an artist) and I went to the Ackland Art Museum at the University of North Carolina. In one of the rooms, I saw a plaque:

> Artists like Cennino Cennini, Leonardo da Vinci, Giorgio Vasari, and others who wrote about artistic training all agreed on one thing: although copying completed works of art was a valuable method for teaching students how to solve artistic problems, they should, above all, consider nature their principal teacher. This recurring theme helped students remember that they should develop their own individual style rather than simply reproduce that of their teacher. It also reinforced a growing conviction among artists at this time: studying nature's underlying scientific principles enhanced the quality of art.

Cennini, an Italian painter of the 14th and 15th century, wrote a handbook for would-be painters.* The relevant point of the handbook to this book is that it describes all aspects of painting at that time, for example, preparing paints, brushes, and canvas, besides how to paint, as processes. Cennini even describes how apprentices should develop a process for learning from masters. As part of that process of learning, he recommended studying scientific principles to enhance their art quality.

Doing art is a process.

Mammography

We all have topics that we might not be capable of understanding. So, we rely on experts. However, there may very well be experts who do not understand but still advise laypeople.

---

* Cennino d'Andrea Cennini and Daniel V. Thompson Jr., *The Craftsman's Handbook: "Il Libro dell'Arte,"* (Mineola, New York: Dover Publications, 1954).

For example, the new USPSTF (U.S. Preventive Services Task Force) mammography guidelines* illustrate how, for the majority of adults on topics critical to our health, we do not learn to understand or to make better decisions.

News media support our disposition for rote learning by providing headlines that summarize complex issues rather than just listing the topic. For example, the *New York Times* headline on November 17, 2009, was "Panel Urges Mammograms at 50, Not 40" rather than neutrally, perhaps, "New Mammogram Guidelines Issued."

The difference is critical. The actual headline gives the impression it is all one needs to know. The actual headline does not state any caveats that *are* included in the guidelines. The actual headline does not provide the reasoning for the recommendations. The new guidelines did provide details— although not the assumptions and analyses and raw data. Did people read them entirely to learn the how and the why along with the what? Most of us learned *by rote* the headline version and interpreted it as no mammograms for any woman until the age of 50 or all women under 50 stop taking mammograms. That is not exactly what the guidelines recommend.

More recently, new guidelines have been proposed based on more research. One doctor opined that "Mammograms help all women between the ages of 50 and 75." Yet, studies do not show that every woman who received mammograms benefited while those who did not receive mammograms did not benefit. Instead, studies show only that the average benefit was greater for the former group relative to the latter. Finer distinctions are therefore necessary. For example, women with dense breasts may not benefit at all or much from mammograms.

The problem rests not just with the layperson. If a doctor also learns the guidelines as simplistically as the news headlines, then this expert is not the expert female patients will expect. This doctor will not be able to develop an appropriate (minimum risk and anxiety) program of screening for every woman. While voters can harm themselves because they are rote learners, rote doctors can harm others by giving the simple mammography advice. The result may "cause" more preventable anxiety and early cancer-related deaths than the advice of a profound doctor.

Will there be rote doctors of mammography screening? Yes, I believe so. There are such doctors with respect to other health issues (based on

---

* U.S. Preventive Services Task Force, Screening for breast cancer: U.S. Preventive Services Task Force recommendation statement, *Ann Intern Med*, vol. 151, no. 10 (2009): 716–726 (accessible at http://annals.org/article.aspx?articleid=745237).

personal experience). One reason is unacceptable. Studies show that some doctors, as other professionals, can be persuaded by financial gain to make certain recommendations.*,†

Another reason is straightforward, understandable, and acceptable: doctors are incapable of keeping current to the level of expert understanding, given all the advances in medicine. It is ludicrous for us to expect that of them—and then blame them for not doing it. Doctors should also admit they cannot keep up and accept the help they need (effort using technology) to provide more profound knowledge-based health care.

The rote learner in the capacity as a doctor will tell his or her patients to follow the guidelines *because those are the new guidelines.* The advice will not be based on the expert's understanding of the theory behind the guideline. This will limit or prevent appropriate modifications alluded to in the guidelines but not even suggested by the headlines.

God Gene

Besides survivability, socialization also helps create a fertile environment for rote learning. Psychologists use the term *cognitive dissonance* to describe another critical element of human thought. When we make difficult decisions or hold strong beliefs, we tend to accept statements that support our decisions or beliefs and disregard contrary statements. Our psyche does not want to engage in a conflict. Often, we do not seek to understand the statements consistent with our decision or belief. We accept it through rote learning.

Unsurprisingly, people do have contradictory beliefs, although they may not be aware of them. Rote learning almost ensures unawareness, and therefore, contradictory beliefs. If the rote learning restricts the application of each contrary belief to areas that do not cross, they will not recognize the inconsistency.

This is not restricted to just politics or medicine or art.

Consider a common example. You have a great meal at a restaurant you patronize for the first time. You tell others how much you enjoyed the meal. They will pass on to others what you told them. The way the message is transferred contributes to it becoming a "truth." "I had a wonderful

---

* Katie Thomas et al., Detailing financial links of doctors and drug makers, *The New York Times*, October 1, 2014, B1.
† E.M. Johnson, Physician-induced demand, *Encyclopedia of Health Economics* (Cambridge, MA: MIT, 2014): 77–82.

meal at X" becomes "X is a wonderful restaurant." It could be that X is a wonderful restaurant. Yet, if it isn't, people believe something false. People who have never been there may falsely believe that X is a wonderful restaurant. A false belief accepted as true *regardless of all evidence to the contrary* becomes a myth as we use the term in everyday language.

Rote learning is accepting and believing without understanding. Faith is a particular kind of acceptance. Faith is the basis for both myth and religion. In one sense, a myth is a false story. In another sense, a myth is an oral or written story about how the world and everything or, at least people, in it came to be, expressed as a narrative and often by a narrator. In this second sense, a myth is sacred to and believed true by the society that holds the myth.*

Myths are built on rote learning. They are beliefs that are accepted as true by a group of people. Believers do not delve into the myths to understand or verify. Myths become the glue that binds the society, ostensibly increasing its chances of survival and health. However, myths can lead to the extinction of a society or group, for example, the Jonestown massacre or mass suicide.

Spirituality comes from or is an extension of myths. Based on studies on spirituality and genes, Dean Hamer in *The God Gene*[†] postulated that spirituality is hardwired via evolution or a "god" gene. This theory is dependent on the ability to define and measure *spirituality*. Hamer uses an operational definition of *spirituality* adopted from work by the psychologist Robert Cloninger.[‡] Hamer's spirituality scale has three personality types: "self-forgetfulness" or self-absorption, "transpersonal identification" or feeling connected to something greater, and "mysticism" or an openness to believe things not literally provable. Using this scale, he found a statistically significant correlation between spirituality and changes in a specific gene, VMAT2.

Setting aside that *Understanding Statistics and Statistical Myths: How to Become a Profound Learner* is about statistical myths and whether Hamer used any of these or others for his analysis and interpretation, there are other questions to be asked. The main one is what the highest (pun intended) level on the spirituality scale or mysticism really means. If *spiritually* means "openness to believe things not literally proven," this

---

* Joseph Campbell, *The Power of Myth* (New York: Anchor, 1991).
† Dean Hamer, *The God Gene: How Faith Is Hardwired into Our Genes* (New York: Anchor Books, 2005). Note: It appears that Hamer no longer ascribes to the point of view that spirituality is hardwired.
‡ C.R. Cloninger, *Feeling Good: The Science of Well-Being* (New York: Oxford University Press, 2004).

sounds very much like rote learning. Openness to believe does not appear to be materially different from accepting things without proof.

We have evidence that rote learning is necessary for survival. If religion is a type of acceptance, as is rote learning, perhaps VMAT2 isn't a god gene but a rote learning gene. For example, if one studied only Chinese-speaking societies and knew nothing of other languages, one might develop a theory that there is a Chinese gene. Once it is recognized that humans are capable of learning any of the human languages, the Chinese gene theory is replaced by the more general language gene theory, a richer theory. Similarly, the god gene of spirituality perhaps can be replaced by the more general rote learning gene: gene of acceptance without requiring understanding or proof. This may be a richer theory, capable of explaining more phenomena.

In the second sense of myth, there is no difference between religion and myths: codification of beliefs that explain things we otherwise cannot or do not explain or sacred narratives that are to be accepted on faith that describe the origins of the world through super beings or god(s). The key commonality is that myth and religion are based on rote learning. Understanding is not the primary point—acceptance is.

STATISTICS AS A RELIGION

In 1985, David Salsburg published a satire on how medical journals use statistics.* Although not his intention (as far as I know), it is also a satire on rote learners of statistics. They were not only the editorial boards and reviewers of scientific journals but also medical professionals. The holy grail of this "religion" was to produce a statistically significant result, quantified by a probability less than 0.05. This probability is known far and wide as p, or, redundantly and commonly, p value (see Myth 10).

Unfortunately, the religion of statistics is not restricted to medical journals or professionals. Statistics is a commonly used discipline in which millions of practitioners are rote learners. The association of statistics to lies and damned lies might be enough to prove this claim. Since statistics is based on logic, as with other branches of mathematics, we might as well

---

* David Salsburg, The religion of statistics as practiced in medical journals, *The American Statistician*, vol. 39, no. 3 (1985): 220–223.

say "lies, damned lies, and logic" or "you can prove anything with logic [statistics]." (You can!)

Besides the search for *p*'s less than 0.05, rote application of statistics proliferates because of the advent of software that allows anybody to enter any information and produce any statistical calculation. Thanks to mistake-proofing in software, this is less true. There are many published examples of the results of rote learning applications of statistical or mathematical analyses. Reporters summarizing statistical results often are rote learners of statistics. Unfortunately, reporter's sources can also be rote learners.

To cite one example, years ago, as female marathoners dramatically reduced the world record, some researchers predicted that women runners would eventually run faster than men. Disbelieving reporters questioned the researchers, who explained that these were not their opinions. They were simply publishing what the data indicated. They published graphs predicting when women runner times would be less than those of men runners, by showing trend lines of both male and female times that eventually crossed.

Their explanation was an example of rote learning *by the researchers*. The researchers had made at least two mistakes of which they were unaware. Presumably, they did not understand the theory. No data in and of itself defines a relationship (see Myth 30).

This analysis did not result in any harm.

However, if rote learning is behind the analyses for the mammography guidelines, then preventable injury and death may result. For what issues might a profound learner seek explanation? The following are just a few questions a profound learner might ask:

- It seems coincidental that the mammography analyses resulted in exactly age 50 and not 47 or 51 as the cutoff; and, exactly age 40 and not 39 or 42. If the studies compared two groups, say 40–49 and 50–59, how can the analysis show that some other ages other than the age dividing the groups would be a better cutoff for when screening should begin?
- It seems coincidental that the interval for screening is exactly 2 years and not 25 months or 19 months. Was this also a case of applying a particular relationship without knowing that one was assuming it? Would other relationships fit the data better—and reach a different conclusion?
- The guidelines divide women neatly into four age groups: 40–49, 50–59, 60–69, and 70–74. The reason for stopping at 74 is that no women age 75 and older were included in the studies reviewed. What about women younger than 40? It seems coincidental that the groups

neatly divide at the decades. What medical or scientific theory to augment the statistical analyses explains the drastic change in recommendations from 39 to 40, from 49 to 50, and from 59 to 60?

- All inferential statistical analyses require assumptions. On what assumptions were the statistical conclusions based? How were they verified or shown to be tenable?
- Reports (e.g., *The New York Times* article) quote panel members as using the phrase "precise figure" when referring to the cutoff age. No statistical inference can provide an exact figure. What is meant by "precise?" What is the actual margin of error and how wide? Did they rely solely on that aforementioned p or could they have calculated confidence intervals or probabilities?
- The guidelines state that "The harms resulting from screening for breast cancer include psychological harms, unnecessary imaging tests and biopsies in women without cancer, and inconvenience due to false-positive screening results." How reliable are the measures of "psychological harms" and "inconvenience?" How are these harms weighed against "unnecessary imaging tests and biopsies?" Is that weighting based on measurements from women themselves or the opinion of the guideline committee? All measurements have some margin of error. What is the margin of error on the assessment of "harm?"
- Just as doctors may "rotely" apply guidelines, were the statisticians doing the analyses of the studies also "rotely" applying statistics, for example, using only p values and only comparing to alpha = 0.05? Were the panel members also "rotely" applying statistics?

We can ask a similar list of questions of Hamer, given that he spoke about a statistical correlation and based that correlation on other statistical analyses by Cloninger:

- While a correlation may be statistically significant, how much does one variable explain about the other?*
- Statistical correlation does not mean causation. Why is the cause–effect relation extrapolated from the analysis in the direction proposed (gene is the cause, spirituality is the effect) rather than the other way around?

---

* Carl Zimmer, Faith-boosting genes: A search for the genetic basis of spirituality, *Scientific American* (October 2004).

- Some correlations exist not because one is the cause of the other, but because there is a third factor that causes both. Is there a rote learning gene that causes both the changes in VMAT2 and the claimed propensity for spirituality?
- All statistical inferences require assumptions. What are the assumptions in this analysis?
- Was the significance based on the *p* value alone?
- One variable in this correlation is the three-point scale of *spirituality*. How is each level operationally defined? What were Cloninger's assumptions for that analysis?
- Since every measurement has error, how repeatable and reproducible is this measurement of *spirituality*? What was the margin of error? What did Cloninger and Hamer use as "truth" to calibrate the measurement or scale?

Six Sigma

There is one reason for the proliferation of rote learners of statistics in the past two decades.

Six Sigma is based on a body of statistical knowledge. Since the 1990s, businesses worldwide have been using Six Sigma to try to improve their financial performance. Consultants, for obvious reasons, have recommended a certain percentage of a company's employees be certified in Six Sigma. As a result, millions of people have taken Six Sigma courses, ranging from a few hours to a few weeks long. These courses cover varying amounts of statistics.

On the basis of my exposure to various materials and experts, I would guess that easily tens of thousands if not hundreds of thousands of people have been exposed to one or more of these myths. From these numerous rote learners have come many rote experts. By an expert, I mean a certified Six Sigma Belt, a person with extensive experience in the role of a Belt, a trainer of Belt courses, or a consultant/facilitator/coach of people who apply Six Sigma.

That is not necessarily bad or wrong. For example, always crossing at the intersection is not necessarily a bad learning to keep throughout one's life. It could be inefficient to always do it and might not be possible in some situations. But, by understanding why the advice is given, one can alter or discard the advice and apply it in more appropriate or effective ways.

Similarly, a rote expert applying tools will be inefficient and ineffective at times. Organizations that hire, unknowingly, a cadre of rote experts each with their own way of doing things may unknowingly find that it takes considerably longer to accomplish things and with results less than anticipated, expected, or needed.

For an organization headed by a rote expert, the undesirable results can be even more severe. This could explain in part why despite so many companies adopting lean or Six Sigma or both or any of the previous approaches (quality circles, Total Quality Management [TQM], reengineering, and so on), few were sustainably successful. They may have started with a rote expert who installed a rote learning program to create a pool of rote learners.

People are not rote learners because they are malicious. As noted previously, there are various reasons why they cannot or choose not to be profound learners *for a specific topic*, for example, statistics. Yet, experts should be profound learners for their area of expertise. While not everyone can have profound knowledge, we can all engage in profound learning.

KNOWLEDGE THEORY

The fields of education and psychological development have developed three-dimensional frameworks of learning—and hence, of teaching. These three dimensions are knowledge types, knowledge stages, and learning types. In educational circles, Bloom's 1956 taxonomy applied to cognition formed the basis for understanding levels of knowledge.* This taxonomy led to a systematic way of creating learning objectives, which in turn structures teaching, which then determines assessment. There are two kinds of assessments: on the student to assess learning and on the teaching including the instructor, materials, and format to assess effectiveness.

Critically, the taxonomy assigns observable behavior to each level of knowledge. The original version had six levels of cognition with some behavioral descriptions for each level:

1. Knowledge: for example, list or repeat, as evidence of ability to remember

---

* B.S. Bloom et al., *Taxonomy of Educational Objectives: The Classification of Educational Goals. Handbook I: Cognitive Domain* (New York: David McKay Company, 1956).

2. Comprehension: for example, classify or explain, as evidence of ability to understand
3. Application: for example, apply or solve, as evidence of ability to use/apply
4. Analysis: for example, discriminate or appraise, as evidence of ability to decompose
5. Synthesis: for example, create or propose, as evidence of ability to create
6. Evaluation: for example, assess or defend, as evidence of ability to judge

In 2001, Anderson and Krathwohl led a revision.* The resulting taxonomy reorders the last two levels, relabels the levels using gerunds, and emphasizes the subcategories over the major categories by focusing on them as processes. The 19 subcategories defined behavioral processes that a person at that level of cognition would be capable of engaging.

The revised levels are remembering, understanding, applying, analyzing, evaluating, and creating. Creating is now the old synthesis, but with the same meaning: to put together in a novel way, to generate something new, to reorganize into a coherent whole.

It might seem at first glance that the profound learner is at the sixth level of cognition: creating. However, in the sense used in this book, that is not completely correct. As illustrated with examples in the previous sections, there are people who have at least the fifth level of cognition (evaluating) but may not be profound learners. In the sense used in this book, a profound learner can be at any of the six levels of cognition, because it is not their knowledge that determines whether someone is a profound *learner*. It is their behavior.

Here, a profound learner is better described by another type of knowledge in the revised taxonomy but not in the original. In the original, Bloom spoke of the levels representing factual (remembering), conceptual (understanding, applying), and procedural (analyzing, evaluating) knowledge. The revision includes a fourth element on this knowledge scale, metacognitive. Metacognition means reflective knowledge or self-knowledge, an understanding of one's own cognitive processes and cognition.

---

* Lorin W. Anderson and David Krathwohl, *A Taxonomy for Learning, Teaching, and Assessing: A Revision of Bloom's Taxonomy of Educational Objectives*. Abridged Edition (Upper Saddle River, NJ: Pearson, 2000).

In this book, a profound learner, *regardless of the cognitive level*, could have metacognition—but not just that. It is someone who has other characteristics described in the next section.

Thus, Bloom's original and revision taxonomy may not completely capture the characteristics of a profound learner because such a person always believes, by definition, that he or she has more to learn. For example, consider the application of the revised taxonomy. Teachers are given lists of verbs and example questions for each cognitive level to use for creating learning objectives. Some examples for the creating level include questions of the form "Can you...?"

A profound learner would immediately ponder the validity of this form of a question. If I cannot, is no the correct answer to "Can you...?" If I can, is the correct answer yes? How many teachers (presumably they are at the creating cognitive level in their subject) have just accepted and use this type of question expecting the answer to be an action either showing or not showing whether the person can?

This reminds me of a class where the instructor asked the participants to be creative. He said to draw nine dots on a piece of paper as below on the left in Figure P.1 and then connect all nine dots with just four straight lines and without lifting the marker off the paper. Of course, the idea was for us to literally move out of the "box" created by the nine dots, as shown in Figure P.1 on the right.

When I told him I had two answers each with fewer than four lines, he refused to believe it. I showed him my solutions. The first one involved folding the paper along the dotted lines so the left and right columns of three dots overlap with the dots of the middle column (Figure P.2). That puts the three rows of dots against each other, so one line connects them.

The second answer involved drawing another set of nine dots inside the center dot of his nine dots so they were all covered by his line, as shown in Figure P.3.

FIGURE P.1

Think-out-of-the-box puzzle. Draw four straight lines without lifting the pencil off the paper to connect all nine dots.

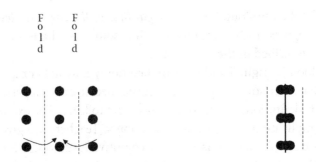

FIGURE P.2
Solution: fold the paper so that all three dots in each row are overlapping, creating one column. Draw one line touching all the dots in that column.

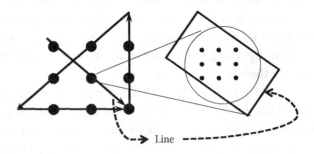

FIGURE P.3
One-line solution: Draw a line and then draw another set of nine dots inside it. One line then touches all nine dots.

He rejected my solutions. Apparently, thinking out of *the* box did not mean thinking out of *his* box.

There is a much older example of thinking out of the box. One version of the Gordian Knot is that Alexander the Great cut the knot when he could not find the ends to untie it.

In both examples, it is not the knowledge but the action taken to learn other ways to solve a problem that is critical. In other words, it is not the level of knowledge, but the type of learning that determines the kind of learner. The earliest classification of styles simply stated that learning occurs visually, auditorily, or kinesthetically. Others, for example, Kolb and the 4MAT group, have expanded on this providing more depth. Both use a four-stage approach that includes two dimensions: feeling–watching and thinking–doing. These can be viewed as incorporating the three modes of receiving information: visually, auditorily, and kinesthetically.

Kolb, for example, uses four types of learning styles called Diverging, Assimilating, Converging, and Accommodating.*,† This model defines the four stages as concrete experience, reflective observation, abstract conceptualization, and active experimentation. The psychological behaviors of feeling, watching, thinking, and doing are embedded pairwise, for example, feeling–thinking and in the order listed in the four learning styles.

The 4MAT group has a variation of Kolb's four types that are based on meaning, concept, skill, and adaptation. Some learners need to have learning that is meaningful, so a key question to answer in teaching them is "Why?" The conceptual learner needs to have lessons answer the question "What?" The learner who is primarily seeking skills wants to know "How?" And, the learner who needs to have learning that is applicable should be able to answer "What if?"

The result is a four-stage learning cycle designed to engage (meaning), share (concept), practice (skill), and perform (adaptation). The engage stage is better suited for those who want to know why they should learn this. Share is better for those who want to learn what they need to do. Practice is better for those who want to know how to use the learning. And, perform is better for those who want to know what is possible (what if) from the learning.

At first glance, one might suspect that Kolb's divergers or Assimilating learners and 4MAT's "what-if-ers" would be candidates for profound learners. However, no other characteristics in these models ensure an attitude or predisposition to learn.

Kolb's model adds another dimension, development. He states that his model is consistent with three development stages of learning: acquisition during preadult, specialization during college/training and early career, and integration during midcareer to later life. This matches previous comments that rote learning occurs in early life and on initial learning of any topic.

Unfortunately, Kolb does not blend the development stages with the four-stage learning cycle and the four learning styles. Thus, it is possible for very young people to be profound learners. While teaching may be rote for young or early-in-their-career people, it doesn't mean that the learner is not a profound learner.

---

* D.A. Kolb, *Experiential Learning: Experience as the Source of Learning and Development* (Englewood Cliffs, NJ: Prentice-Hall Inc., 1984).
† D.A. Kolb, *The Kolb Learning Style Inventory—Version 3.2.* (Philadelphia: Haygroup, Experience Based Learning Systems Inc., 2013).

In fact, a more recent study of learning by Dietrich and Kanso* suggests that the right-brain versus left-brain dichotomy on which the learning types are based is a myth. In which case, every type of learner regardless of the learning style would be capable of being a profound learner.

So far, I have made a simple distinction between a rote learner and a profound learner. Daniel Willingham, professor of psychology at the University of Virginia, has written extensively on learning based on empirical research. He makes a distinction between rote knowledge and inflexible knowledge and identifies a third type of knowledge called flexible knowledge.†

Rote knowledge is the bane of education and, according to Willingham, is merely parroting not what was said but what was heard. He gives an example of a teacher, in defining *equator*, saying "imaginary line" and the student hears "menagerie lion." A person with this rote knowledge will use what they heard even if it makes no sense. This learning is based on mondegreens—misinterpreting what is heard as a result of homophony. When I was a child, the famous mondegreen was the hymn line "Gladly, the cross-eyed bear" for "Gladly, the cross I'd bear."

Clearly, this type of learning is worthless.

However, Willingham defines inflexible knowledge as a necessary first step toward any kind of complex knowledge. Inflexible knowledge is that knowledge that parrots what was said rather than heard and understands sufficiently to apply at perhaps the first level or two of Bloom's original or revised taxonomy. It cannot go beyond that since such knowledge is not mastery of the underlining structure. Using Deming's terms, inflexible knowledge is knowledge without a theory.

Flexible knowledge occurs when, besides facts, there is understanding of a theory or deeper structures. Studies supporting Willingham's distinction between flexible and inflexible knowledge include one of classifying objects or concepts. Experts (flexible knowledge) classify according to theory, for example, principles of physics, rather than superficial differences as nonexperts (inflexible knowledge) did. "Similar experiments, using knowledge of dinosaurs, have likewise shown that experts' memories are

---

* A. Dietrich and R. Kanso, A review of EEG, ERP, and neuroimaging studies of creativity and insight, *Psychological Bulletin*, vol. 136, no. 5 (2010): 822–848.

† Daniel Willingham, Ask the cognitive scientist inflexible knowledge: The first step to expertise, *American Educator*, vol. 26, no. 4 (Winter 2002): 31–33.

organized differently than novices', whether the experts are *children or adults* [italics added for emphasis]."*

He emphasizes that inflexible learning is an essential passage between rote learning and flexible learning. To convert inflexible learning to more flexible learning requires repetition, examples, more facts, and clearer distinctions between examples and theory. This empirically based advice is from the instructor's or teacher's perspective. From the student's perspective, there are also requirements of a desire and capability to learn theory.

THE PROFOUND LEARNER

Socrates is perhaps the epitome of a profound learner. Plato, a student of Socrates, illustrated in his dialogues how Socrates sought knowledge. In these dialogues, Socrates asked questions of those he and others thought were wise people. Sometimes, the wise people Socrates questioned were self-anointed wise. His questions sought to understand, then develop theories, and thereby acquire knowledge. This is now known as the Socratic Method.

A profound learner is a philosopher in the original sense of the word. From the Greek *philo*, meaning "to love," and *sophia*, meaning "wisdom or knowledge," *philosophy* is the love of knowledge. A philosopher is a lover of wisdom. It is this sense of knowledge that I use to interpret Deming's statement that there is no knowledge or learning without theory.

To go from a rote learner to an inflexible learner to a profound learner, one must have a love of knowledge: a desire to understand, to explain, and to have a theory. This book is not just to debunk statistical myths I encountered when working with Six Sigma experts and "industry standard" training materials. It is also to provide theories for the topics related to the myths and suggestions on how to move beyond inflexible learning to profound learning. The key to traversing this path is to question what one hears or reads to understand the theory and assumptions behind the statements.

The idea of a profound learner and the process of profound learning started with John Dewey at the turn of the 20th century. For Dewey, deeper understanding, knowledge, and learning depended on being able

---

* Daniel Willingham, Ask the cognitive scientist inflexible knowledge, 33.

to distinguish between information and knowledge, focus on understanding, and use reflection as the critical steps to acquire deeper learning.*

A key method of focusing on understanding comes from Socrates. He asks for definitions of words. Today, the term for doing this is *operational definition*. An operational definition is a measurable definition of a term or concept. But, having operational definitions is not all there is to the Socratic questioning. There is also the testing of the definition, challenging it by applying it to see what limitations it has and whether it serves its purpose. Such challenging is part of Dewey's third component of reflective thinking: "(a) a state of perplexity, hesitation, doubt; and, (b) an act of search or investigation directed toward bringing to light further facts which serve to corroborate or to nullify the suggested belief."†

Numerous educators and psychologists have expanded on Dewey's ideas. For example, Bloom's revised taxonomy has metacognition as the fourth element to the knowledge hierarchy. This level is Dewey's reflection: awareness of one's own cognitive level.

Bowring-Carr et al. use the terms *effective learners and effective teachers*‡ while West-Burnham and Coates§ use the terms *deep and profound learning* and *learners* to get at this reflection level. They provide a list of strategies that could develop deep and profound knowledge. Their use could be evidence that someone is a profound learner:

- Understanding one's learning style to increase learning effectiveness
- Using all levels of knowledge from memorizing to analysis to theory development
- Understanding the value of and using as many sources and support to increase effective learning
- Constant review and reflection

Again, we see not only the ability to reflect but the habituation of reflection as a critical component of the profound learner.

Leon Festinger developed the term *cognitive dissonance* to indicate the human tendency or need for consistency among one's cognitions.

---

* John Dewey, *How We Think* (New York: D.C. Heath & Co.: 1910): Chapter One.
† Dewey, *How We Think*, 9.
‡ Christopher Bowring-Carr et al., *School Leadership in the 21st Century*. 2nd ed. (London: Routledge, 2005).
§ John West-Burnham and Max Coates, *Personalizing Learning: Transforming Education for Every Child* (Network Continuum, 2006).

Inconsistencies motivate one to alter something to retain the consistency. One can alter different things: the importance of one's dissonant cognitions, the weighting of the dissonant cognitions, or the dissonant cognitions themselves. A common example of the second altering is when a person makes a choice and then is especially attuned to evidence supporting the "correctness" of the decision and close-minded or dismissive about contrary evidence.

It is this particular alteration that a profound learner must overcome through reflection. It is much easier to reflect on others' viewpoints than one's own. It is far easier to question others, especially if what they say is contrary to one's beliefs. Nevertheless, the profound learner has the capacity to even question one's own beliefs and take the third altering approach.

> He who loves practice without theory is like the sailor who boards ship without a rudder and compass and never knows where he may cast.
>
> **Leonardo da Vinci**

ROTE AND INFLEXIBLE LEARNING BEHAVIORS

Years ago, as I started my career in industry, someone gave me this advice: In a decade, you had better have 10 years of experience and not 1 year of experience 10 times. Rote and inflexible learners become the inflexible expert with repetition of the same year's experience without realizing it. Like da Vinci's sailor, they steer without rudder and compass.

To borrow from the comedian Jeff Foxworthy, you might be a rote or inflexible expert if

- Your ultimate explanation is "that was what I was taught"—and are proud of it
  - In discussing one of the myths, a Six Sigma consultant eventually admitted "That was what I was taught"—and continued to believe it.
- You say "I've done it that way for years"—and don't plan to change
  - The company leader of a lean program heard the same criticism from all his direct reports about an approach they had tried. When I suggested an amendment to the tool to address the

weakness everyone else saw, he replied, "In the 13 years I've been using this tool, I've never used it that way." He didn't then either.

- You teach and preach—but don't practice what you teach and preach
  - One external trainer required all lean projects to reduce the process cycle time by 70% regardless of current performance or type of process. When I told him I did his 10-day course in 3 days, he replied, "Well, you must not cover everything." Given that even in 10 days, it would be impossible to cover "everything," I let the comment go.
- You are not a statistician—but believe everything you know about statistics is absolutely true
  - The head of a company's Six Sigma program offered that she was not a statistician. When we got to discussing statistics, I stated that capability is just percent good. She said it wasn't. When I asked what it was, she simply cited a formula for process capability index (C_p).
- You know only one way to do things—and believe it's the correct way
  - One Master Black Belt (MBB), when challenged about how to do something, defended himself by saying "I've been doing it this way for 15 years. I don't understand how others can do their projects if they don't do it this way." The fact that people had successfully done projects other ways was not, for him, contrary evidence.
- You support your view—by quoting others (only if they have the same view)
  - When a friend sent a comment I made about control charts to a Deming blog, several responses were *ad hominem* attacks, for example, "It's clear he's never read Shewhart (or Wheeler)." I have. To quote da Vinci: "Anyone who conducts an argument by appealing to authority is not using his intelligence; he is just using his memory."
- You know how to use many tools—and use them religiously whether needed or not
  - When interviewing for a position, a Black Belt asked me about when I would use a control chart. I described the basic purpose of the tool and said it could be used in the first and last steps of their five-step methodology to deliver their purpose or it could be modified and be used in the third step. He asked why I would not use it in the second step. I said I've never had a reason to use it. Later, I learned his review of me included that he wasn't sure I really understood control charts.

Since inflexible learning does not require understanding, inflexible experts will not see inconsistencies among their beliefs. As a consequence, they may be inefficient in their applications. They may also be ineffective, for example, by insisting a tool be used at a certain step, when other tools may work as well or when the purpose of the tool has already been accomplished.

Inflexible experts are change agents because they are experts. However, by being inflexible, they themselves have not changed in years and will be unwilling to change in those situations where they are experts. Inflexible experts were told something and they accepted it without question and understanding. They may say "I was taught this" or "This is the way I have done it for *x* years" but not recognize that their comments do not explain why. Inflexible experts quote other experts as if, because someone else said it, it makes it true or they think experts are infallible.

Inflexible experts can undermine the ability of the organization to develop beyond "industry standards." For example, almost every employee needs computer skills. Most skills they need are rote and inflexible skills— they do not need to know computer science, IT, and other disciplines (theories) to know how to use software. The profound knowledge does not and need not reside with all users of computers or technology. That profound knowledge can successfully reside within an IT function, for example.

This is not to say inflexible experts are not valuable and good problem solvers, facilitators, coaches, or trainers. The above behaviors caution companies and others who rely on inflexible experts on the following three points:

1. Inflexible experts believe that there is a correct way of doing things (their way), which limits the flexibility required for solving different kinds of problems and addressing the needs of different audiences.
2. Inflexible experts pass on their false beliefs as gospel. These beliefs can be benign in most situations, but can have severe consequences in a few. Inflexible experts will be susceptible to seeing all situations as the same.
3. Inflexible experts may not practice what they preach. They will not see inconsistencies in their beliefs, so they will unknowingly provide contradictory advice. Developing a common and consistent language will be difficult as terms will have different meanings in different applications and they will be inflexible in how terms are defined and used. They will not heed Humpty-Dumpty: "which is to be master [you or the words]—that's all?"

PROFOUND LEARNING BEHAVIORS

West-Burnham and Coates* identify three types of learning based on the type of question the learner can answer: shallow learners answer what, deep learners answer how, and profound learners answer why. With this foundation, they identify six scales on which to differentiate shallow, deep, and profound learning: means, outcomes, evidence, motivation, attitudes, and relationships. The key scales for the purposes here are evidence, attitude, and relationship.

As a learner proceeds from shallowness, to deep, to profound learning, the evidence used would be replication, understanding, and meaning.

The attitude would change from compliant to interpretation to challenge.

The relationship between the learner and the teacher would change from dependent to independent to interdependent.

LESSONS FOR ORGANIZATIONS AND INDIVIDUALS

What is the impact on organizations and others who rely on inflexible experts? For one, if they have been spreading myths and yet that doesn't stop projects from being successful, it raises the question of the value of the learning on the myth-related topic. For example, for centuries, people believed that the earth was flat. This was an inflexible learning. However, it did not stop human progress: cities and civilization, democracies and philosophies, libraries and bridges, sciences and religions were all created while believing the world was flat. Since the information of a flat earth was not used for this progress, then why teach it?

At least a few experts believe one or more of the myths described in this book and yet they are competent problem solvers and contributors to their employer's success. If the myths in this book were not used to solve problems, why teach them?

Consider Myth 10. Probabilities called *p* (see Statistics as a Religion) are used to make decisions when testing statistical hypotheses. There is a decision rule that says to accept or fail to reject the null hypothesis when

* West-Burnham and Coates, *Personalizing Learning* (see also http://www.johnwest-burnham .co.uk/index.php/understanding-deep-learning?showall=&start=1 and http://www.nwlink.com /~donclark/hrd/development/reflection.html).

$p \geq$ alpha and accept the alternative hypothesis when $p <$ alpha. People can successfully use this rule and make the (statistically) correct decisions according to the rule without knowing what p (or alpha) is. They certainly don't need to know how p is calculated.

Many people do use the rule successfully—even when they mistakenly believe Myth 10.

Knowing or not knowing what p is, is irrelevant to applying the decision rule for statistical tests. This suggests that, for rote learners, rather than spend time teaching in great detail what p is (especially when a myth is taught), courses could teach a general concept related to p and spend more time on other knowledge that is used. For example, a lesson on p could be simply on the procedure: p is a probability used to decide which hypothesis to accept using the decision rule....

Because of the propensity and need for inflexible learning, perhaps the guidance for adult teaching is to take advantage of it. One could combine inflexible learning with a concept from another quality movement. Lean, or the Toyota Production System, is a complementary movement to Six Sigma. Its objective is to remove waste. One approach is to create standard work. Standard work is a documented standard operating procedure that is the best known way of doing something. People who need to do that specific work are trained on the standard work. Understanding why is not essential. Following the procedure is.

This approach can be used for course development and hierarchy. Basic courses would teach standard work, to take advantage of our inherent tendency for inflexible learning. Standard procedures would be taught and practiced with little explanation of theory or why. Organizations that simply create standard work that is diligently adhered to will make substantial improvements in the performance of their processes by reducing variation and costs. However, to instill a culture of continuous improvement, there must be a reservoir of understanding based on theory. There must be profound learners.

More recent work on adult learning indicates that the neural connections people develop as they go from rote learning as children to inflexible learning to flexible learning become stagnant when adults specialize and narrow their interests and attention. Rather than reverting to rote and inflexible learning, such as teaching facts like the myths debunked later in this book, the adult brain should be challenged. This challenging of assumptions makes the brain grow, some scientists believe. The profound learner—and older adults, if this adult learning theory for those beyond 40 years of age is applicable—will flourish in an educational environment that encourages the stretching of the mind through challenges of established assumptions.

So, for advanced courses, prospective candidates can be assessed on their ability and interest and compatibility with the characteristics of a profound learner. If they qualify, then they would take advanced courses that develop their understanding and require them to demonstrate their understanding by testing their theories. These courses would provide practice on the profound learner sequence from theory to understanding. Practice leads to habituation.

Benefits to organizations and individuals include the following:

- Opportunities to relearn
- Opportunities to deepen one's knowledge
- Sources of and culture for profound expertise
- Reduction of the time and money spent on knowledge and skill transfer
- More qualified employees/consultants
- Greater flexibility—able to go beyond industry standards
- Creation of a learning organization with experts who are profound learners

I cannot teach anybody anything; I can only make them think.

Socrates

STRUCTURE OF BOOK

Six fictional characters have 30 discussions on various topics taught in the course covering seven statistical areas: data, estimation, measurement system analysis, capability, hypothesis testing, statistical inference, and control chart. Each discussion is about a statement one of the characters makes, as if it was universally true. Because each is not universally true, each is a myth.

Each discussion is like a Platonic dialogue. The purpose of a Platonic dialogue is to analyze a concept, statement, hypothesis, or theory through questions, applications, examples, and counterexamples, to see if it is true, when it is true, and why it is true when it is true. The dialogues lead to understanding why the statement is not always true under all conditions and when it contradicts other myths.

Socrates, one of the characters, is a student in a Six Sigma Master Black Belt course. Socrates has a background in mathematics but no industrial experience. The course objective is to teach students how to apply statistics to improve processes. The students apply what they learn to a project in which they are team members.

Socrates is a profound learner. He learns by asking questions of instructors, experts, and fellow students. He uses four approaches to profound learning: soliciting definitions of words and phrases, identifying assumptions, determining what questions can be answered if the statement is true, and determining the purpose of the statement. These understandings provide opportunities for development of other concepts, tools, and applications that are richer, more powerful, and more generally applicable or better applicable for specific cases.

The reader may wish to answer this question: Which character reminds you of yourself? To assist the reader, below is a way to distinguish among the three learning types based on two critical behavioral characteristics. One characteristic is what the learner perceives and the other is what action the learner takes based on those perceptions:

- Rote learning: believe what you perceive (e.g., hear, read, touch) without understanding what you perceived; when corrected, you will change your belief even if it is incorrect but still without understanding.
- Inflexible learning: check whether you perceived correctly to adjust your belief to match the correct perception even if the understanding of the perception is wrong, but won't change your belief once the perception is accepted.
- Profound learning: check whether you perceived correctly and adjust your belief to match the correct perception; check whether the belief is correct by seeking to understand why.

CHARACTERS OF THE DIALOGUES

Instructor, consultant, mentor	Daskalos
Master Black Belt	Mathitis
Black Belt	Peismon

Science researcher	Kokinos
Science journalist	Agrafos
Student (math background)	Socrates

Scenes—various rooms (hall lounge, classroom, breakroom, project meeting room) of a building where the course is being taught

1

Myth 1: Two Types of Data—Attribute/ Discrete and Measurement/Continuous

Scene: *Hall lounge before class discussion on data types*

BACKGROUND

Socrates: Our instructor Daskalos said there were two types of data, measurement or sometimes called continuous, and attribute.

Peismon: That's right. For example, having only a finite number of choices, such as pass–fail, is attribute data. Other attribute data include colors and counts and proportions, while measurement data are like temperature, length, and weight.

Socrates: Are you just grouping things into two arbitrary categories that you called attribute and measurement, or do the labels attribute and measurement mean something specific?

Peismon: Of course, they mean something. I find that using these labels helps people understand the types of data better.

Socrates: Can you explain what you mean by the labels or types besides giving examples?

Kokinos: I find examples help me understand better than definitions.

Peismon: That's what I mean. People understand better with examples.

Socrates: What is the difference between attribute and measurement?

MEASUREMENT REQUIRES SCALE

Peismon: Measures require a scale. Attribute data do not require a scale.

Socrates: What do you mean by a scale? Can't we have a color scale? For example, you want to buy white paint to repaint a wall in your house. The person at the store gives you a palette containing a range of shades of white. Isn't that a scale?

Mathitis: I think it is. That scale also occurs when we want to choose the color of a font or touch-up photos. Software menus show palettes or ranges of shades.

Peismon: Those aren't scales.

Socrates: What is your definition of scale?

Kokinos: I looked it up. It has different meanings in different disciplines. It can be a progression of steps or degrees, marks at fixed distance, or a balance for weighing things.

Agrafos: Maybe not the last.

Peismon: A progression of degrees or fixed distances. That's what I mean. But the last definition of needing an instrument also applies.

Socrates: Then, both attribute and measurements have scales. Even the pass–fail measurement has a scale: pass, fail. Colors have scales—the range of shades. Each are degrees or steps.

Peismon: Colors aren't scales since they are not a fixed distance apart.

Mathitis: We see colors because of wavelength. Wavelength is a scale with fixed distances.

Socrates: Counts clearly have scales, as they are a fixed distance apart. But are nonlinear measures using instruments not scales, e.g., the Richter scale?

Agrafos: Proportions are fixed distances apart. Aren't they on scales?

Socrates: I'm not sure I understand. Aren't these scales: pass/fail, five car colors, 1 through 5, and 0% to 100%? Is it a progression of steps or degrees or a fixed distance apart that indicates a scale?

Mathitis: It seems so to me.

GAUGES OR INSTRUMENTS VS. NO GAUGES

Peismon: Attribute data mean that there isn't an instrument or gauge like a balance or meter to get the measurement. Remember, I said that was also a critical part of measure types of data.

Mathitis: You use instruments—hardware and software—to determine font colors.

Peismon: That just lists the possible choices. It doesn't measure like a balance or measuring stick does. Colors, for example, are determined by just looking. There is no measurement. Attribute data need to be observed or counted while measured data must be measured.

Socrates: When I say that the car is red, what am I counting?

Agrafos: What difference does it make?

Socrates: I am trying to understand whether I always must count to get discrete data or that whenever I count I automatically have discrete data. This would help me know which I have.

Peismon: If you need to count, then you have discrete data.

Mathitis: You count to get percentages and proportions.

Socrates: Okay. What about identifying colors or when I give a score of 4 on a satisfaction survey, what am I counting?

Peismon: Perhaps not for those examples. But for continuous data, you need a measurement instrument.

Socrates: So not all discrete data involve counting.

Peismon: Perhaps not.

Socrates: Don't we count with continuous data? Have you ever measured the length of a room by counting the number of steps it takes to cross it? Even using a yardstick to measure something, you are actually counting how many inches or centimeters there are from one end to another.

Agrafos: You don't have to count since you can read the amount on the ruler.

Socrates: True, you don't have to. The reason is that the ruler has already made the count for you and labeled the counts. The ruler has the numbers 1, 2, 3, which represent counts of whatever the measurement unit is, such as centimeters or inches. If the ruler just had the inch markings on it without the numbers written on it, you would have to count them.

Peismon: The real difference is that you don't use an instrument to measure with attribute data. As I said, you can count or just observe, like colors.

Socrates: What do you mean by measure or use an instrument?

Mathitis: What he means is that you need a gauge to determine the value—the number.

Peismon: We call the instrument a gauge, like a ruler, thermometer, or bathroom scale.

Socrates: How do I tell what color something like a car is if I am blind?

Peismon: You may not be able to, but others may tell you or it may be written in Braille.

Socrates: How does the person who determined the color do it?

Mathitis: That's easy, Socrates. If they are not blind or color blind, they look at the car and determine it.

Socrates: They need an eye to see the color.

Mathitis: Of course.

Socrates: Then the eye is the instrument or gauge.

Peismon: No.

Socrates: What is the difference between a thermometer gauging the temperature and an eye gauging the color?

Kokinos: I was taught that attribute data did not need a gauge, so I would conclude that the eye is not an instrument or gauge.

Mathitis: I agree, Socrates. Yes, the eye is an instrument. However, we typically mean that a gauge is something separate from the person.

Socrates: Let's suppose that is true—the gauge or instrument must be external. We do have pass–fail or go–no go gauges. Does that make pass–fail a measurement?

Agrafos: Pregnancy tests are pass–fail. They measure the presence or absence of a hormone.

Mathitis: There are many medical instruments that are equivalent to pass–fail. They report positive or negative with respect to a disease or condition, such as HIV or tuberculosis.

Kokinos: There are instruments to measure motion, light, or sound.

Mathitis: Also, to measure colors as wavelengths. Astronomers use wavelength to determine the distance of stars from earth.

Agrafos: We also have counting instruments. Machines that count coins or bills—I've used them to return jars of pennies for dollars. Or bottles or cans at recycling centers.

Socrates: These are examples of external gauges that measure what you said were attribute data. Are you saying that data are attribute type until an instrument exists to measure it? Before we had rulers to measure length, length was attribute type?

Mathitis: That would be odd. I'm not sure that attribute versus measurement is a good distinction nor is the requirement or use of a gauge.

DISCRETE, CATEGORICAL, ATTRIBUTE VERSUS CONTINUOUS, VARIABLE: DEGREE OF INFORMATION

Peismon: But we also use other terms to distinguish data types: discrete versus continuous.

Kokinos: That's what I learned. By discrete, we meant attribute, categorical, or qualitative. These other terms are either synonyms for discrete or ways of being discrete.

Peismon: That's correct. Sometimes, we use the term *variable* for *continuous*.

Socrates: I see that pass–fail or colors are discrete. But I don't completely see how proportions and percentages are not variable. By having more than one choice, aren't all these examples variable because there is more than one option?

Peismon: One way of looking at the difference is that continuous data are more informative. Pass/fail is not as informative as temperature.

Socrates: How are continuous data more informative?

Peismon: Continuous data provide a degree of goodness. If we have specifications, it is better to know how far off you are than just whether it is good or bad.

Socrates: I can't always have continuous data. If the color is important to me when buying a car, I can't convert that to a number.

Peismon: No, but if you have a choice, use continuous data.

CREATING CONTINUOUS MEASURES
BY CHANGING THE "THING" MEASURED

Agrafos: That's what my mentor for my Green Belt project told me. I should have quantitative measures. She said that I can convert my discrete data to quantitative data by using counts. So, you do always have a choice.

Socrates: What do you mean?

Agrafos: My project involved reducing complaints from customers who called and were transferred to the wrong person. Either it was transferred correctly the first time or it wasn't. I had a binary measure—which is discrete.

Socrates: How was that converted into a continuous or quantitative measure?

Agrafos: She said to use the number of calls transferred correctly the first time.

Socrates: But that's a different measure.

Agrafos: Yes, it is a continuous measure, not discrete.

Socrates: Sorry. I meant that you are now measuring something else.

Agrafos: Yes, because the measure is changed.

Socrates: Originally, correct or incorrect transfer was a measure on what "thing?"

Agrafos: On each call.

Socrates: What is the number correctly transferred on a single call?

Agrafos: Oh, I see: it's either 0 or 1.

Socrates: So when your mentor suggested changing to counts, what is the "thing" you were counting? Is it still a single call? To have a count greater than 1, how many calls are needed?

Agrafos: No, it's not a single call, so you need more than one.

Socrates: When do you stop counting? Is it the number per hour or shift or day?

Agrafos: I see. Now the "thing" would be the hour or the shift or the day.

Socrates: Yes, this makes that measure quantitative but not the original measure quantitative. It is not a measure of correct or incorrect transfer but of the number of correct or incorrect transfers *per some period of time* rather than per call.

Agrafos: But, if it is more informative, as Peismon said, what's wrong with making that change?

Socrates: There are three possible problems.

Agrafos: What are they?

Socrates: You now have to sample a certain number of days, say 25 days and count all the calls for that day, while before you would sample 25 calls, which could be spread over several days.

Agrafos: So more work is involved and it will take longer.

Socrates: The second issue is that you now need specifications for each day and a goal for all the days. For example, you might say we want each day to have no more than 5 incorrect transfers but then you have to set a goal of what percent or how many of the days are to meet this specification of no more than 5 incorrect transfers per day.

Agrafos: That certainly would make it a lot harder and more complicated.

Socrates: The third issue is that in either case, you have to look at each call to determine why they aren't being transferred correctly.

Agrafos: But it could be something systematic on particular days.

Socrates: Yes, it could be, like peak hours. But you won't know that unless you investigate each call to determine whether it's systematic or not. But, what is the issue? Too many hours, shifts, or days with incorrect transfers, or too many calls incorrectly transferred?

Agrafos: Too many calls incorrectly transferred.

Mathitis: So, you didn't make a discrete measure continuous but created a different measure on a different thing which isn't what you wanted to address.

Agrafos: I understand that now.

DISCRETE VERSUS CONTINUOUS: HALF TEST

Socrates: What do you mean by discrete and continuous?

Peismon: Let's use examples, like in Table 1.1.* Discrete means there is a gap between options while continuous means there is no gap.

Socrates: What does continuous mean mathematically?

Peismon: That it passes the half test.

Kokinos: What is the half test?

---

* The contents of tables like Table 1.1 have occurred in Six Sigma training material developed by and used by numerous companies including Goal/QPC, GE, Microsoft, and Merck.

TABLE 1.1

Examples of Continuous and Discrete Data Types[a]

Discrete (Attribute, Categorical, Qualitative)	Continuous (Variable, Quantitative)
Binary, e.g., pass/fail	Length, e.g., inches, centimeters
Colors, e.g., list of five colors	Weight, e.g., pounds, grams
Rating, e.g., 1 to 5	Time, e.g., seconds, days
Percentages or proportions	Temperature, e.g., Fahrenheit, Centigrade

[a] http://www.thequalityweb.com/measure.html.

Peismon: That you can divide an amount by half and it makes sense. Using this definition, counts are not continuous because you can't have half a person. For example, if the temperature is 30° then we can take half of that and it is a valid temperature, and half again, and half again. The same can be said for lengths: 1 m, 500 cm, 250 cm, 125 cm, and so on.

Mathitis: Mathematically, it means that between any two numbers there is another. Between 0 and 1 there is 1/2, between 0 and 1/2 there is 1/4, and so on.

Socrates: Then, the difference between discrete and continuous is that continuous measures are infinite and discrete are finite?

Peismon: Yes, that's the major difference. By being able to take half, then you get the concept of an infinite number of choices.

Socrates: Then, the difference between continuous and discrete is not that one has a scale and uses gauges and the other does not. It is that one passes the half test and the other does not.

Peismon: The scale and gauge help, but yes, the key is passing the half test.

Socrates: You don't think that colors pass the half test?

Peismon: No they don't.

Socrates: You agree that there is white and black.

Peismon: Yes, and white/black would be attribute or discrete measures.

Socrates: Can we get an infinite number of gray shades?

Mathitis: I think we can. We just mix different combinations of white and black.

Socrates: Let's see if we have an infinite number of gray shades. Mix equal proportions of white and black. That would give you a gray halfway between the two. You now choose whatever direction you want—lighter toward white or darker toward black.

Mathitis: I see. If we choose to go lighter, we take equal proportions of that half-mix with all white, then equal proportions of that mix (one-quarter white and three-quarters black) with all white, and so on. We "cut" each shade of gray in half.

Peismon: Yes, you can do that but those grays do not represent a quantity like temperature or length or weight.

Mathitis: You're right, they don't. But the issue was whether there were an infinite number of choices. Attribute scales can have an infinite number of choices, making those scales continuous. We can create an infinite number of colors by combining any two colors just as we created an infinite number of gray shades.

NOMINAL, ORDINAL, INTERVAL, RATIO

Peismon: Okay. But you can't do that with counts. Besides, as is noted in Table 1.1, continuous is quantitative while attribute or discrete is qualitative.

Socrates: Then, the distinction is not between measuring or not, scale or not, gauge or not, infinite or not, but quantitative or not.

Peismon: That's a critical issue, a distinguishing feature.

Socrates: Proportions and percentages are quantitative. Doesn't that contradict the column heading of your table—it says "qualitative?"

Mathitis: They are a special case. I often explain data types not just as discrete versus continuous but also as nominal, ordinal, interval, or ratio (Steven 1946).

Agrafos: What are those?

Mathitis: Nominal refers to names or labels. Even if numbers are attached, they only represent labels, like identification numbers—your social security number, for example.

Agrafos: So colors are also nominal.

Mathitis: Yes, they can be. Ordinal refers to labels that can be ordered by magnitude. In the discrete column, you see ratings. They are ordinal data. Satisfaction surveys, rankings, and so on. There is a sense of magnitude but no exact quantity. Interval data are ordinal with the additional fact that a unit difference represents the same amount everywhere.

Peismon: On a rating scale, the difference between 1 and 2 is not the same as the difference between 3 and 4. A one-unit difference does not represent that same quantity everywhere.

Kokinos: You mean, for example, that the difference between "Very satisfied" and "Satisfied" is not equal in satisfaction as between "Neutral" and "Satisfied" even using numbers from 1 to 5.

Peismon: Exactly. But for interval—that's why the name—one unit difference is the same everywhere on the scale.

Mathitis: However, there is no absolute zero. The common examples of interval scales are Fahrenheit and Centigrade. Zero on these scales does not mean lack or absence of.

Peismon: But the ratio scale does contain an absolute zero, like length. A length of zero means it has no length, it lacks length.

Agrafos: What is the advantage of the four types over the two types?

Peismon: As you move from nominal to ordinal to interval to ratio, more information is provided and finer distinctions can be made.

Socrates: Arithmetic can be applied when the numbers represent quantities. For example, even if we label colors 1 = red, 2 = blue, 3 = green, and 4 = black, it doesn't mean we can add and subtract the data to represent a quantity. Subtracting, 4 − 3 = black − green doesn't represent anything and doesn't equal 3 − 2 = green − blue. But with interval scales they do: 43°C − 36°C = 7°C and that is the same as 70°C − 63°C.

Mathitis: That's correct. But we can't say that 40°C is twice as hot as 20°C.

Agrafos: Why isn't it twice as hot?

Mathitis: Because there is no absolute zero. When we change the scale, the ratio is not the same. If we change to Fahrenheit, the temperatures are 104°F and 68°F. Clearly, 104 is not twice 68.

Socrates: Interval data can be added and subtracted but not multiplied and divided while all four arithmetic operations apply to ratio data.

Peismon: That's right.

Kokinos: Then, does discrete mean nominal and ordinal while continuous mean interval and ratio?

Peismon: Exactly, because of the calculations we can do.

Socrates: Do these four categories cover all possibilities?

Peismon: Yes.

Mathitis: I believe they do.

MEASUREMENT TO COMPARE

Socrates: In any case, isn't the point of measuring to make distinctions?

Peismon: Yes. If everything is the same amount or quantity, then why measure? We couldn't differentiate between things nor would we need to. That's why continuous data can be compared.

Kokinos: Can't we compare black to white or red to not-red? Or, pass to fail?

Peismon: I can't say that red is twice not-red or seven less than not-red.

Socrates: But there are other ways of comparing. We do in fact compare colors. If you didn't or couldn't compare, you wouldn't be able to say that the car you got was the color you wanted.

Mathitis: Peismon is saying that with continuous data you can compare quantitatively. You can't do that with discrete data.

Socrates: Then, it isn't that discrete data can't be compared. After all, we do it all the time—with our five senses, such as, this tastes good or bad, or sweet or sour; it's too bright or too dim; it's hot or cold; loud or soft; rough or smooth. We are comparing in those cases.

Mathitis: You are right, Socrates. We do comparisons even with discrete data.

Socrates: Let me see if I understand then. Both attribute and measurement data are measures on a scale; they both use instruments although one is external and the other is the person (but not always); both are used for comparing, although with continuous data the comparison is quantitative, and some attribute/discrete data can pass the half test while not all quantitative data pass the half test.

Mathitis: I think you have it.

SCALE TYPE VERSUS DATA TYPE

Socrates: Is it the data that are discrete or continuous?

Agrafos: I don't understand.

Socrates: Is the number 1.34 discrete or continuous?

Agrafos: Continuous because of the decimal places.

Socrates: What if that is a label or a percent?

Agrafos: Then it wouldn't be.

Mathitis: So, we can't tell if a number is discrete or continuous until we look at the scale.

Socrates: Did we not agree that the purpose of measuring is to make distinctions? We do that with choices. Scales are just lists of choices: {Pass, Fail}, {Red, Green, Blue}, or {1, 2, 3, 4, 5}.

Peismon: Yes, we did.

Socrates: Then, the measure or the scale is simply the choices available. There could be a finite number of choices either quantitative or qualitative, or there could an infinite number of choices either quantitative or qualitative.

Peismon: When the choices are finite, we have discrete data; otherwise, continuous data.

Agrafos: But we saw that the gray scale is infinite—and we could do the same mixing of any two other colors, e.g., green and blue—but you have *color* as discrete.

Socrates: Is each individual choice quantitative or qualitative or is it the scale?

Mathitis: I say it is the scale, list of choices.

SCALE TAXONOMY

Socrates: It seems that we are discussing things at different levels then.

Agrafos: What do you mean?

Socrates: Are you familiar with the biological classification system?

Kokinos: Yes. Kingdom, Phylum, Class, Order, Family, Genera, Species.

Socrates: When we say discrete or continuous or measure or attribute or nominal or interval, we need to make sure we are talking at the same hierarchical level. Saying counts and colors are attribute or discrete and different from continuous or measure, we may be confusing levels.

Agrafos: How is that?

Socrates: The first clue is that given the examples, we see that attribute/ discrete contain both quantitative and qualitative scales. We saw that values for an attribute scale can be produced by an external instrument (e.g., pass/fail gauge or counting machine) and pass the half test (e.g., shades of gray). All these ambiguities suggest that we are not classifying scales appropriately.

Kokinos: It sounds similar to the issue of classifying a platypus. It has mammary glands, so it should be a mammal. Yet, it lays eggs, so it shouldn't be.

Mathitis: But we saw that the four categories of nominal, ordinal, interval, and ratio occur without overlap under quantitative and qualitative.

Kokinos: Does that mean that the highest classification is qualitative versus quantitative?

Mathitis: I would say yes and then place nominal and ordinal under Qualitative.

Kokinos: And interval and ratio under Quantitative.

Agrafos: Where do discrete, attribute, continuous, and measure occur?

Socrates: That depends on what we mean by these terms. But let's put our examples in this hierarchy. Where does pass/fail occur?

Agrafos: It's Qualitative and Nominal.

Mathitis: Temperature—specifically Fahrenheit and Centigrade—occurs under Quantitative and Interval since temperature except for the Kelvin scale, does not have an absolute zero.

Kokinos: Satisfaction survey scales, for example, ratings from 1 to 5, would be Qualitative and Ordinal.

Agrafos: Distances, speed, weights would be Quantitative and Ratio.

Socrates: What about counts?

Kokinos: They are Quantitative and have an absolute zero. We can add, subtract, multiply, and divide them. They must be Quantitative and Ratio also.

Mathitis: But if division results in a fraction when only whole numbers are possible, then we can't divide. Otherwise, they meet all the criteria for Interval scale.

Peismon: But counts do not pass the half test.

Socrates: Then, to be continuous, do they have to be quantitative and pass the half test?

Peismon: Yes, I think so.

Mathitis: I agree that counts don't pass the half test.

Socrates: That suggests that Continuous as a classification is below the Interval and Ratio or Nominal–Ordinal–Interval–Ratio taxonomic level. But these do not distinguish for all types of scales.

Agrafos: Why not?

Socrates: Because if counts meet all the criteria for Ratio scale but are not continuous, we have two choices. We can have Ratio–Continuous or Ratio–Noncontinuous (or Discrete, if we mean not continuous) scales. Although counts are infinite in number.

Peismon: If we don't accept that?

Socrates: Counts are clearly Quantitative. If they are not Interval or Ratio, then there must be a third option.

Mathitis: We could name it Discrete Ratio and others Continuous Interval and Continuous Ratio.

Kokinos: What about the infinite number of shades of gray? That scale passed the half test.

Mathitis: We have infinite counts that are quantitative but don't pass the half test and we have infinite gray shades that are not quantitative but do pass the half test.

Socrates: We need to distinguish not only between finite and infinite but also different kinds of infinite.

Agrafos: There are kinds of infinite?

Socrates: There are countable infinites—like counts. But there are also uncountable infinites like the real numbers. We consider a length scale as the last: uncountable infinite.

Mathitis: Then uncountable infinites pass the half test. That's why the shades of gray pass.

Kokinos: But they don't have to be quantitative.

Socrates: Then, instead of saying discrete versus continuous, we should be saying finite versus countable infinite versus uncountable infinite.

Kokinos: That helps me understand why we had counts as discrete when they are quantitative.

Socrates: The taxonomy looks like in Table 1.2 where the highest level is Qualitative versus Quantitative.

Mathitis: I guess the important thing is to note that attribute, discrete, and continuous attributes do not distinguish between scales until after we have classified them as Quantitative–Qualitative and Nominal–Ordinal–Interval–Ratio and further distinguished between finite, countable infinite, and uncountable infinite.

Kokinos: And, that measure applies to all scales.

Agrafos: So do gauges—whether external or internal. Technology can replace internal ones with external ones.

TABLE 1.2

Scale Taxonomy with Examples[a]

Qualitative			
Nominal		**Ordinal**	
Finite/Countable Infinite	**Uncountable Infinite**	**Finite/Countable Infinite**	**Uncountable Infinite**
Pass/fail	Colors Grayness	Ranks Ratings	
Quantitative			
Interval		**Ratio**	
Finite/Countable Infinite	**Uncountable Infinite**	**Finite/Countable Infinite**	**Uncountable Infinite**
	Centigrade Fahrenheit	Counts	Length

[a] Uncountable infinite means it passes the half test—between any two choices, there is another, so there are no "gaps" in the scale. Finite or countable infinite means it does not pass the half test.

PURPOSE OF DATA CLASSIFICATION

Socrates: Why do we even need to know what type of data or scale we have?

Mathitis: It determines the type of statistical analysis you will use.

Socrates: Is discrete versus continuous or attribute versus measure a distinction that helps us determine what analysis is applicable?

Peismon: Yes, you can only do certain analyses if the data are continuous, not discrete.

Socrates: For example?

Peismon: Regression can only be done with continuous data, not on discrete data.

Kokinos: We've done analyses on counts, which we originally classified as discrete.

Mathitis: Even with a nominal scale: logistic regression converts the data into proportions.

Socrates: Then, isn't the issue whether it is quantitative rather than continuous?

Mathitis: Yes, of course. Regression produces an equation, say Y is a function of X. We need both Y and X to be quantitative.

Socrates: Whether we have an internal gauge or external one, if the scale is quantitative, we can do the same analyses as if it were continuous. So, measure versus attribute is not the issue either.

Mathitis: I agree. Only quantitative versus qualitative determines the appropriate statistical analyses.

Socrates: But what if there are only two or three choices when Quantitative? Would you do regression analysis on a quantitative scale that has only the choices {0, 1} or {0, 1, 2}?

Mathitis: Then it is a combination of Quantitative and degree of "discreteness" or number of choices. But not, as we thought at first, discrete versus continuous. It can be finite, countably finite, or uncountably infinite as long as it is quantitative.

Agrafos: I like that distinction better.

2

Myth 2: Proportions and Percentages Are Discrete Data

Scene: *Classroom discussion on data types*

BACKGROUND

Agrafos: Where do proportions and percentages belong in the taxonomy table (Table 1.2)?

Socrates: Earlier, you said that your mentor recommended changing your measure from qualitative to quantitative by using counts.

Agrafos: But we saw that doing that changed what we were measuring.

Kokinos: However, that is the same thing we do with binary logistic regression. We form groups and then calculate a proportion passing for each group. Then, compare those proportions across time or some other quantitative factor.

Socrates: Exactly. Now, initially we said that counts, proportions, and percentages were not continuous and that, therefore, they would have the same problem of not being as informative as possible. But if quantitative scales are more informative, then aren't counts, proportions, and percentages just as informative?

Mathitis: I see what you are saying. The issue is whether the data are quantitative or qualitative.

Peismon: But counts, percentages, and proportions scales are not as informative as continuous scales because of the half test rule. They fail it.

Socrates: If regression involves two or more quantitative factors and we classify counts, percentages, and proportions as attribute factors, then can we do binary logistics regression?

Kokinos: Not using the original table (Table 1.1) of data classification. But now that we have reclassified these scales, we can.

Peismon: But counts, proportions, and percentages are discrete. That's why I disagree with the way you classified these scales.

Socrates: I agree that counts do not pass the half test even though there is an infinite number of counts or integers. But proportions and percentages can.

Peismon: No, there are not an infinite number of choices.

Socrates: Let's use the taxonomy we created. Are proportions and percentages quantitative?

Kokinos: Yes, of course.

Socrates: Do they have an absolute zero? Does zero proportion or percent mean lack of?

Kokinos: Yes.

Socrates: Can we add, subtract, multiply, and divide proportions and percentages? Are they not therefore ratio data?

Mathitis: Yes, you can do all that.

Peismon: Also with counts, you can add, subtract, multiply, and divide if it produces another whole number. But they are not continuous or uncountably infinite.

Socrates: That was why I asked whether the categories nominal through ratio covered all cases. We saw they didn't because we could have discrete (not pass the half test) and continuous (pass the half test) scales for each of these. The key difference was not discrete versus continuous but finite and countably infinite versus uncountably infinite.

Mathitis: But we agreed that counts were discrete types of ratio scales even though they are countably infinite.

Socrates: Yes, but are proportions and percentages always discrete ratio, i.e., finite or countably infinite as Peismon says?

DENOMINATOR FOR PROPORTIONS AND PERCENTAGES

Peismon: But the key issue for proportions and percentages is that there are a finite number of choices—not infinite, whether countable or not. Therefore, they don't pass the half test.

Kokinos: Good point, Peismon. For example, I take a sample of 30 meals delivered at a restaurant. I count how many of them are good and calculate the percent that are good. That percent or proportion good is discrete.

Socrates: Why?

Mathitis: Well, since there are only 30 meals, then the proportions can only be 0/30, 1/30, 2/30, and so on. The scale has a finite number of proportions.

Socrates: Is 10/30 twice 5/30?

Mathitis: Of course it is.

Socrates: Is 0/30 an absolute zero—does 0/30 mean lack of meals?

Mathitis: Yes. I agree with what you are saying. So we classify them as Quantitative and Ratio, but Finite/Countably Infinite (Table 1.1). But there must be an uncountably infinite number of choices to be continuous. That's what we said was the mathematical definition.

Kokinos: Even if we use counts, there are only a finite number of meals: 0 to 30.

Peismon: They are not infinite and so not continuous.

Socrates: Good. I agree. My question is "Are all proportions and percentages discrete?" Are there no proportions or percentages that meet the definition of continuous?

Peismon: No. By definition, it is dividing one quantity by another. They must be finite.

Socrates: Is miles per hour continuous?

Peismon: Yes.

Socrates: That's a proportion achieved by dividing two continuous measures.

Peismon: But those are different scales.

Socrates: Still, is it continuous?

Mathitis: Yes it is. I'm beginning to change my mind.

Socrates: Why is length continuous?

Peismon: Because between any two centimeters there is another. It passes the half test. There are an infinite number of choices. Uncountably infinite.

Socrates: We agree that there are an infinite number of values from, say, 0 cm to 1 cm. Let's use 1 cm as the denominator. If we express each of those infinite numbers between 0 and 1 cm as a percentage of 1 cm, don't we have a continuous scale? They meet the definition of ratio scale and there are an infinite number of choices.

Peismon: But now they are in percentages.

Socrates: Either way. But let's just keep them as proportions. Each number between 0 and 1 cm when expressed as a proportion of 1 m is exactly the same number.

Agrafos: Yes, because 1 cm means one-hundredth of a meter.

Kokinos: So, if centimeters are continuous, then they are continuous when the centimeter unit is removed and expressed as a unitless proportion or as a percent.

Socrates: Have we found cases of proportions that are continuous?

Peismon: I'm still not convinced. That's what I was taught, Socrates, proportions and percentages are discrete. Take the sample of 30 meals. Count how many are hot or entrees or some other characteristic. Then, calculate the percent with that characteristic. You have a finite number of possible results.

Socrates: Isn't that because the original table you showed us (Table 1.1) had misleading examples? It shows proportions as discrete and attribute and categorical when they are quantitative ratio data by the definition. As you said, they are one number divided by another. If they are ratio and pass the half test, then they are continuous. There is a contradiction between the listing and the definitions.

Mathitis: At the very least they are quantitative, which means they don't belong in the first column of Table 1.1.

Peismon: But that's my point—they do not pass the half test.

Socrates: Consider this. There are an infinite number of portions of a race one can complete. These would be continuous proportions since we're dividing by a continuous value.

Mathitis: Or, another example is an eclipse or phases of the moon. Does the moon go from full to new in jumps or is it continuous?

Kokinos: Continuous. Then, the proportion of area that is dark is a continuous scale.

Mathitis: Proportions and percentages are continuous when the denominator is continuous and discrete when the denominator is discrete. I see that now. Proportions depend on the denominator. If the denominator is discrete, limiting the number of proportions possible, then proportions are discrete. If the denominator is continuous, then the proportion is continuous.

Kokinos: The same for percentages.

Peismon: All the examples we see with real data, we have discrete proportions and percentages.

PROBABILITIES

Socrates: Let's try one more way—especially since we are discussing statistics. Agrafos, draw a normal distribution.

Agrafos: (draws the curve [Figure 2.1a])

Socrates: How much is the area under the entire curve?

Agrafos: 100%.

Socrates: Now, draw a vertical line anywhere dividing the curve into two sections. They don't have to be equal.

Agrafos: (draws the line [Figure 2.1b])

Socrates: How much area is to the left of the line?

Agrafos: I don't know, but it isn't 100%.

Socrates: Is the normal distribution a probability distribution?

Agrafos: Yes it is.

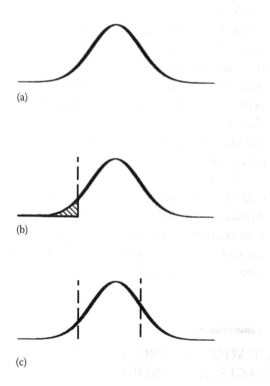

FIGURE 2.1
Normal continuous probability distribution showing that probabilities, proportions, and percentages can be continuous scales. (a) Normal probability distribution. (b) Probability = shaded area. (c) Normal probability distribution and half test.

Socrates: So, what does that area to the left represent?

Agrafos: It represents the probability of getting a result to the left of the line.

Socrates: Is the normal distribution a continuous or discrete probability distribution?

Agrafos: Continuous.

Socrates: So, are those probabilities continuous or discrete?

Agrafos: Continuous, of course.

Socrates: Draw another line anywhere. Can I draw a third line between them (Figure 2.1c)?

Agrafos: Yes. I see—it passes the half test. You can draw a fourth line between the third and the first, and so on, as we did with shades of gray. There are an infinite number of probabilities.

Kokinos: Is this why bar charts for discrete data have the bars separated, not touching. But a histogram, which is a bar chart for continuous data, has the bars touching each other?

Mathitis: That's right.

Socrates: Mathitis said that all probabilities are proportions and percentages, and Peismon said all proportions and percentages are discrete and discrete data do not pass the half test. Therefore, all probabilities are discrete—no probability scale passes the half test.

Agrafos: Why would they be called continuous probabilities if they were discrete?

Kokinos: I didn't fully realize that probabilities are just proportions and percentages.

Mathitis: Nor I. I think I was so focused on each topic that I never realized that I was accepting contradictory statements. Of course, it makes sense that a continuous probability distribution would have continuous probabilities and so it can't be discrete. And, therefore, not all proportions and percentages are discrete—some are, some are not.

CLASSIFICATION OF PROPORTIONS, PERCENTAGES, AND PROBABILITIES

Kokinos: Then proportions and percentages, and also probabilities, fit in the taxonomy as Quantitative–Ratio, and either Discrete or Continuous.

Mathitis: Or, more accurately, they are Quantitative–Ratio–Finite/Countably Infinite.

Socrates: Recall the purpose of classifying scales and the resulting data.

Agrafos: It was to determine the kind of analyses that can be used.

Kokinos: Given that they are quantitative, we know we can use quantitative analytical tools.

Agrafos: That's why, for example, binary logistic regression works.

Kokinos: But we have changed the measure from, say, pass–fail, to counts, proportions, or percentages of pass or fail.

Socrates: If we rely on the categories discrete and continuous, can we say that all and only data from continuous scales can be analyzed with quantitative tools?

Mathitis: Initially, I would have said yes. But since discrete scales can be either quantitative or qualitative, then no.

Socrates: What about the dichotomy attribute versus measurement?

Kokinos: If counts, proportions, and percentages are attribute, no for the same reason.

Agrafos: By starting at the first taxonomic level of Quantitative versus Qualitative scales, we don't have that problem and confusion.

3

$$Myth\ 3: s = \sqrt{\left[\sum_{1}^{n}(x_i - \bar{x})^2/(n-1)\right]}$$

Is the Correct Formula for Sample Standard Deviation

Scene: *Break room (after-class lesson on variation)*

BACKGROUND

Socrates: I was thinking about the material we covered in class the other day on statistical analyses. The instructor described the difference between descriptive and inferential statistics.

Mathitis: Yes, there is a difference. For example, statistics calculated for sports events, like a hitter's batting average or a basketball player's free throw percentage, are descriptive statistics. These statistics only summarize what actually happened. They are the truth.

Socrates: What about inferential statistics? We discussed two kinds the other day: confidence intervals and hypothesis testing.

Mathitis: In the case of descriptive statistics, you are calculating statistics on the population. These are the exact values or descriptions of the population. Inferential statistics is used when you take a sample and make an inference to the population.

Socrates: In that case, you don't have all the data, so you infer from the sample value what the population value might be.

Mathitis: That's right. In descriptive statistics, we have a known population—you have all the data or facts. No inference is required. You count or enumerate. Unless you count or do arithmetic wrong, you can't make a mistake. But for inferential statistics, you have a sample, so there is a chance of being wrong. Confidence intervals estimate the population value with a margin of error. Hypothesis testing tests whether the population value is a certain amount with a probability of being wrong.

Socrates: When calculating the population standard deviation, we are calculating the true value according to the definition.

Mathitis: Yes.

Socrates: What is the formula for calculating the population standard deviation?

Mathitis: It depends on whether the population represents a continuous random variable or a discrete random variable.

Socrates: But isn't there a common definition that makes them both standard deviations?

Mathitis: Yes, that definition is the square root of the variance. The variance is defined as the average squared deviations from the population mean. With a continuous random variable, you use calculus to integrate across the entire range of possible values.

Socrates: For discrete random variables?

Mathitis: Then, you can calculate all the squared differences and divide by the population size N. The formula where μ is the population mean and x_is are the individual population values is

$$\sigma^2 = \sum_{1}^{N} (x_i - \mu)^2 / N \qquad (3.1)$$

Socrates: Our instructor said that the formula for the sample standard deviation is the same as for the population standard deviation except he replaced N with $n - 1$, where n is the sample size, and replace μ with the sample mean $\bar{x}$. We have this formula for the sample variance:

$$s^2 = \sum_{1}^{N} (x_i - \bar{x})^2 / (n-1) \qquad (3.2)$$

And then we take the square root to get the sample standard deviation:

$$s = \sqrt{\left[\sum_{1}^{n}(x_i - \bar{x})^2 / (n-1)\right]} \qquad (3.3)$$

Mathitis: That's right.

Peismon: That's the correct formula for the sample standard deviation.

Socrates: But that can't be the standard deviation of the sample according to the definition Mathitis gave.

Peismon: Why not?

Socrates: Because it is not the square root of the average. The average requires dividing by n, not $n - 1$.

Peismon: If you use n, then you are treating it like the population.

Socrates: Yes, because that's the definition.

Peismon: But it is a sample not a population. Each has its own formula.

Socrates: Are you saying that we are not calculating the standard deviation of the population but of the sample?

Mathitis: Look at it this way. We are using the sample standard deviation to estimate the population standard deviation σ.

Socrates: When we use $n - 1$, is it to infer what the population standard deviation is?

Peismon: That's right. Otherwise, the data wouldn't be a sample but the population.

Socrates: Then we aren't in fact calculating the standard deviation of the sample—that would use n, not $n - 1$. But when we are calculating the sample standard deviation to infer what the population standard deviation is, we are estimating.

Mathitis: Yes, we use it as a point estimate or in a confidence interval estimate.

CORRECTNESS OF ESTIMATIONS

Socrates: But estimation formulas can't be correct or incorrect.

Mathitis: Why not?

Socrates: There are certain characteristics that do not apply to estimates just as there are certain characteristics that do not apply to bicycles or sounds or people.

Kokinos: I don't understand. What characteristics are you talking about?

Socrates: Does color apply to sounds? Would it make sense to ask what color was that sound?

Kokinos: No.

Socrates: Does it make sense to ask how tall an idea is?

Kokinos: No—although we might say "that's a big idea."

Peismon: How does that apply to what we are talking about?

Socrates: If correctness applies to estimates, then what does it mean to be correct?

Peismon: That it provides the right answer, the correct answer.

Socrates: Consider the equation for calculating the circumference of a circle.

Peismon: Yes, it's pi times the diameter.

Socrates: Are you saying that is the correct equation?

Peismon: Yes.

Socrates: Is it a loud or quiet equation? Is it a tall or short? Is it a green or blue?

Peismon: No.

Mathitis: I see. Equations don't have those features, so it doesn't make sense to ask about them.

Socrates: That's my point. Not all characteristics apply to all things.

Mathitis: Obviously—now.

Socrates: What is the correct formula for estimating someone's age?

Mathitis: There are different ways of estimating. I think I see what you are saying. If there are different ways to estimate, then there is no correct way to estimate. Is that it?

Socrates: In part. What's the correct way to estimate the distance to a planet or a star, or the length of room, or the size of building?

Kokinos: Except for the distance of a planet or a star, we can get the right measure.

Socrates: But if we are estimating, what is the correct way to estimate? Have you ever estimated the length of a room, by counting how many paces from one end to the other?

Kokinos: Yes, almost everyone has.

Socrates: Was your answer correct—exactly correct?

Kokinos: No, because I was estimating.

Socrates: Is that the only way to estimate?

Kokinos: No.

Socrates: Would you then say it is not the correct way but it is one way?

Kokinos: I suppose so.

Peismon: How is this related to statistics?

Socrates: If you are estimating, isn't it because you either do not know the correct answer or you do not need to know the exact answer?

Mathitis: Sure. We don't know the population standard deviation. But we estimate it from a sample standard deviation. Just as in hypothesis testing, we don't know which hypothesis is true, we just accept or reject one based on the data.

Socrates: Are we estimating the circumference of a circle when we multiply the diameter by pi?

Mathitis: No.

Socrates: Does the circumference equation always give the correct answer?

Mathitis: As long as we have the correct diameter, yes.

Socrates: When we are talking about the data being all the population, then using *n* gives us the correct answer every time—because that's the definition.

Mathitis: Yes.

Socrates: Let's say we have a fair die with the numbers 1, 2, 3, 4, 5, and 6. They are the population. What is the standard deviation?

Mathitis: It's 1.708, to three decimal places.

Socrates: Will the sample standard deviation from 25 tosses equal 1.708?

Mathitis: Probably not.

Socrates: Then how is the formula correct if the value it produces is incorrect?

Agrafos: I don't see how it could be.

Socrates: What do we mean by correctness when we are estimating?

Agrafos: I would imagine it means that the estimate gives the true value.

Socrates: What are we estimating with the sample standard deviation?

Agrafos: The population standard deviation.

Socrates: Does *s* using *n* − 1 always give the correct answer?

Mathitis: Every sample will produce a different result because of sampling error or variation. So, no, it is not correct in that sense.

Socrates: What other sense is there? What is the population mean for the fair die?

Mathitis: 3.5.

Socrates: Is this the same formula for the population mean?

Mathitis: Yes.

Socrates: The sample mean and the population mean are calculated the same way. Because the same formula is used, the issue of whether the sample formula is correct does not arise. But does it apply? In what sense is it correct?

Mathitis: Maybe because it gives the right result.

Socrates: Do you think the sample mean formula will always give the correct mean, of 3.5?

Mathitis: No.

Socrates: If my sample size n is an odd number, I will never get an average of 3.5.

Mathitis: Why not?

Socrates: The sum of any amount of the numbers 1 through 6 will always be an integer, right?

Mathitis: Yes, if I add whole numbers, I get whole numbers.

Socrates: The average is the sum divided by n. Let's say sum/n = 3.5. Then, the sum equals 3.5 times n. If n is odd, say, 3, will 3.5 times an odd number ever be a whole number?

Mathitis: No, because of the half: 3.5 times 3 is 10.5. We need n to be an even number.

Socrates: More than that.

Mathitis: Why?

Socrates: To be an even number means it is divisible by 2. If n is even, we can write it as $2k$, where k is an integer. Then, the average is the sum divided by $2k$. For that to equal 3.5, the sum must equal 3.5 times $2k$. What is that?

Mathitis: That's $7k$. I see. Not only must n be even but also the sum must be a multiple of seven.

Socrates: Even the equation for the sample mean will not produce the correct answer or the population mean most of the time. What can we mean when we say it is the correct equation?

Mathitis: If I take a larger and larger sample, then the calculation gets closer and closer to the population mean. But I can see that that won't help because it won't guarantee the correct result until I take the entire population.

Socrates: Is there another way that it is correct?

Mathitis: I can't think of any.

ESTIMATORS AND ESTIMATES

Socrates: When calculating the population standard deviation, the formula always produces results that are correct—barring any calculation error. Just as the formula for the circumference always produces results that are correct.

Mathitis: That's right.

Socrates: We say it is correct because the results are always correct.

Mathitis: Yes.

Socrates: For estimation, let's make a distinction between the formula and the result of using the formula with some specific data. Let's call the formula an estimator and the resulting number an estimate. Is the population standard deviation a number?

Mathitis: Yes, for each population of numbers, its standard deviation is also a number.

Socrates: Is the formula a number?

Agrafos: It produces a number, but it is not itself a number.

Mathitis: So, the equation is correct if it produces numbers that equal the number it is estimating.

Socrates: Using the estimator with $n − 1$, does the sample standard deviation equation always produce a correct estimate?

Mathitis: No, rarely if ever.

Socrates: Then why do we want to say it is the "correct" formula?

Mathitis: Ok, so correctness does not apply to estimation formulas or estimators.

Socrates: Can we say a particular sample standard deviation is correct? That is, can a particular sample standard deviation equal the population standard deviation?

Kokinos: Yes, it could coincidentally.

Socrates: Particular estimates can be correct, but the formula, the estimator, cannot be correct.

Mathitis: Each sample produces a different value and most will not equal the population standard deviation.

Kokinos: Unlike the equation for the circumference of a circle, where regardless of the circle and diameter, it always gives the correct answer.

Mathitis: Unless we come up with a different definition or meaning of *correct*, we cannot say estimators are correct, although particular estimates are. But we won't know which ones.

PROPERTIES OF ESTIMATORS

Socrates: If correctness doesn't apply to estimators, what characteristics or properties do apply?

Mathitis: I'm not sure.

Socrates: What would we like an estimator to do given that it won't give the correct answer?

Mathitis: We'd like it to produce approximately correct results.

Socrates: How might we determine that?

Mathitis: I'm not sure.

Socrates: You already mentioned one way.

Mathitis: You mean that if I take a larger and larger sample, it becomes more accurate.

Socrates: Yes, and you just said another characteristic.

Mathitis: Accuracy.

Socrates: But not correctness.

Mathitis: Accuracy means unbiased. We want estimators that are unbiased.

Socrates: You mentioned sampling error.

Mathitis: Small sampling error; small variation or precision would be another. This is just the way we evaluate measurement systems.

Socrates: Doesn't that make sense? Isn't a measurement system just a way of estimating?

Mathitis: I hadn't thought of that, but yes. We never say all measurements are correct. Measurements have error. Yes, it does make sense.

Socrates: Do we necessarily want an estimator that is accurate, or unbiased?

Mathitis: Yes, of course, why not?

Socrates: Supposing the true value is 10 and I estimate it to be 9.2. Is that difference attributed to inaccuracy or imprecision?

Mathitis: I don't know with only one value. I need more to know how much variation there is.

Socrates: We could combine accuracy and imprecision and just say we want the difference—perhaps the average difference—between estimates and the true value to be small.

Mathitis: We know the sum of the differences between each value and the mean is zero.

Agrafos: Why?

Mathitis: It's easy to prove. The sum of the differences is $\sum(x_i - \bar{x}) = \sum x_i - \sum \bar{x}$. But these two sums are equal: $\sum x_i = n\bar{x}$ and $\sum \bar{x} = n\bar{x}$.

Socrates: We can always use absolute difference to get around that.

Mathitis: That's true. How does that eliminate the need for accuracy?

Socrates: That difference could be the same amount because (1) the precision is small enough to compensate for the bias or (2) the lack of bias compensates for the imprecision.

Mathitis: I see we could have either of these two scenarios (Figure 3.1). The curve to the left is centered, so no bias. The curve to the right is not centered but has less variation.

Kokinos: Look at this. I've done an Internet search on estimators. These are various properties or characteristics of estimators. Unbiasedness, precision... those we talked about... minimum variance, consistency, efficiency, maximum likelihood.... I'm not sure what those are.

Mathitis: Yes, I see... I don't see correctness on the list.

Socrates: Estimators produce estimates. We can ask whether one estimator is better in some way than another, but not whether it is correct.

Mathitis: We didn't study any of that in my Six Sigma courses.

Kokinos: So, instead of correctness, what do we say?

Socrates: We ask which is better. To find estimators that are better requires defining *better*.

Mathitis: I think I get it.

FIGURE 3.1
Distribution with bias and less variation versus distribution with no bias and more variation.

Agrafos: It seems to me that statistics, at least inferential statistics, is different from other types of mathematics. Unfortunately, we weren't taught that difference.

Kokinos: I wasn't. How could I or others know? Instead, we learn the wrong things.

Socrates: In other branches of mathematics, there are correct formulas.

Mathitis: But we also have correct formulas in statistics. There is a correct formula for the normal probability distribution or for any probability distribution.

Socrates: That's true. But we are not making inferences when talking about that formula. Those formulas merely describe mathematically the bell-shape curve. When we make inferences, we use inferential statistics, which is the discipline of estimation. We could be estimating the distribution of the population, the mean of a population, the relationship between two factors. But we are estimating—even when testing hypotheses. There are no correct formulas for estimating; only formulas that have more or less desirable characteristics than other formulas. Or, we could use a formula that better meets our purpose for estimating.

Kokinos: How do we choose such formulas?

Peismon: That's what I thought was why we used these formulas.

Socrates: For example, do you need to know the exact width of your car to determine whether it fits in your garage given that you have things stored in the garage?

Kokinos: No. I could stretch a rope from one side to the other and then lay it inside the space in the garage. So, comparisons can be made without knowing the exact or even approximate sizes.

Mathitis: We should make this clear in our classes and when mentoring others. We should say inferential statistics is about estimating. Estimates have error in them and we can't know how much it is.

Kokinos: I was taught that there are sampling errors. But I still understood that there were correct formulas for the population standard deviation and for the sample standard deviation. Something else needs to be said than just estimates have errors. People understand that.

Socrates: I agree. Perhaps we should teach more about the characteristics of good estimators.

4

Myth 4: Sample Standard Deviation

$$s = \sqrt{\left[\sum_{1}^{n} (x_i - \bar{x})^2 / (n-1) \right]} \; \text{Is Unbiased}$$

Scene: *Break room (continuation of previous discussion)*

BACKGROUND

Agrafos: If the formula for the standard deviation s is not correct, why do we use $n - 1$ when the sample size is n? Why don't we use n when calculating the average?

Kokinos: Because that's the formula. You don't need to take my word, look at all the statistical software. There are two standard deviations, one for the population and one for the sample.

Socrates: Yes, but the question isn't *what* is the formula but *why* that formula.

Peismon: The reason is because that is the definition of the sample standard deviation.

Socrates: There is no definition of a sample standard deviation. There is a definition of a standard deviation.

DEGREES OF FREEDOM

Peismon: The reason has to do with degrees of freedom.

Agrafos: What is that?

Peismon: Since the differences are between each data and the mean, you need to calculate the mean. That uses one degree of freedom.

Agrafos: What is a degree of freedom?

Kokinos: The easiest way to explain it is the example used in my class. Suppose we have 10 chairs and 10 people. The first person can sit wherever he or she wants. The next person is restricted by one chair but still has choices, the next has fewer choices, and so on until there is one person and one chair left. The last person has no choice—no freedom—and must sit in the remaining chair. Thus, there are 9 degrees of freedom.

Peismon: When we said that the sum of the differences was zero, we can do the same thing. If I have a sample of 10 and calculate the mean and then calculate the differences, I can stop after nine differences are calculated—thus 9 degrees of freedom. Since the sum has to be zero, the last difference is determined.

Agrafos: I don't see the connection.

Peismon: You can't use the data to calculate the mean and also to calculate the differences. You are using an extra degree of freedom.

Kokinos: That would be cheating.

Socrates: What was the cheating? Whether you calculate 10 differences directly from the mean or calculate 9 and determine the last to get a sum of zero, you are still calculating 10 differences.

Peismon: I was taught that you can't use the sample mean twice.

Socrates: I don't know what sense of *can't* you are using because we do use it. But let's look at the formula for the variance.

Peismon: It isn't that you aren't using it. By *can't*, I mean that there is a cost if you use it again. The correct variance is based on using the population mean. We don't know what it is, so we guess what it is. Our best guess is the sample mean. But since that is not exact, then we adjust the variance calculation by dividing by $n - 1$. This gives us a better estimate.

Socrates: What if we don't use the sample mean?

Peismon: You have to since that's the definition. Remember that the equation for the variance uses the sample mean, $\bar{x}$.

Socrates: In class, the instructor showed us another formula because he wanted us to calculate it by hand for a sample he gave us. Do you remember it, Agrafos?

Agrafos: Yes, it's this and I don't see the sample mean $\bar{x}$:

$$\frac{\left[n \sum_{i=0}^{n} x_i^2 - \left(\sum_{i=0}^{n} x_i \right)^2 \right]}{[n(n-1)]} \tag{4.1}$$

Kokinos: Can you do that—rearrange the equation so it doesn't require calculating the sample mean?

Socrates: Clearly, we can do it since we just did. Does it make the argument for $n-1$ based on degrees of freedom invalid?

Kokinos: Why not?

Socrates: If using the mean twice determined the degrees of freedom but we can calculate the variance without using the mean even once, isn't the degrees of freedom n? Or, if we use each value twice as in the second equation: once to get the sum of the squares—the first summation, and again to get the square of the sums—the second summation, does that mean we have zero degrees of freedom?

t DISTRIBUTION

Peismon: But there is another reason why $n-1$ is used. It is the degrees of freedom used for the sample standard deviation in estimating or hypothesis testing. If you use the wrong standard deviation for the *t* distribution, you'll get the wrong answer for tests and confidence intervals because the *t* value will be wrong.

Socrates: What is the equation for the *t* value? Assume you are testing the mean μ of a population.

Peismon: If the null hypothesis is H$_\text{o}$: μ = some value, say 10, then you need the sample mean $\bar{x}$ and the sample standard deviation s:

$$t = \frac{\bar{x} - \mu}{s/\sqrt{n}} \tag{4.2}$$

Socrates: You are saying that if *s* is calculated using *n*, then the *t* value is incorrect.

Peismon: That's right.

Socrates: But this equation can also be rewritten.

Peismon: How?

Socrates: What is the denominator of *s*?

Peismon: Square root of *n* − 1.

Socrates: Let's label that standard deviation using *n* − 1 as s_{n-1} and label the standard deviation using *n* in the denominator s_n. Then, your equation with these labels is this:

$$t = \frac{\bar{x} - \mu}{s_{n-1}/\sqrt{n}} = \frac{\bar{x} - \mu}{\sum_1^n (x_i - \bar{x})^2 / \sqrt{(n-1)n}} \tag{4.3}$$

Peismon: That's right.

Socrates: By switching *n* and *n* − 1, the equation can be written as this:

$$t = \frac{\bar{x} - \mu}{s_n / \sqrt{n-1}} \tag{4.4}$$

Kokinos: That's a neat trick, Socrates! I like that better because it shows why the *t* distribution has *n* − 1 degrees of freedom—you can see the *n* − 1 in the equation.

Peismon: I think that's exactly what it is—a trick.

Socrates: In fact, we can separate the denominator from *s*, as we did before. We can write the equation using only the x_i's, the individual values of the sample—even for the sample mean. The denominator can be simply the square root of the sum of squares (SS):

$$\frac{\text{Sum of Squares}}{[n(n-1)]} = \frac{\text{SS}}{[n(n-1)]} = \frac{\left[n \sum_{i=0}^n x_i^2 - \left(\sum_{i=0}^n x_i \right)^2 \right]}{[n(n-1)]} \tag{4.5}$$

Kokinos: I guess that makes it irrelevant for this calculation whether you use *n* or *n* − 1.

Mathitis: In either case, it is the best guess of the population variance, and the square root of the population standard deviation.

Socrates: What does best guess mean?

Peismon: These formulas give us unbiased estimates of the population variance and standard deviation. Remember, that's what you said was one desirable characteristic of estimators.

Socrates: You are saying that using $n - 1$ gives us both an unbiased estimate of the variance and of the standard deviation?

Peismon: Yes. If one is unbiased, so is the other.

DEFINITION OF BIAS

Mathitis: The reason the sample variance equation uses $n - 1$ is to make it an unbiased estimate of the population variance. Since the square root of that is the standard deviation, then s is an unbiased estimate of the population standard deviation.

Agrafos: What does unbiased mean?

Mathitis: It means that the expected value equals the population value.

Agrafos: What does expected value mean?

Mathitis: Expected value is the average of a statistic for all possible samples of the same size. If I were to take all possible samples of size n, calculate the standard deviation for each using the same formula, and then calculate the mean of those standard deviations, it would be the expected value of the sample standard deviation.

Kokinos: If the expected value—the way you described it—equals the population value, then it is unbiased?

Mathitis: Yes. The sample mean is unbiased. If I take samples of size 4 or 40 or 400—it doesn't matter what size as long as all the samples are of the same size—and calculate the sample mean for each sample, then calculate the average of the averages, I will get the expected value of the sample mean. It turns out that it will equal the population mean. So, it is unbiased.

Kokinos: You don't need every sample mean to equal the population mean?

Mathitis: No. That wouldn't make sense. A random variable has variation called sampling error, so taking samples would not be using all the information from the population. We would not expect each sample mean to equal the population mean. The mean of a fair die is 3.5, the average of the numbers 1 through 6. If I toss the die twice, I might not get an average of 3.5. Yet, if I took all possible samples of size 2, calculate their means, and then the mean of those means, I will get 3.5. That's true for any sample size. If I calculate the mean of each possible combination of seven tosses of the die, the mean of those means will equal 3.5.

Socrates: Is that something you can show through simulation methods or is it a mathematical calculation, for example, a theorem?

Mathitis: In the die example, we can show directly since there are only 15 possible samples of size $n = 2$. For a larger population, I suppose you could check through Monte Carlo simulations. But it can be proven mathematically. I've seen it done for the mean.

Peismon: Great. We are saying that since the variance is unbiased using $n - 1$, then the standard deviation is also. That's why it is the right equation to use.

Socrates: Here's my concern. Taking the square root is not a linear transformation, like addition and subtraction are.

Peismon: So?

Socrates: Not all properties are preserved when the transformation is not linear. Bias, because it is an average, might not be preserved by taking the square root. If using $n - 1$ for variances makes it unbiased, then the square root transformation will not preserve that bias, and vice versa. What makes the standard deviation unbiased will not make the variance unbiased.

Peismon: But everyone, including my instructors, have said that $n - 1$ makes both unbiased.

Socrates: We can do a simple exercise to show that the square root transformation does not preserve unbiasedness.

Mathitis: How?

Socrates: Let's use a die and its six values. What is the average of the numbers 1 through 6?

Mathitis: 3.5.

Socrates: Now, square the numbers. What is the average of 1, 4, 9, 16, 25, and 36?

Mathitis: 15.167.

Socrates: If unbiasedness is preserved, then the square root of 15.167 should equal 3.5, the average of the unsquared numbers. Does it?

Mathitis: No, the square root of 15.167 is 3.89.

Socrates: If we took all possible samples of size *n* from a population and calculated both the standard deviation and the variance for each sample using *n* – 1, the square root of the average variance would not equal the average standard deviation. If the average variance using *n* – 1 is unbiased, then the average standard deviation using *n* – 1 would not be.

Mathitis: It makes sense with the example of the numbers 1–6. How did you know that?

REMOVING BIAS AND CONTROL CHARTS

Socrates: I knew that from my math classes. But it also occurred to me when we learned about control charts. The instructor made us calculate control limits using tables and formulas. Have you looked at the formulas for calculating control limits?

Mathitis: Yes, most software packages include the formulas and I suspect every book on control charts has some of the formulas. But before software packages, people used tables to determine the factors used to calculate the limits.

Socrates: In calculating the limits for the standard deviation, there is a factor called c_4. Do you know what that factor is?

Mathitis: Yes, it is a factor to adjust the average standard deviation for the standard deviation control chart or S chart.

Socrates: How is an S chart constructed?

Mathitis: All control charts are based on subgroups. For each subgroup, a statistic is calculated. The S chart is a control chart that monitors standard deviations. The standard deviation, using *n* – 1, is calculated for each subgroup. Those values are plotted on the chart in time sequence.

Socrates: How are the limits computed?

Mathitis: The average of the subgroup standard deviations, $\bar{s}$, is calculated and used as the center line for the chart. The control limits are determined by multiplying $\bar{s}$ by an appropriate factor. The upper control limit, UCL, is $\bar{s}$ times a factor B_4 and the lower control limit, LCL, is $\bar{s}$ times the factor B_3.

Socrates: What are the values of B_3 and B_4?

TABLE 4.1

Factors for S control chart limits[a]

Subgroup Size, n	B_3	B_4	c_4
2	–	3.267	0.7979
3	–	2.568	0.8862
4	–	2.266	0.9213
5	–	2.089	0.9400
6	0.030	1.970	0.9515
10	0.284	1.716	0.9727
25	0.565	1.435	0.9896

Note: Center line $= \bar{s}$, LCL $= \bar{s} \times B_3$, and UCL $= \bar{s} \times B_4$.

[a] A search for control chart constants will find numerous downloadable pdf tables, for example, http://onlinelibrary.wiley.com/doi/10.1002/0471721212.app9/pdf.

Mathitis: They depend on the subgroup size. Let's find a table on the Internet. Here is one (Table 4.1). You can see that if the subgroup size is less than 6, $B_3 = 0$, while B_4 varies from 3.267 when $n = 2$ to 1.435 when $n = 25$, and keeps getting smaller.

Socrates: Presumably, it will equal 1 when you have the entire population.

Mathitis: Yes. And B_3 keeps increasing and will equal 1 when the sample is the entire population.

Socrates: $B_3 = B_4$ when the sample size equals the population size.

Mathitis: Yes, because when you have the entire population, you don't need to estimate. You can calculate the standard deviation exactly.

Socrates: Why is $B_4 = 3.267$ when $n = 2$?

Mathitis: It's to determine the 3 sigma limits.

Socrates: Then, why isn't $B_3 = -3$ and $B_4 = +3$?

Mathitis: I'm not sure. But now I suppose it's because of the bias.

Socrates: Let's use $B_4 = 3.267$ when $n = 2$ as an example. What is the sampling error?

Mathitis: That would be the standard deviation divided by the square root of the sample size times a factor, C, representing the confidence level: $C \times s/\sqrt{n}$.

Socrates: To make the math easy, let's use $s = 1$.

Mathitis: Okay, so then, the UCL should be 3 times 1 divided by the square root of $n = 2$. That's 2.1213—which doesn't equal $B_4 = 3.267$.

Socrates: No, it doesn't. Now divide your result by c_4.

Mathitis: For $n = 2$, $c_4 = 0.7979$ and 2.1213/0.7979 = 3.267. That's B_4!

Socrates: That's why c_4 is used.

Mathitis: But doesn't this just remove the bias when you are taking an average standard deviation? Isn't each standard deviation unbiased and only the average is biased?

Socrates: The average is a linear transformation. If each component is unbiased, then the average is unbiased. But we can show that using your definition of unbiased.

Mathitis: How is that?

Socrates: You said that unbiased means the expected value equals the population value, or the average of all the sample values equals the population value.

Mathitis: That's correct.

Socrates: Then, using $E[\ ]$ to denote the expectation operation, $\bar{s}$ for the average of the sample standard deviations, and σ to denote the population standard deviation, $E[\bar{s}/c_4] = \sigma$.

Mathitis: Okay.

Socrates: But $\bar{s}/c_4 = ((s_1 + s_2 +...+ s_k)/k)/c_4$. Substituting, $\sigma = E[\bar{s}/c_4] = E[((s_1 + s_2 +...+ s_k)/k)/c_4] = E[\Sigma s_i/(k \times c_4)] = \Sigma E[s_i]/(k \times c_4)$. Since the expectation of each s_i is the same, then this reduces to $E[s_i]/c_4$. This means that each s_i is biased by the c_4 factor.

Mathitis: I see that c_4 is always less than 1, so the sample standard deviation using $n - 1$ always underestimates the population standard deviation. For small sample sizes, the bias can be large: it's 25% [1/0.7979] for $n = 2$ and more than 6% [1/0.9400] for $n = 5$.

Socrates: The bias depends on the population distribution. The c_4 factor used for control charts is based on a normal distribution. For other distributions, the bias would be different.

Mathitis: I didn't know that. So, when we don't have a normal distribution, then it is even worse.

Socrates: It could be.

Kokinos: How is c_4 calculated?

Socrates: Let's do a search. Here it is:

$$c_4 = \left(\frac{2}{n-1}\right)^{1/2} \Gamma\left(\frac{n}{2}\right) \Gamma\left(\frac{n-1}{2}\right) \qquad (4.6)$$

where $\Gamma()$ is the gamma function.

Agrafos: What's the gamma function?

Socrates: It's a complicated equation, but to give you an idea, $\Gamma(n) = n!$ and $\Gamma(1/2) = \text{sqrt}(\pi)$.

Kokinos: I guess I would rather just use the c_4 factor.

Agrafos: Why were we told that s using $n - 1$ is unbiased?

Peismon: I think it was to let people know that it was the best guess.

Kokinos: Since we weren't taught any other way of calculating s and software automatically use s with $n - 1$, it doesn't really make any difference.

Mathitis: I agree. When calculating confidence intervals or doing hypothesis testing, whether the standard deviation is biased or not is irrelevant. The same is true for control charts, as they account for the bias in determining the limits.

Socrates: Then, when would it be important to know whether it is biased or not?

Mathitis: If we are simply estimating the population standard deviation, then I would teach that the divisor $n - 1$ makes the variance unbiased but not the standard deviation. To remove the bias from the sample standard deviation formula, you need to use the c_4 factor. I would give an idea of how much bias based on sample size by showing how c_4 changes with n. But I think it is important also to note whether removing the bias is essential.

Kokinos: What do you mean?

Mathitis: I would explain—depending on the level of the class—that, for some uses, the bias is not an issue. In fact, except for a point estimate of the population standard deviation, the formulas seem to be self-correcting.

Agrafos: I don't understand the point about self-correcting.

Mathitis: When Socrates rearranged the formulas for the t test, he showed that it didn't matter whether you used n or $n - 1$ because both factors are included in the formula. You have to be sure that you adjusted the formula if you switched. But, if you use some other way of calculating the standard deviation, you need to adjust the formula accordingly.

Socrates: The important thing to note about the t distribution is that it is the ratio of the mean to the sum of squares. Those two factors contain the random variables, not the sample size n, or $n - 1$. That is why we can rewrite the equation using s_n or s_{n-1}.

5

Myth 5: Variances Can Be Added but Not Standard Deviations

Scene: *Project meeting room (team discussing MSA)*

BACKGROUND

Socrates: Mathitis, can you help us?

Mathitis: Sure, Socrates. What do you need help with?

Socrates: We need to check that my data are reliable. So, I need to do a measurement system analysis or MSA. Perhaps you can describe the MSA you did in your project.

Mathitis: We had continuous data, so we did a Gauge R&R.

Socrates: That's to identify the reproducibility and repeatability of the measurement system.

Mathitis: Yes. The Gauge R&R is an analysis of variance (ANOVA) that breaks down the total variability into various parts. It starts with the total variance of the data and breaks into two components: variance of the process and variance of the measurement system (MS):

$$\sigma^2_{\text{total}} = \sigma^2_{\text{process}} + \sigma^2_{\text{measurement system}} \tag{5.1}$$

Socrates: Why do you use variances and not standard deviations since they are in the same unit as your original measurement?

Peismon: The issue is in how these quantities are used. The variances are used to determine the contribution of each source of variation. The equation that decomposes the total variation into that from the process and that from the measurement is further decomposed into other components, like this:

$$\sigma^2_{\text{total}} = \sigma^2_{\text{parts}} + \sigma^2_{\text{repeatability}} + \sigma^2_{\text{reproducibility}} \tag{5.2}$$

Then, the contribution of each can be determined by representing each variance, σ^2, as a percentage of the total variance σ^2_{total}.

Kokinos: You can add variances but not standard deviations.

SUMS OF SQUARES AND SQUARE ROOTS: PYTHAGOREAN THEOREM

Socrates: What is one standard deviation plus two standard deviations?

Kokinos: Three standard deviations, of course.

Socrates: Isn't that adding standard deviations?

Kokinos: I was taught you couldn't add standard deviations.

Mathitis: Well, yes, you can add standard deviations in that way but not when separating different sources of variation.

Socrates: I don't understand. When can I add variances but not standard deviations?

Peismon: It's like the hypotenuse equation for right triangles. In fact, we use that as an example to explain variance decomposition. You remember the Pythagorean theorem for triangles?

Socrates: Yes.

Peismon: Suppose the hypotenuse is c and the other two sides are a and b. Then, $a + b$ does not equal c. However, $a^2 + b^2$ does equal c^2.

Socrates: Isn't the reverse true also?

Peismon: No.

Mathitis: What do you mean?

Socrates: If a plus b equals c, then $a^2 + b^2 \neq c^2$. For example, what's $3 + 4$?

Mathitis: Seven.

Socrates: What's $3^2 + 4^2$?

Mathitis: Twenty-five.

Kokinos: That's not 7^2.

Socrates: That's my point. Why don't we argue that while standard deviations can be added variances cannot? How do we know which is the case here? The problem with using the Pythagorean theorem to illustrate your point is that you are saying that squares can be added but not square roots. Are you saying that squares of numbers can be added but numbers cannot? We add numbers all the time—distances, costs, people, dollars—we don't add their squares.

Peismon: No, this is just to illustrate why we add variances. The advantage of using variances is that the sum of the components equals the total, so the sum of the percentages will equal 100%. That isn't true for the standard deviations.

Mathitis: They do not add up to 100%. My understanding is that it is because the standard deviations are derived from trigonometric functions (Wheeler 2006).

Socrates: I don't know if they are. But the reason the standard deviations expressed as percentages do not sum to 100% is simple. If $c^2 = a^2 + b^2$, then clearly a^2 and b^2 as percentages of c^2 will sum to 100%. Since $c \neq a + b$, then a and b as percentages of c will not sum to 100%.

Kokinos: Even I see that.

Socrates: However, if we start with $c = a + b$, we can show that the percentages of the parts will total 100%. But then, $c^2 \neq a^2 + b^2$ and the sum of these percentages will not total 100%. If $10 = 3 + 7$, then $100\% = 30\% + 70\%$. But after squaring each number, $100 \neq 9 + 49$, so $100\% \neq 9\% + 49\%$.

Mathitis: I hadn't thought of that.

Socrates: Consider the right triangle example. The Pythagorean theorem explains why squares of numbers can be added to produce an equality, and when that is true, then the square roots do not add. But if you had used the perimeter equation, you can show the opposite.

Kokinos: How is that?

Socrates: What is the perimeter of a triangle?

Kokinos: It's the sum of the sides. If a, b, and c are the lengths of the three sides of a right triangle, then the perimeter = $a + b + c$.

Mathitis: I see what you are saying. Then, it's false that (perimeter)^2 = $a^2 + b^2 + c^2$.

Kokinos: Then, it depends on what equation you start with.

Peismon: But we *do* start with the equation of the variances.

FUNCTIONS AND OPERATORS

Socrates: In reality, however, the examples I gave and those of the Pythagorean theorem use constants or scalars. But the variance is none of these. It's an operator.

Kokinos: What do you mean by an operator?

Socrates: By operator, I mean a function like a trigonometry function, for example, sine or tangent of x and not just x. In this case, we are talking about variance of x not x. With the Pythagorean theorem, we are referring to x and not a function of x. The same is true with the perimeter equation: it is an equation of a, b, and c, and not a function of a, b, and c.

Kokinos: How does that apply?

Socrates: Then we must look at properties of functions or operators.

Kokinos: What properties?

Socrates: One property is being additive. Is the function of $(a + b)$ equal to the function of a plus the function of b? Consider the absolute value operator, $|\ |$. Is $|a| + |-a| = |a + (-a)|$?

Kokinos: No. The left side is $2a$ while the right side is 0.

Mathitis: I see. The sine function is not additive: $\text{sine}(X) + \text{sine}(Y) \neq \text{sine}(X + Y)$. Similarly, the logarithm function is not additive: $\log(X) + \log(Y) \neq \log(X + Y)$.

Peismon: How does that apply to variances?

Socrates: Let's look at the Pythagorean theorem first. There is no function hypotenuse, say $\text{hyp}(x)$. The Pythagorean theorem is not $\text{hyp}(a + b) = \text{hyp}(a) + \text{hyp}(b)$. It is $a^2 + b^2$ equals c^2.

Mathitis: I see what you are saying. Variance is an operator. So, we aren't saying $x + y = z$. If we let $\text{Var}(x)$ be the operator, we are saying $\text{Var}(X) + \text{Var}(Y) = \text{Var}(X + Y)$. For MSA, we are not saying part + measurement = total, we are saying $\text{Var}(\text{part}) + \text{Var}(\text{measurement}) = \text{Var}(\text{total})$.

Socrates: Exactly. Those were the Equations 5.1 and 5.2 you wrote using σ_2 with subscripts rather than $\text{Var}(x)$. However, $\text{Var}(x)$ or variance of x is not the same as the square of x. So, using the Pythagorean theorem is not an appropriate analogy.

Mathitis: But we also know that the operator standard deviation is not additive: $\text{STDEV}(X) + \text{STDEV}(Y) \neq \text{STDEV}(X + Y)$.

Peismon: That's why we say you can add variances but not standard deviations. We are saying that Var(X) + Var(Y) = Var($X + Y$).

Mathitis: But *Socrates* is saying that using Sq(x) as the squaring operator, we know that Sq(3) + Sq(4) ≠ Sq(3 + 4). So, neither square root or square works.

RANDOM VARIABLES

Socrates: Let's look at that claim from another perspective. There is another discrepancy with using the Pythagorean theorem as an example. The sine function applies to degrees or radians only. It doesn't apply to lengths such as meters or weights. In other words, the question "What is the sine of 38 ounces?" doesn't make sense because the units to which the sine function applies are degrees or radians, not ounces. The Pythagorean theorem applies to lengths of the sides of triangles. It does not apply to the colors of the sides of a right triangle.

Peismon: I don't see the point.

Socrates: The operator Var() applies to random variables not to numbers.

Kokinos: Why doesn't it apply to numbers? The variances are numbers.

Socrates: The variance of a single number is what?

Kokinos: Oh, I see. It's zero.

Socrates: To do the ANOVA to decompose the sources of variation, we are assuming that the sources are random variables, not single numbers. Using the right triangle to illustrate this is not a good example since it doesn't use operators and doesn't use random variables.

Mathitis: Perhaps not, although people seem to understand it.

Socrates: It seems to me they are not understanding correctly.

Mathitis: Why do you say that?

Socrates: The point you are making is that the operator Var() is additive while the operator STDEV() is not. But we know neither is. That is also different from saying that the square of the hypotenuse of a right triangle equals the sum of the squares of the other two sides. In the right triangle example, you are not saying that an operator is additive, you are just providing an example of the summation of squares when discussing the sides of right triangles.

Kokinos: Maybe. But operators and their properties may be too difficult for most people to understand. They can understand the Pythagorean theorem.

Socrates: Are you saying that explaining with an irrelevant or wrong example that is "understood" is better than not explaining at all?

Kokinos: If they understand the example, then yes.

Socrates: If I don't think you can understand gravity I can explain it by saying it's a giant at the center of the earth inhaling and drawing you toward it. Do you understand what gravity is then?

Kokinos: No, but I get a sense for it.

Agrafos: I would not want an incorrect understanding even if it made sense. Aren't we scientists?

Kokinos: Good point, Agrafos.

Socrates: Do variances add because the squares of the shorter sides of a triangle add?

Mathitis: No, we use it as an example.

Socrates: So what is understood from the example?

Mathitis: They understand that you can add variances.

Socrates: Are you sure? The question is what do you want them to understand about the Pythagorean theorem? I showed you that you can add standard deviations and I showed you that squares do not add in the sense that the sum of the two numbers does not mean that the sum of the square of the numbers equals the square of the sum. The Pythagorean theorem doesn't explain why you add variances.

Kokinos: Maybe they don't need to know why.

INDEPENDENCE OF FACTORS

Socrates: Then, why even explain with the Pythagorean theorem? Let's go back to the equation. Are you saying that the Var() operator is always additive?

Kokinos: I'd say yes because that was what I learned, but I get the impression that you are going to tell me that it isn't true.

Socrates: The expectation operator, which is the mean, is additive: $E(X + Y) = E(X) + E(Y)$. Remember we talked about this when we discussed what it means to be unbiased.

Kokinos: It's the expectation or average of all possible samples.

Socrates: The Var() operator can be expressed as an average or expectation operator $E()$.

Kokinos: I didn't learn that.

Mathitis: We did. We can use the equation: $Var(X) = E(X^2) - [E(X)]^2$. In other words, the variance of a random variable X is the expectation of the random variable X^2 minus the square of the expectation of the random variable.

Kokinos: And $E(X)$ is just the mean?

Mathitis: That's right. So, we are saying the variance of X is the mean of X^2 minus the square of the mean of X.

Socrates: What does the right side of the equation, $Var(X + Y)$, equal?

Mathitis: That is the same as $E[(X + Y)^2] - [E(X + Y)]^2$.

Kokinos: What is that, after expanding?

Mathitis: $E(X^2 + 2XY + Y^2) - [E(X) + E(Y)]^2$. That's the same as $E(X^2) + E(2XY) + E(Y^2) - [E(X)^2 + 2 \times E(X)E(Y) + E(Y)^2]$.

Socrates: Now, rearrange the terms so the Xs are together and the Ys are together.

Mathitis: Like this: $[E(X^2) - E(X)^2] + [E(Y^2) - E(Y)^2] + 2 \times [E(XY) - E(X E(Y)]$.

Socrates: What is the first term in brackets?

Mathitis: Oh, I see. That is simply $Var(X)$. The next term in brackets is $Var(Y)$.

Kokinos: What is the third term? It's a combination of X and Y.

Mathitis: That's the covariance of X and Y. In other words, $Var(X + Y) = Var(X) + 2Covar(X,Y)2 \times Covar(X,Y) + Var(Y)$.

Socrates: But you were saying that $Var(X + Y) = Var(X) + Var(Y)$. What happened to $2 \times Covar(X,Y)$?

Mathitis: I get it. The variances add only when the covariance = 0; $Var(X + Y) = Var(X) + Var(Y)$ only when $2 \times Covar(X,Y) = 0$.

Socrates: When does that occur?

Mathitis: When X and Y are independent.

Socrates: Let's return to the hypotenuse equation. Why isn't the square of the long side of any triangle equal to the sum of the squares of the two shorter sides?

Mathitis: It has to be a right triangle.

Socrates: Do you remember the law of cosines? It states the relationship between any two sides of a triangle, the angle between them, and the third sides.

Mathitis: No.

Kokinos: Nor I.

Socrates: Assume we have any triangle with two short sides a and b and the longest side c. Then, the law of cosines says $c^2 = a^2 + b^2 - 2 \times a \times b \times \cos(ab$ angle$)$. Except for the $-2 \times a \times b \times \cos(ab$ angle$)$ term, it looks like the Pythagorean theorem formula.

Mathitis: For it to work for a right triangle, $-2 \times a \times b \times \cos(ab$ angle$)$ must equal zero. But a and b are not zero—otherwise there wouldn't be a triangle.

Kokinos: I'm guessing that the cosine of 90° is 0. When the ab angle is a right angle or 90°, we get the Pythagorean theorem instead of this equation.

Socrates: Exactly.

Mathitis: So, for the variances to add, that extra term must equal zero. The only way for the variance of the sum to equal exactly the sum of the variances is for the variables to be independent. The covariance term is not zero when the variables are correlated.

Socrates: So, do variances add?

Mathitis: Not necessarily. Even variances of random variables cannot "add" unless the random variables are uncorrelated.

Mathitis: That means for the ANOVA used in the MSA, we must assume that the factors are uncorrelated.

Kokinos: I wasn't taught that.

Mathitis: And the Pythagorean theorem did not reveal this to me.

Socrates: You see, *Kokinos*, that using an analogy but not explaining what is the same and what is different between the actual thing and the analogy may not lead to understanding.

Kokinos: I see that now. We haven't explained why variances add.

Agrafos: Nor when they add nor when and why they do not add.

Mathitis: Nor that variance is a mathematical operator, which is not what the Pythagorean theorem is about.

OTHER PROPERTIES

Socrates: We might want to know other things about the variance operator.

Kokinos: What else?

Socrates: For example, you should know that $\text{Var}(XY) \neq \text{Var}(X) \times \text{Var}(Y)$. If you have a nonlinear relationship for the random variables, that relationship may not hold for the variances.

Mathitis: But we don't typically have that situation.

Socrates: Perhaps not. But you may have a linear relationship that is slightly but critically different from $X + Y$ and that will occur more often.

Mathitis: For example?

Socrates: If you have a regression equation from historical data or develop an equation from a designed experiment, you will have a linear, additive relationship of the variables. But the coefficients will not all be the same or one.

Kokinos: How does that affect the additive property of the variance operator?

Socrates: Remember what the variance of a single number, say, a, is? What is $\text{Var}(a)$?

Kokinos: It's zero because there is no variability, it's a constant.

Socrates: Now, what is the variance of a constant times a random variable? What is $\text{Var}(a \times X)$?

Kokinos: Isn't $\text{Var}(a \times X) = a \times \text{Var}(X)$?

Mathitis: No, it is $a^2 \times \text{Var}(X)$.

Socrates: In other words, $\text{Var}(a \times X) \neq \text{Var}(a) \times \text{Var}(X)$ because we know that $\text{Var}(a) = 0$.

Kokinos: Okay, but how does that apply to a regression equation?

Socrates: If you have an equation of the form $Y = aX_1 + bX_2$, what is the variance of Y?

Kokinos: I see. It isn't $\text{Var}(a) \times \text{Var}(X_1) + \text{Var}(b) \times \text{Var}(X_2)$ if X_1 and X_2 are uncorrelated.

Mathitis: No, it is $\text{Var}(Y) = a^2 \times \text{Var}(X_1) + b^2 \times \text{Var}(X_2)$.

Kokinos: We were not taught this in class. Just simply that you can add variances.

Socrates: Then, for the MSA, what is the equation we use to apply ANOVA and decompose the total variability?

Mathitis: It has to be the $Y = \text{Process} + \text{Measurement}$, where process and measurement are random variables.

Kokinos: And they have to be uncorrelated or independent of each other.

Mathitis: The key is understanding the assumptions and conditions of the analyses.

6

Myth 6: Parts and Operators for an MSA Do Not Have to Be Randomly Selected

Scene: *Break room (after Measurement System Analysis lessons)*

BACKGROUND

Agrafos: Kokinos, can you help us with our measurement system analysis (MSA)?

Kokinos: Sure. What's the issue?

Agrafos: We've had problems with two parts not always fitting together. But we suspect that our gauge for measuring their diameters is suspect. So, we want to do an MSA.

Kokinos: That's great. You should always check your gauges before you start any improvement projects. It might solve all your problems.

Agrafos: I was telling Socrates that we should select the parts but he says we should take a random sample.

Kokinos: There are different criteria you will look at to evaluate your measurement system, so you want to pick parts that cover the expected range. That's what I was taught.

Socrates: Why do you pick the parts?

Kokinos: We do it to make sure we cover the range of parts produced by the process.

Agrafos: What analysis will we do?

Kokinos: The first analysis separates the sources of variation, starting with separating the process variation from the measurement system variation. Process variation is the variation of the parts. You need two or more parts to determine that source of variation.

Agrafos: And measurement variation? We need several operators to measure each part several times.

Kokinos: That's right. The several measurements from the same operator and from different operators provide the variation of the measurement system.

Mathitis: You can't evaluate purely the measurement system unless you have more than one measurement on the same part.

Socrates: Do you assume that the part does not change while being measured?

Kokinos: Yes, you have to. There may be some situations, for example, chemical reactions, where delays between measurements actually change the "true" value of the part. In those cases, there is typically a time limit in which all the measurements must be taken to be useful.

Peismon: Repeatability is the variation within each operator and reproducibility is the variation between operators.

Agrafos: Great! So, we need to make sure we select the parts that cover the production range to ensure we get a good estimate of the process variation.

Peismon: That's right. Select parts that are in spec and parts that are outside of specifications.

Agrafos: So, both good and bad parts.

Peismon: Yes.

Kokinos: I was told to select parts that cover the specification range.

Peismon: And also select your operators.

TYPES OF ANALYSES OF VARIANCE

Socrates: Are you sure it is appropriate to select parts and operators?

Peismon: Why not?

Socrates: Look at the equations you wrote (Equations 5.1 and 5.2). How are these variances determined?

Peismon: You use analysis of variance or ANOVA.

Agrafos: We used ANOVA to test whether the means of two or more groups were the same. How do we use it to analyze the variances?

Mathitis: ANOVA is used for both testing means and estimating variances. Remember when we talked about variances adding but not standard deviations?

Agrafos: Yes.

Mathitis: That addition is what ANOVA does or, in this case, separates the sum into its parts.

Peismon: ANOVA produces this equation:

$$\sigma^2_{total} = \sigma^2_{part} + \sigma^2_{repeatability} + \sigma^2_{operator} + \sigma^2_{operator \times part\ interaction} \qquad (6.1)$$

The total variation is the sum of the variation from the parts and the measurement system (Equation 5.1). The MS variation is decomposed into more sources: repeatability and reproducibility (Equation 5.2), and reproducibility is decomposed into operator and the interaction of operator and part, as shown in this Equation 6.1.

Socrates: Are we interested in only the parts that we sample?

Agrafos: No. We are interested in all the parts, perhaps even those we haven't produced.

Socrates: How are you proposing getting the parts?

Kokinos: We will select them as we said to cover the range and mix of good and bad.

Socrates: Remember when we talked about adding variances?

Kokinos: Yes.

Socrates: It applied to random variables. How much is the variance of a constant?

Mathitis: It's zero. I see. If we assume that parts are fixed, then it has zero variance.

Socrates: You can only calculate the variance components for a random variable.* Using Equation 5.1 where the total variance equals the variance of the process or parts plus the variance of the measurement system, we see that we want to estimate these

---

* There are three types of ANOVA: fixed-effects, random-effects, and mixed-effects. The type of analysis depends on the factors. A factor is either a random variable or a constant. Fixed-effects means all the factors are fixed in terms of the choices for each factor. They are fixed because all the choices make up the population so the factor is constant. A random-effect factor is a random variable and a sample from a population is collected. If all the factors are random-effects, then the analysis is random-effects. Mixed-effects ANOVA means that some factors are fixed and others are random.

variances. If the parts are fixed, then there is no variance component to estimate.

Mathitis: Then, to have variances for parts and operators, they must be random variables.

Agrafos: Doesn't that mean we should also select operators randomly?

Socrates: True.

Mathitis: Even though we don't.

Kokinos: But we shouldn't deliberately select the parts.

Socrates: Nor the operators.

Kokinos: What if we only have two or three operators? They always do the measuring.

Socrates: In a strict sense, then operators are a fixed effect and there is no variance of operators.

Peismon: I still don't see why. We've done this many times and it works. We pick the parts and we pick the operators. Why does it seem to work when we select parts—and when we have no choice on the operators?

Socrates: By "works," you mean you get numbers when you do the calculations?

Mathitis: It could be you just think it works, but it's only the proverbial garbage in–garbage out.

Socrates: You can convert colors into numbers and calculate a standard deviation, but what does it tell you? You enter numbers into a software program and it produces results. What do the results mean if the numbers aren't appropriate?

Peismon: Yes, you get results but also our decision on whether the measurement system is okay or not seems to be correct.

Socrates: Now, remember that ANOVA is a statistical model used to ostensibly give us meaningful information. Even with violations of assumptions, it can provide us meaningful and useful answers. However, we need to understand what it is telling us when we violate the assumptions. We need to know how to interpret those results.

Mathitis: I guess we can think of it this way. Sometimes, we use normal theory analysis even when we know a distribution is not normal. Socrates is recommending that you randomly select parts and try to randomly select operators if you can.

Socrates: There are at least a couple of issues raised if select parts and operators.

Mathitis: What are they?

Socrates: Do you use the measurement system you are evaluating to choose the parts?

Kokinos: Yes, because that's what we have.

Socrates: Isn't that circular reasoning?

Peismon: Why is that? We don't have another way to measure the parts.

Socrates: If you use the measurement system that is unacceptable to get a "good" part, how do you know it was a good part?

Kokinos: I hadn't thought of that.

Mathitis: But if you randomly select parts, then you don't need to measure them to select them.

Socrates: True.

Kokinos: How do I ensure I have covered the range?

Socrates: If you have already produced parts, then randomly sample from that production.

Peismon: That still doesn't make sense to me. I want to choose the parts.

Socrates: The question is "Do we want to make an inference from the parts we chose to all parts or just to those chosen parts?"

Peismon: To all parts, of course.

Mathitis: Then, you have to randomly select the parts. Otherwise, the inference is invalid.

Socrates: If you wanted to estimate only process variation or test whether the mean was a certain amount, would you select the parts or randomly sample?

Peismon: I would randomly sample.

Mathitis: Randomly sampling is an assumption for all statistical tests.

Socrates: In the equations for the MSA, is one component process variation?

Mathitis: Yes. I see. We do want to estimate process variation. That is represented by the variation of the parts.

Peismon: I'll have to think about this.

Socrates: Would you also select the measurements you want?

Peismon: That's different.

Socrates: Why? You want to estimate the variation of each, so if you think it is wrong to select the measurements, why are you selecting the parts?

Peismon: If we select the measurements, we can make the system look better than it actually is.

MAKING A MEASUREMENT SYSTEM LOOK BETTER THAN IT IS: SELECTING PARTS TO COVER THE RANGE OF PROCESS VARIATION

Socrates: I agree. Can't that occur by picking the parts?

Peismon: Why?

Socrates: By selecting them, you make the distribution of parts different from what it may actually be.

Peismon: How?

Socrates: What distribution do you typically assume for an MSA?

Peismon: Normal distribution.

Socrates: That's for both parts and measurement system results?

Peismon: Yes.

Socrates: What is the probability of getting a part that is far away from the mean?

Peismon: Not very high. It could be 5%, 2.5%, or 1% or less.

Socrates: Yet, if you select 10 parts and pick one at each end of the range, you make it appear as if each extreme value occurs 10% of the time when in reality it occurs 1% or less often.

Mathitis: That would give the false impression that the parts are distributed somewhat equally across the range.

Socrates: If you select the parts to be equally distributed across the range, what kind of distribution is that?

Mathitis: A uniform distribution, which has no mode. That's quite different from a normal distribution.

Socrates: How does the variance of a uniform distribution compare to that of a normal?

Mathitis: A uniform distribution that extends from A to B has a variance equal to $(B - A)^2/12$.

Socrates: Assume that 99.73% of the assumed normal distribution covers the range A to B. What's the variance of that normal distribution?

Mathitis: There would be six standard deviations from A to B. Dividing $(B - A)$ by six and squaring gives us the variance: $[(B - A)/6]^2$ or $(B - A)^2/36$. That's one-third the amount of variation as the uniform.

Kokinos: Or, we made the process variation three times larger using a uniform distribution rather than a normal distribution.

Peismon: What if only 95% was covered in the range A to B?

Kokinos: Then, the variance would be $[(B - A)/4]^2 = (B - A)^2/16$.

Mathitis: That's still approximately 25% more variation for the uniform distribution.

Socrates: Why is it important to estimate the part variation?

Peismon: We want to know how much of the total variation comes from the process and how much comes from the measurement. We want the measurement system variation to be a small contributor to the total variation. That's how we assess the measurement system.

Mathitis: When selecting the parts to cover the range of production uniformly, we make the part variation larger than it really is. That makes the measurement system seem better than it really is.

Agrafos: No wonder Peismon's MSAs always work!

SELECTING BOTH GOOD AND BAD PARTS

Socrates: Kokinos, you said that you were taught to use parts that are good and bad.

Kokinos: That's right.

Socrates: If you select five good parts and five bad ones, how many modes would you have?

Kokinos: It would be bimodal: one mode for the average of the good parts and another mode for the average of the bad parts. But the normal distribution only has one mode. Okay, then I would be changing the assumed distribution making the analysis invalid. I hadn't realized that.

Mathitis: Or three modes if you have two specifications and you select some bad parts below the lower specification limit and select others above the upper specification limit.

Socrates: If you select parts that cover the range of the specifications or extend past that range, you would be inflating the variation estimate of the process. Again, if you are comparing measurement variation to part variation or total variation, you have changed the distribution of the process by selecting half bad and half good.

Mathitis: Or one-third bad versus lower spec, one-third good, and one-third bad versus upper spec.

Socrates: Does your process produce 50% bad or two-thirds bad?

Kokinos: No. If it did, we would have another problem.

Mathitis: That would make the measurement system look better than it really is. Then, we might search for causes of not meeting specifications in the process when it could be the measurement system.

Kokinos: It seems that selecting good and bad parts or selecting the parts to cover the range of process variation can make the measurement system look better than it is—a bad one look good.

Mathitis: Or, the reverse: a good one look bad.

Socrates: How can you reduce the probability of these false impressions?

Mathitis: One way is to ensure that we meet the assumptions of the statistical analysis we use. That means randomly sample the parts and operators for the MSA.

7

Myth 7: % Study (% Contribution,
Number of Distinct Categories)
Is the Best Criterion for Evaluating
a Measurement System
for Process Improvement

Scene: *Break room (continuation of Measurement System Analysis discussion)*

BACKGROUND

Socrates: Peismon, you said that the reason for selecting parts that cover the process range or selecting good and bad parts was to determine whether the measurement system (MS) variation was small compared to the process variability.

Peismon: That's right. We need to separate the process variation from the measurement variation as the equation (Equation 5.1) we wrote showed for total variability and its components (Myths 5 and 6).

Socrates: Then what?

Peismon: We can then evaluate the MS in a variety of ways. The best way to evaluate the MS for improvement is to use the standard deviations of the MS and process and calculate the percent of the total variation attributable to the variation of the MS. This is called the % study ratio.

Mathitis: The formula is

$$\%study = 100 \times \frac{\sigma_{measurement}}{\sqrt{\sigma^2_{measurement} + \sigma^2_{process}}} \qquad (7.1)$$

If this is less than 30%, it's acceptable.

Socrates: Is this the only criterion?

Mathitis: The MS variability consists of gauge repeatability and reproducibility or total gauge R&R. You could compare the measurement variation to the tolerance, using the *P/T* ratio, or precision-to-tolerance ratio. A third criterion is the percent contribution: the variance from R&R as a percentage of the total variance. And, a fourth criterion compares the standard deviations of the process and MS to create a number of distinct categories.

Socrates: What are the formulas and acceptance levels for these criteria?

Peismon: They vary but these are the typical criteria:

1. $P/T = 6 \times \sigma_{measurement\ system}/(USL - LSL)$: excellent if <10%, marginally acceptable if <30%.
2. $\%contribution = 100\% \times \left(\sigma^2_{measurement\ system}/\sigma^2_{total}\right)$: acceptable if <10%.
3. $\%study = 100\% \times \sigma_{measurement\ system}/\sigma_{total} = 100\% \times \sqrt{contribution}$: acceptable if <30%, prefer <10%.
4. Number of distinct categories $= \sqrt{2}\left(\sigma_{process}/\sigma_{measurement\ system}\right)$: acceptable if >4, marginal if 3 or 4.

Mathitis: The MS standard deviation is the combined standard deviation of repeatability and reproducibility, the process standard deviation is the standard deviation of the parts, and the total standard deviation is the sum of measurement and parts standard deviations.

Socrates: What are distinct categories?

Kokinos: I was told that it is approximately the number of times the measurement error range can fit in the range of parts used in the study. It is acceptable if it is at least five times.

Socrates: Only the first criterion compares the measurement variation to the tolerance and the other three compare the measurement variation to something other than the tolerance.

Mathitis: That's right.

Socrates: What's the purpose of these other three that they don't compare to the tolerance?

Peismon: It's what I said before—to check that the measurement variation is small compared to the process variation.

Kokinos: We want the contribution of the MS to the total variation to be small.

Socrates: Of these four criteria, why is the % study the best for process improvement?

Mathitis: To show that you have improved your process, you need the measurement error to be small compared to the total variation. Otherwise, you will not be able to see whether the changes you made improved the process.

% CONTRIBUTION VERSUS % STUDY

Socrates: If the three criteria including % study compare measurement error to process or total error, why is % study the best for showing improvement?

Mathitis: The % study criterion uses the standard deviation, which is what is used to determine process capability. The % contribution uses the variance.

Socrates: And you would use that even though previously you said that standard deviations can't be added and variances can?

Kokinos: That's what I learned.

Socrates: And didn't we show that if we take the square root of the variances and express them as percentages, then the sum of all the components will not be 100%?

Mathitis: We did, but you also showed that we can sum them to get a total and use that as the basis for the percentages.

Socrates: Is that the total that is used for % study?

Mathitis: No, not exactly.

Socrates: If % study uses the standard deviations and % contribution uses the variances, then isn't % study just the square root of the % contribution?

Mathitis: I'm not sure.

Socrates: Is the acceptance criterion for the % study just the square root of the acceptance criterion for the % contribution?

Mathitis: I hadn't thought of that, nor checked it.

Socrates: The acceptance criterion for % contribution is 10%, correct?

Mathitis: Yes.

Socrates: What is the square root of 10%, or 0.1?

Peismon: It's 0.316.

Mathitis: That's 31.6%.

Socrates: What is the acceptance criterion for % study?

Peismon: 30%.

Mathitis: That's very close. I hadn't realized that.

Socrates: Then, one can't be better for evaluating the MS if they are the same thing.

Peismon: Well, the square root was not exactly 30%.

Mathitis: Okay, but % study is the slightly more rigorous test.

Socrates: If your MS passed % study, it must pass % contribution because if % study <30%, squaring % study to get % contribution will be less than 10%.

Kokinos: Then, either one can be used and both are preferred to the other criteria.

Mathitis: Better yet, only use the more rigorous one—% study.

Peismon: That's what I said.

P/T RATIO VERSUS % STUDY

Socrates: But now we have a reason for choosing it. You mentioned process capability when explaining the *P/T* ratio. How is process capability determined?

Mathitis: That's the extent product is meeting the customer requirements.

Socrates: How are customer requirements represented?

Mathitis: By the tolerance, the upper specification limit (USL) minus the lower specification limit (LSL).

Socrates: When you say for improving the process, do you mean with respect to meeting customer requirements?

Mathitis: Yes, showing that performance is better after you implement a solution than before you implemented it.

Socrates: What ratio compares the measurement error to customer requirements or the tolerance?

Mathitis: The *P/T* ratio.

Socrates: So, wouldn't the *P/T* ratio be better than % study or % contribution to evaluate the MS against the tolerance for showing process improvement?

Mathitis: Not if the measurement error is a significant contributor to the total process variation.

Kokinos: Why can't we do both? Isn't that why we have four criteria and not just one?

Socrates: We can. But that is different from saying that % study is the best criterion, isn't it?

Kokinos: I guess. If we have to do both, shouldn't we say one is better?

Socrates: I'm asking.

Peismon: I think % study does both. Before we can check *P/T*, we need to be sure that the measurement variation is small enough. We prioritize the criteria—% study must be done first.

Socrates: We agreed that the tolerance tells us when a unit meets specifications or is good. If measurement error is small compared to the tolerance, what does it allow us to do?

Mathitis: It helps us distinguish between good product and bad product.

Socrates: Does the *P/T* ratio tell us whether or to what extent we can distinguish between good and bad?

Mathitis: Yes.

Socrates: Do we need to do that to know whether we have improved? Do we need to be able to distinguish between good and bad to do three things: (1) determine the baseline performance, (2) determine the performance after a change, and (3) determine whether there is a difference?

Peismon: Yes, we need all that. But, again, if the measurement error is large compared to the total variation, then we won't be able to do any of that.

Socrates: That's possible. How are % study and % contribution determined?

Peismon: We've gone through that several times, Socrates.

Kokinos: Both are the measurement divided by the total variation. But for % study, the standard deviation is used, and for % contribution, the variance is used in the ratio.

Socrates: And, the tolerance, which defines what is good, is not in either equation?

Mathitis: No, it isn't.

Socrates: The total variation is composed of what?

Mathitis: That is the measurement variation plus the process variation, which is the part-to-part variation (Equation 5.1).

Socrates: Then, % study and % contribution depend on the part-to-part variation only and never on the tolerance.

Mathitis: Yes. I see what you are getting at. If % study and % contribution do not include the tolerance or specifications, how can they tell us about good or bad?

Kokinos: It's like saying my car is faster than your car, but it doesn't tell me whether either one is going the speed limit—which would be the specification.

Socrates: If we use similar parts, won't the part-to-part variation be small?

Kokinos: Yes.

Socrates: If we use dissimilar parts, will it be large?

Kokinos: True.

Socrates: If the part-to-part variation is small, % study will be large and possibly make the MS unacceptable, and vice versa, large part-to-part variation is likely to indicate acceptable MS.

Kokinos: Yes and yes.

Mathitis: Now, I see why the discussion of selecting parts (Myth 6) can give false information about the MS variation relative to the process variation.

Socrates: Before, we thought that selecting the parts to cover the range of production, tolerance, or good and bad units was appropriate.

Kokinos: But now we know it isn't right.

Socrates: We now have another reason. If that selection makes the process variation larger than it actually is, it could give the false impression that the MS is acceptable, right?

Kokinos: Yes—maybe that's why we tended to think we didn't have measurement problems.

DISTINGUISHING BETWEEN GOOD AND BAD PARTS

Socrates: And, does having a wide range of parts affect whether the MS can tell whether a single part is good or bad?

Mathitis: I don't understand.

Socrates: Let's say I have an MS that can reliably tell whether a part is good or bad. The measurement variation relative to the tolerance is

5%. Now, I check the % study using this same, reliable measurement system. I choose very similar parts and conclude the % study is 50%, so it is unacceptable. Do I now conclude I cannot tell whether the parts are good or bad?

Peismon: You have to because the measurement variation is a large contributor.

Socrates: Now, I redo the study using parts that are very different and the % study is acceptable. Does the ability to tell whether a single part is good or bad change simply because I chose very different parts to evaluate the % study?

Peismon: It has to because that's its purpose.

Agrafos: That doesn't make sense to me. How can choosing parts that are alike or different determine whether I can tell each part individually whether it is good or bad?

Mathitis: It can't. Identifying good or bad is only relative to the specifications. It has nothing to do with how much measurement variation there is relative to similar or dissimilar parts.

Socrates: Process improvement requires knowing whether more good parts are made after a change in the process: % study won't tell us that whether it is acceptable or not.

Mathitis: I have to agree with you now.

Kokinos: Then why do we use the % study?

Agrafos: Maybe when we don't have specifications.

Socrates: We agree that the P/T ratio is to tell good parts from bad parts.

Mathitis: Yes.

Socrates: And that % contribution and the % study tell us whether we can distinguish parts, but if we can select them, then we can manipulate the results.

Mathitis: It appears so. Now, I see why there is another reason why the parts and operators must be randomly selected.

Socrates: What about the number of distinct categories. But what does it tell us?

Mathitis: It is similar to % study and % contribution in that they compare the measurement variation to the total variation.

Kokinos: The total variation consists of the measurement variation and the part variation.

Socrates: We saw that the part variation can be manipulated if parts are selected deliberately rather than randomly.

Kokinos: Yes.

Socrates: Then, for all three to be valid, the parts must be randomly selected.

Mathitis: Yes.

Socrates: None uses tolerance or the specifications in the calculations, correct?

Kokinos: That's correct.

Mathitis: % study and % contribution use the total variation and number of distinct categories uses the part variation, specifically the standard deviation.

Socrates: Then, none of these three helps us directly in informing us whether we can distinguish between good and bad parts.

Mathitis: That's right. They don't include the tolerance and therefore they can't tell us whether we can separate the good units from the bad ones.

Socrates: If the purpose of process improvement is to increase the percent of good parts, these three criteria are not useful to us in assessing our MS.

Peismon: I disagree. They do tell us relative to the process variation whether the measurement error is large or small. That is useful to know.

Socrates: Then, the information it provides is only indirectly useful.

Peismon: What do you mean?

Socrates: We are assuming that less than 30% for % study is acceptable.

Mathitis: That's the worst-case acceptable. We prefer to be less than 10%.

Socrates: Whether 30% or 10%, it applies equally to a tolerance of 10 units or 1000 units.

Mathitis: Well, yes.

Socrates: Since the percent is not relative to the tolerance, then the 30% could be 100 times greater for the tolerance of 10 than for the tolerance of 1000. We would reach the same conclusion that it's acceptable for both.

Mathitis: I see what you're saying.

Peismon: But we use this all the time. It works. I believe it's useful to know that if 90% of the variation is attributed to the MS, then it probably isn't very useful.

Socrates: For distinguishing good from bad?

Peismon: Yes, why not?

Mathitis: What Socrates is saying is that it can't do that if the criterion doesn't include the specifications or tolerance. The same measurement variation contribution to the total variation may be acceptable as a small contributor but it may be unacceptable for a specific tolerance.

Kokinos: That's why I said that we need to do all four—or at least three since either % contribution or % study can be used.

Socrates: Let's see if it is possible that even with randomly selected parts, % study, % contribution, and number of distinct categories is misleading.

Mathitis: How do you propose doing that?

Socrates: Consider this example. There are three businesses, each producing hinges. The hinges come in two sizes: 4 cm long for cabinet doors and 10 cm for regular doors. Business A produces only the 4-cm hinges, business B produces only the 10-cm hinges, while business C produces both. All three businesses use the same MS and the same tolerance = USL − LSL for the length of all hinges.

Mathitis: I see where you're going with this. A and B, because they don't have a wide range of hinges, will conclude that the MS is unacceptable, while C will conclude that the same MS is acceptable because it has a wider range of parts. The % study will be large for A and B but small for C.

Socrates: Do you think that conclusion is correct?

Mathitis: It doesn't appear so since all three companies use the same MS.

Socrates: It looks like we have a paradox. If I make hinges that are very different in size, then I can measure the same hinge more precisely than if I have hinges that are very similar in size. Do you think that the variation in hinges affects whether I can measure an individual hinge more precisely?

Mathitis: No, that doesn't make sense. The MS should somehow be independent of the range of sizes. Otherwise, A and B could just periodically produce a huge hinge, include it in their measurement system analysis, and conclude that the MS is acceptable— even though they didn't change it. It could have gotten worse, but % study would not recognize that.

Socrates: Also, isn't that one way of making the MS "better" without changing it: increase the variation of the parts?

Mathitis: Yes, but that isn't right.

Socrates: Kokinos, you said that you were told to select good and bad parts. Do we want more variation or less variation in our processes?

Kokinos: Less process variation.

Socrates: Do you find it odd that doing the opposite—increasing process variation—makes your MS better? Without doing anything to the MS.

Kokinos: That can't be right.

Socrates: It is if we deliberately select parts to cover the range of production. Do you see that a business with lots of problems, producing parts that are really good and parts that are really bad will conclude that the same MS is better than a business that consistently produces really good parts?

Peismon: If the specifications were the same.

Mathitis: That's the point. These criteria do not use the specifications, only measurement variation and part variation.

Kokinos: I hadn't thought of that.

Agrafos: That means that if all we are interested in is how much of the total variation comes from the MS, we can use % study, % contribution, and number of distinct parts.

Kokinos: Not even for that if we can manipulate the results by the parts we select.

DISTINGUISHING PARTS THAT ARE DIFFERENT

Socrates: And, that leads to your second question. Why is 10% the acceptability criterion for all three hinge companies?

Mathitis: I'm thinking that it shouldn't be the same.

Socrates: What do % contribution, % study, and number of distinct categories have in common?

Mathitis: For one, they do not use the specifications or tolerance. Also, they both include the part-to-part variation. So, that makes me think that one use for them is to see the part-to-part variation. Given our discussion that similar and dissimilar parts change the calculations and conclusions, then we must somehow determine what is acceptable.

Socrates: If we are comparing measurement variation to part variation, then consider this. If I have two parts that are different, when would I want the measurement variation to be small compared to the part variation and when does it not matter?

Peismon: It should always matter.

Kokinos: Perhaps not when the difference is negligible or inconsequential.

Mathitis: Ok, I get it. The three criteria comparing MS variation to process variation or total variation are used to determine whether I can tell those parts apart. They supposedly answer the question "Can I distinguish between them?"

Kokinos: In a perfect world, all the companies would make hinges that are identical.

Agrafos: Then, we couldn't tell them apart—the MS would not—should not—be able to distinguish among them because there would be no difference.

Socrates: Do we then want to say that the MS is unacceptable? If all parts are identical, the MS would fail all three criteria: % study, % contribution, and number of distinct categories.

Mathitis: Of course, we would not want to say it is unacceptable.

Socrates: So, how should you select parts then?

Agrafos: You have to start with parts that are different.

Mathitis: I could determine what difference I want to recognize and then see whether the MS can measure that difference.

Kokinos: So, maybe we don't need to do variance decomposition?

Socrates: It seems we have two questions: (1) Can I distinguish between good and bad parts? (2) Can I distinguish between two different parts that I want or need to recognize as different?

Mathitis: ANOVA is useful for the first to separate the sources of variation. But to compare the measurement variation to the tolerance because the tolerance is the range of what is good, I need to use the *P/T* ratio criterion.

Socrates: And, for the second question?

Mathitis: I need to deliberately select two parts whose difference I need to recognize and see whether I can.

Kokinos: Then, the ANOVA based on random variables is not applicable since you are deliberately selecting the parts.

Socrates: What analysis would you do if you select the parts deliberately?

Mathitis: If I can't use ANOVA, I'm not sure what analysis to do. I could just measure each part and see if I recognize them as different.

Socrates: How might you do that?

Kokinos: Could we use confidence intervals?

Mathitis: Okay. We measure each part several times, calculate a confidence interval for the means, and if they do not overlap, then the MS can distinguish between them.

Socrates: But for small sample sizes, the confidence intervals will be large.

Mathitis: I could determine the sample size needed so that the confidence interval width is the size of the difference I want to measure.

Socrates: That's one way. What is required for that approach?

Mathitis: I would need to have two parts that are different by that amount.

Kokinos: How would you know they were that different? You can't use the MS you are evaluating. That was the problem Socrates noted in selecting good and bad parts.

Mathitis: We would have to use another MS. Or, get those parts, for example, from the National Institute of Standards and Technology.

Socrates: What if you didn't have to find or create the two parts that are exactly the smallest difference we want to recognize?

Mathitis: That would be great.

Agrafos: Can you do that?

Socrates: Reverse engineer the fourth criterion of number of distinct categories.

Kokinos: How do you do that?

Mathitis: I think I know how. Measure a part several times to determine the MS standard deviation. Then, using five as the number of distinct categories, solve the equation for the standard deviation.

Kokinos: Show me. What's the equation?

Mathitis: Here's the equation where c = number of distinct categories.

$$\frac{\sqrt{2}\sigma_{parts}}{\sigma_{measurement}} = \text{number of distinct categories} = c \geq 5 \qquad (7.2)$$

Let $c = 5$ and solve for the measurement standard deviation:

$$0.2828 \times \sigma_{part} \geq \sigma_{measurement} \qquad (7.3)$$

Peismon: We still have to determine the parts variation. You can't get around that.

Mathitis: We can if we replace the parts standard deviation by the smallest difference we want to recognize. If Δ is the difference we want to be able to measure, then $\Delta = 0.2828\sigma_{parts}$.

Peismon: That's not a valid assumption.

Socrates: If we have only two parts whose difference is Δ, what is the standard deviation?

Mathitis: It's half of the difference, with only two parts.

Socrates: Then, we can substitute 0.5Δ for σ_{parts} in the last equation.

Mathitis: Then, $\sigma_{measurement} \leq 0.1414\Delta$ for the MS to recognize a difference of Δ.

Socrates: Okay, there are two ways to do this. One, get parts that have a difference of Δ or, two, calculate the measurement standard deviation σ. Then use either the equation. Which is easier and direct?

Mathitis: It seems like I only need one part and several operators because I'm no longer interested in process variation.

Agrafos: Why not?

Mathitis: Because I can replace the other three criteria with just the difference equation and I only need the measurement variation, $\sigma_{measurement}$.

Kokinos: Then, to answer question 1, you use P/T to tell you whether you can distinguish between good units and bad units. To answer question 2, replacing parts variation with this last delta equation, you only need measurement variation, which tells you whether you can measure the smallest difference you need or want to.

Peismon: But what if you don't have specifications? Then P/T can't be used.

Agrafos: Then, there are no good or bad parts.

Kokinos: But I can still determine the smallest difference my MS can measure.

Peismon: We are also assuming that the variation of the MS doesn't change for difference amounts.

Mathitis: That's true. But we can check that with three parts representing different sizes that span an appropriate range. We can estimate what differences can be measured by interpolating if the standard deviations are dependent on size.

Peismon: I thought we shouldn't choose the parts.

Socrates: In this case, it is okay because we are not estimating the process variation using the parts but are estimating the measurement variation. Think of it this way. When doing a designed experiment, we select the factors and their levels to study. Originally, it was thought that they should be randomly selected. But the design is more efficient when we select factors and levels.

Kokinos: And we are not using a random-effects model that required that assumption.

Agrafos: Again, understanding the assumptions is critical.

8

Myth 8: Only Sigma Can Compare Different Processes and Metrics

Scene: *Project meeting room (team discussing project charter)*

BACKGROUND

Agrafos: Can we now discuss our project objective?

Socrates: Perhaps Peismon can help us? Agrafos and I were talking about our problem statement for a project we are working on.

Agrafos: The instructor insisted that the problem statement needed to state the current sigma level. He said that if everybody made their problem statements in terms of sigma, it would make it easy for management to prioritize projects. Why is that?

Peismon: If everyone proposes a project and we can't support them all, then we need to determine which ones to support now, which ones to support later, and which ones not to support at all. We need to prioritize the possible projects. To do that, management uses criteria for evaluating the projects. One criterion is the severity of a problem. This can be determined by the performance of the process. The worse the performance, the more severe the problem. Keep in mind that that is only one criterion.

Agrafos: Our instructor told us to use process sigma as a measure of that performance.

Peismon: That's right.

Socrates: Why?

Peismon: Because different processes use different metrics. The sigma scale allows us to compare their performances regardless of the metrics.

Socrates: Can you give me an example?

Peismon: For example, I might be comparing cycle time for an HR process, yield for a manufacturing process, and conversion rate on sales calls for a third process. If each team assigned to these projects uses these metrics, how do I know which process is doing worse? The sigma scale allows me to make that comparison.

Socrates: Setting aside why you would want to make that comparison, what is needed to compute sigma?

Peismon: If you have defects, you can calculate defects per million opportunities and convert that to a sigma value. If you have normally distributed data, you can compute the area within the specifications and that determines the sigma value.

Agrafos: You would have to use software for that.

Peismon: It's easier with software, of course. But the end result is a number typically between 0 and 6.

SIGMA AND SPECIFICATIONS

Socrates: To determine whether something is a defect, do you also need specifications?

Peismon: Or, at least a definition of what a defect is.

Socrates: Can I calculate sigma without a definition of a defect or specifications?

Peismon: No. Sigma tells you how many defects or defectives per million opportunities you have.

Socrates: Then, when you said you could compare different processes on different metrics, you can only do so if you are comparing defect or defective rates. Without specifications or a definition of defect, you cannot compare them even with sigma. Is that correct?

Peismon: I'm not following.

Agrafos: Nor I.

Socrates: You used cycle time as an example. If I don't have a specification for cycle time, will I be able to convert to sigma?

Peismon: No you won't.

Socrates: So, even with the sigma scale, I won't be able to compare just cycle time for any process to yield for any other process because I can't convert just cycle time to sigma.

Peismon: I see what you mean.

Agrafos: I don't.

Peismon: I can only compare processes that have specifications using the sigma scale because I want to know how often I'm meeting the specifications. The specifications are either limits or definitions.

Agrafos: Are you saying I can't compare two processes' performances if I don't have specifications?

Peismon: No, only that I can't use the sigma scale. I can compare two average cycle times to see which one is faster. But that is on cycle time. I can't convert the average to the sigma scale.

Mathitis: In the three examples Peismon gave of cycle time, yield, and conversion, they cannot be compared unless we have specifications so we can convert to sigma values.

SIGMA AS A PERCENTAGE

Socrates: What does one sigma mean?

Peismon: It's approximately 70% defects.

Socrates: Three sigma?

Peismon: Approximately 6%–7% defects.

Socrates: So each sigma value is just telling me the frequency of or percent of defects or defectives? In other words, for each sigma value, there is exactly one percentage and vice versa—there is a one-to-one relationship just as there is between Centigrade and Fahrenheit.

Peismon: Yes.

Socrates: Then, doesn't the percentage scale also allow me to compare different processes on different metrics?

Peismon: No, it's not the same.

Socrates: Not everyone knows what a sigma level is, correct?

Peismon: That's right.

Kokinos: I didn't before I took the class.

Socrates: How do you explain what sigma and sigma levels are? When I asked you what was one sigma, what did you say?

Mathitis: I see what you mean. He said a certain percentage. If the only way to explain is by stating the sigma value in terms of percentages, then it is simply another way of stating percentages. It's like translating one language into another. It means the same thing.

Agrafos: Then why convert?

Peismon: It makes clearer the upper end of the percentage scale. It used to be that 99% was acceptable. Now, we are aiming for a higher percentage meeting specifications.

Mathitis: Or, using the normal distribution, 99.73% was acceptable, as that is 3 standard deviations around the mean.

Peismon: People may think that 99% or 99.73% is good enough. But if you are producing millions of units, that still leaves a high number of defective units: 1% of 1 million is 10,000 bad ones and 0.27% out of 1 million is still 2700 bad units.

Mathitis: That's why the upper sigma scale is critical—the values 4 and greater. Four sigma is 99.379%, five sigma is 99.977%, and six sigma is 99.99966% good.

Peismon: The scale is not linear to make clearer that upper end.

Socrates: Why not just state the percentage—that's how it is explained?

Peismon: But there is another reason for using sigma. There is short-term and long-term sigma. The sigma level is different depending on which one you use.

Socrates: How are each calculated or determined?

Peismon: Short-term sigma uses an estimate of the best variation the process is capable of.

Socrates: So, we could just calculate the percentage using that best variation and stop there.

Mathitis: Actually, that occurs in the software we use when we calculate process capability. We get the actual percent of good units and then a short-term percent of good units.

Kokinos: Then, we could just stop there and not convert to the sigma scale.

Agrafos: I think everyone would understand percentages while that would not be true if we use sigma levels.

Kokinos: Since they mean the same, then that's a good reason for using percentages in your charter rather than sigma.

Peismon: However, you don't know whether you are using short-term or long-term variation.

Socrates: Would you know if we use sigma, say sigma = 2.37, which is it?

Peismon: Not if you write it that way. But that's why we use the subscripts "st" and "lt": sigma_{st} and sigma_{lt}.

Mathitis: We could do that with percentages or proportions.

9

Myth 9: Capability Is Not Percent/ Proportion of Good Units

Scene: *Project meeting room (continuation of project charter discussion)*

BACKGROUND

Socrates: In class, we were shown a decision chart to help us develop our project charter. One question was "Is the process capable?" Does this mean that a process is capable? Or does this mean that it is not? It's all or none.

Peismon: No, there are degrees of capability. You calculate a number.

Socrates: Now I am more confused.

Peismon: Why?

Socrates: Previously, we had discussed that when we have a choice between continuous and discrete data, one must use continuous ones because they were more informative. So why are we asking for a discrete answer of yes or no when we can quantify it?

Mathitis: We often ask the question "Is the process capable?" but a better question is "What is the process capability?" Process capability is not all or none, but degrees of capability. Process capability indices measure how capable the process is.

Agrafos: Then, there is an inconsistency between the question "Is the process capable?" and the advice to use the more informative measures, the quantitative and continuous ones.

Mathitis: I see your point.

Peismon: You have to calculate the process capability first. You can use one of the capability indices like C_p or C_{pk}. Then, you check to see whether it is good enough. That gives you the "yes" or "no" answer.

Socrates: Then, there is a minimum amount that is necessary to be capable.

Peismon: Yes.

Socrates: What amount is that?

Mathitis: Traditionally, it was $C_p = 1$.

Peismon: More often now, we want C_p to be at least 1.33 or 1.67, or even 2 in the R&D phase.

Socrates: What values can C_p have?

Peismon: C_p can be any value ≥ 0. In practice, it's hard to get above 2.

Socrates: Why 1 or 1.33 or any other number?

Peismon: That shows the process performance is great. The higher the value, the better the performance of the process.

Socrates: With respect to what?

Mathitis: We want the process output to satisfy customer needs.

Kokinos: Daskalos was my mentor for my project. We had a process capability goal. When I showed that the process was capable after improvements, I meant that the process performance met that goal.

Socrates: How did you show the capability changed?

Kokinos: We used C_{pk} and showed that it was higher after the change than before. We knew after the change that more of the process output would meet the specifications the customer gave us.

Socrates: And those outputs that meet specification are good units?

Mathitis: Yes, from the customer's perspective, as in Kokinos' project.

Socrates: Then, isn't process capability just the percent of good units? I asked the instructor that, but he said no.

Peismon: Your instructor is correct. We aren't calculating the percent good when we calculate process capability.

Kokinos: I was also taught that process capability measures process performance. Besides, we use continuous data when calculating process capability and percentages are not continuous.

Mathitis: Except that we had that discussion the other day (Myths 1 and 2) and we showed percentages and proportions can be continuous—especially when applied to continuous distributions like the normal distribution. If normal distribution probabilities are continuous, then so are the proportions and percentages representing these probabilities. For C_p and C_{pk}, we assume a normal distribution.

Peismon: It's the performance the process is capable of. Look at the formula for C_p:

$$C_p = \frac{(USL - LSL)}{6\hat{\sigma}} \qquad (9.1)$$

The numerator is the tolerance or upper specification limit (USL) minus the lower specification limit (LSL). Like the specifications that Mathitis and Kokinos mentioned. The denominator is the natural tolerance of the process. The ratio is not a percent. It's an index used to make comparisons.

CAPABILITY INDICES: FREQUENCY MEETING SPECIFICATIONS

Socrates: Why is six times the standard deviation the natural tolerance?

Mathitis: Ideally, we would like to include all that the process produces. But C_p assumes a normal distribution. Since it extends to infinity in both the negative and positive directions, then to include everything we would have a natural tolerance of infinity. Traditionally, to get at the vast majority of that distribution, $\pm 3\sigma$ range was used.

Socrates: If $6\sigma = USL - LSL$, then $C_p = 1$. What does a C_p of 1 mean?

Mathitis: It means that 99.73% of the normal distribution can fit within the specifications because we assume a normal distribution.

Socrates: Then, C_p and C_{pk} are just percent or proportion of units that meet specifications.

Mathitis: That's one way of looking at it.

Peismon: But it's not a percent or proportion. It's an index.

Socrates: It is a ratio. But you are right that not all ratios are proportions. Let's look at it another way. Rather than focusing on the equation, let's focus on what the result means. Does the Fahrenheit scale measure temperature?

Kokinos: Yes.

Socrates: If I change the scale to Centigrade, does it still measure temperature?

Peismon: Yes, but I don't see the connection.

Socrates: Is an English sentence converted to another language mean the same?

Kokinos: We hope so.

Socrates: If I use a code, like da Vinci did to write about his scientific studies, does the coded version mean the same as the uncoded version?

Kokinos: If the decoding is done right.

Socrates: Then, isn't C_p just a coded version of percentage? Yes, C_p is a different scale, but they mean the same thing: the percent of units that meet specifications.

Kokinos: I guess I would agree. Just as the sigma scale.

CAPABILITY: ACTUAL VERSUS POTENTIAL

Peismon: I disagree. Capability does not mean actual process performance. It is the potential of the process. For example, you are capable of walking 10 miles or walking to the nearest store even if you haven't done either.

Mathitis: But not always. For example, IT people call what software can actually do software capabilities. C_p means potential capability but P_{pk} means actual capability.

Peismon: C_p is about variation. It tells you how many multiples of six standard deviations fit between the specifications, regardless of the average. Look at the formula. You can see that it is not a percent. But also it doesn't matter what the mean is, so it can't be actual. Part or all of the distribution could be outside of the specifications, like in Figure 9.1.

Mathitis: But the C_p for this distribution is one because *if* the curve was centered, then 99.73% of the distribution would be within the

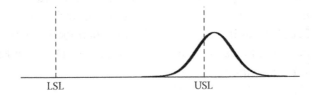

FIGURE 9.1

Normal distribution with $C_p = 1$ with respect to specifications.

specifications. But C_{pk} does include the mean. While $C_p = 1$ in this example, $C_{pk} < 1$.

Socrates: For C_p, it isn't actual percent good, but is it the potential percent good?

Mathitis: I'll accept that. Especially since the standard deviation that is used is the within-group or theoretical best. What we called short-term when discussing the sigma scale (Myth 8).

Socrates: Then, there is a one-to-one correspondence between a C_p value and a percentage value.

Mathitis: Yes, so while Peismon is right that the ratio does not yield a percentage, Socrates is right that the C_p scale is equivalent to and has the same meaning as percent of good units or units meeting specifications.

CAPABILITY INDICES

Kokinos: What about all the other process performance metrics that are used, like yield, DPMO (defects per million opportunities), C_p, C_{pk}, P_p, P_{pk}, and sigma short-term and sigma long-term.

Mathitis: And there are other capability indices: C_{pm}, C_{pkm}, C_{pU}, and C_{pL}.

Socrates: Well, let's start with yield. What is it?

Peismon: There are three kinds of yield. There is the classical yield, which is the percent of good units after rework. If you produce defective units and you rework them to make them good, then it is the percent that are good after the rework. First-time yield is the percent of good units before rework. There is rolled-throughput yield, which is the yield after several process steps, each with their own yield.

Socrates: How is that calculated?

Peismon: You multiply the yields of each step as proportions to get the final yield. For example, if steps 1–3 have yields of 99.3%, 97.8%, and 95.9%, respectively, the rolled-throughput yield is $0.993 \times 0.978 \times 0.959 = 0.931$ or 93.1%. The idea is to show that the final yield will always be no greater than the worse step.

Socrates: Is the rolled throughput yield before or after rework?

Peismon: I don't know—no one has asked that question.

Mathitis: As we said before, we could always add subscripts or other notations to indicate whether it is before or after rework.

Socrates: What about DPMO?

Peismon: DPMO is for defects not defectives. So, it is the proportion of defects given the number of opportunities for defects and then multiplied by 1 million. For example, if I have 10 units, each unit has 2 opportunities for defects, and there are 7 defects total, then my proportion of defects is $7/(2 \times 10) = 0.35$. This gets multiplied by 1 million to get 350,000 defects per million.

Socrates: Then, 0.35 is the proportion of defects and $1 - 0.35 = 0.65$ is the proportion of nondefects or good units.

Peismon: Well, yes, but this is for defects not defectives.

Socrates: Understood. Is DPMO calculated before or after rework?

Mathitis: No one has ever asked that question either. Again, we can clarify with notation.

Kokinos: It makes sense to me that if we distinguish between before and after for yield, we should also make that distinction for DPMO and for rolled throughput yield.

Socrates: Do you make that distinction for the other capability indices?

Peismon: No. I wonder why not?

Socrates: What about C_p and C_{pk}?

Mathitis: We don't distinguish between before or after rework but I imagine it is before.

Socrates: What are the assumptions for C_p?

Peismon: The data come from a normal distribution and there are two specifications, LSL and USL.

Socrates: So, C_p cannot be calculated for nonnormal distributions?

Peismon: Some software calculate C_p for nonnormal distributions.

Socrates: Why do you need two specifications?

Peismon: Because the idea is to determine how much of the normal distribution is within the tolerance. Similar to what Mathitis said about the normal distribution being infinite in size, the tolerance, which is in the numerator, could be infinite with only one specification. Dividing by a finite number still gives a C_p equal to infinity.

Socrates: Then, every process would be capable. What about C_{pk}? You said that takes into account where the mean is relative to the specifications?

Peismon: In the previous example, $C_p = 1$ but $C_{pk} < 1$ because the mean is not in the center of the specifications. $C_{pk} = 1$ when we have this case (Figure 9.2). C_{pk} uses the specification limit closest to the

FIGURE 9.2
Normal distribution with $C_{pk} = 1$ with respect to specifications.

mean of the distribution. C_{pk} equals the distance between the specification limit closest to the mean and the mean, divided by three times the standard deviation:

$$C_{pk} = \min\left[\frac{(USL - mean)}{3\hat{\sigma}}, \frac{(mean - LSL)}{3\hat{\sigma}}\right] \qquad (9.2)$$

In other words, C_{pk} tells you how close your mean is to your specifications.

Mathitis: Another way of calculating C_{pk} is to calculate the capability with respect to each specification limit, like this:

$$C_{pL} = \frac{(\bar{X} - USL)}{3\hat{\sigma}} \qquad (9.3)$$

$$C_{pU} = \frac{(USL - \bar{X})}{3\hat{\sigma}} \qquad (9.4)$$

Then, we can combine them in this manner to get C_{pk}:

$$C_{pk} = \text{Min}(C_{pU}, C_{pL}) \qquad (9.5)$$

Socrates: In this case, is the percent of units that is actually within the specifications 99.73% when $C_{pk} = 1$?

Mathitis: Yes, in this case. But it uses the best or short-term standard deviation. So, C_p, C_{pk}, and sigma short-term are all capable in the sense of potential. They're not the actual percent good.

Socrates: But the potential percent meeting specifications. What about P_p and P_{pk}?

Peismon: P_p has the same formula as C_p in that it only uses the variation and ignores the mean while P_{pk} is like C_{pk} in that it uses both. The difference is P_p, P_{pk}, and sigma long-term all use the total variation.

Socrates: Total variation is what?

Peismon: It's the actual variation of the data, not the best as with C_p and C_{pk}.

Socrates: So, is P_{pk} the actual percent good—the percent of the units within the specifications?

Peismon: It's an index.

Socrates: What about the sigma scales?

Peismon: Sigma short-term and C_{pk} are equivalent. Both are the process potential using the inherent process standard deviation, while sigma long-term is like P_{pk} or the actual process performance.

Socrates: We said before that sigma is simply a percentage.

Mathitis: Yes. A 4.3 sigma short-term equals 99.73%.

Socrates: So, does Table 9.1 describe these capabilities and performance metrics?

Peismon: Yes, but the table doesn't state the assumptions. For example, C_{pk} and C_p assume a normal distribution and require two specifications.

Socrates: Are these (Table 9.2) the assumptions for each measure?

Peismon: Yes, but there is another assumption that the process is stable to get these estimates.

Socrates: Let's postpone that discussion, okay? To review, does the percent within specifications change as the spread changes, as measured by the standard deviation?

Peismon: Yes, of course. And, P_{pk} and C_{pk} change as the mean of the distribution changes while that's not true for C_p and P_p.

Socrates: And, for each value of these indices, there is exactly one value for the percent within specifications?

Mathitis: I would have to say yes, but each index is calculating the percent meeting specifications under different conditions related to actual versus inherent centering and variation.

Kokinos: That's what the assumptions state.

TABLE 9.1

Capability Indices Descriptions

Index	Value	% in Specs	Mean	Standard Deviation[a]	Interpretation[b]
P_{pk}	1	99.73	Included	Actual	Actual performance
P_p	1	99.73	Excluded	Actual	Actual performance as if process was centered
C_{pk}	1	99.73	Included	Inherent "best"	Potential best performance
C_p	1	99.73	Excluded	Inherent "best"	Potential best performance as if process was centered
Sigma short-term	4.3	99.73	Excluded	Inherent "best"	Potential best performance as if process was centered
Sigma long-term	2.8	99.73	Included	Actual	Actual performance
Classical yield	99.73%	99.73	Included	Actual	After rework (actual or potential performance)
First-time yield	99.73%	99.73	Included	Actual	Before rework (actual or potential performance)
DPMO	2700	99.73	NA	NA	Actual for defects not defectives

[a] "Best" in six sigma terminology is referred to as short-term or also common cause variation.
[b] Actual refers to long-term, total, or common cause variation plus special cause variation. Potential refers to performance estimated from the "best," inherent, short-term, or common cause variation.

TABLE 9.2

Capability Indices Assumptions

Index	Assumptions	
	Distribution	Specifications[a]
P_{pk}	None	LSL or USL or both
P_p	None	LSL or USL or both
C_{pk}	Normality	LSL and USL
C_p	Normality	LSL and USL
Sigma short-term	None or normality	LSL or USL or both
Sigma long-term	None or normality	LSL or USL or both
Classical yield	None	LSL or USL or both
First-time yield	None	LSL or USL or both
DPMO	None	LSL or USL or both

[a] LSL, lower specification limit; USL, upper specification limit; each alone is also referred to as one-sided specification, and when both exist, it is a two-sided specification.

PROCESS CAPABILITY TIME-DEPENDENT

Socrates: Which does the customer "feel?" That is, which measures what the customers get?

Peismon: The actual performance. P_{pk}, sigma long-term, and classical yield for defectives and DPMO for defects.

Socrates: Which does the business "feel?"

Peismon: These same ones.

Socrates: If the process changes, will what the customer and business feel change?

Peismon: Of course.

Socrates: When we calculate P_{pk} for example, do we use all the data representing the entire life of a process?

Peismon: No it is for a period.

Socrates: If we calculate P_{pk} using data from this past week or month, is that P_{pk} the same for the entire life of the process?

Peismon: No, because the process can and will change and the customer requirements may change. Besides, P_{pk} is the actual performance or capability.

Socrates: So, P_{pk} is the actual capability for only a specified period defined by when the data were collected.

Peismon: I'm not sure.

Mathitis: I think I understand, Socrates. We should state the period for which P_{pk} is applicable. And, that will also apply to the others: sigma long-term, classical yield, and DPMO—the period corresponding to the data should be stated.

Socrates: Is that also true for P_p?

Peismon: Yes. But I don't think it's necessary for the potential capabilities like sigma short-term, C_p, and C_{pk}.

Socrates: Let's discuss that in a moment too.

MEANING OF CAPABILITY: SHORT-CUT CALCULATIONS

Kokinos: Then, why do we use the C_p formula or these other indices if all we want is the percent within specifications?

Mathitis: For one, because they are easier to calculate if certain assumptions are made.

Kokinos: How?

Mathitis: We know that for all normal distributions, 99.73% of the area under the curve is within ±3 standard deviations of the mean. Using that as a basis, we automatically know that $C_p > 1$ yields more than 99.73% and $C_p < 1$ yields less. We can use these indices when we set goals. If our goal is $C_p = 1.33$ (someone has already determined that percentage is 99.9934%), then we just make our calculation simple by using the formula for C_p and check whether it is ≥ 1.33. Otherwise, we need to calculate the percent within the specifications assuming a normal distribution and using an estimate of the best standard deviation.

Peismon: We set C_p goals in R&D, when developing products or processes.

Socrates: We could have used the formula $C_p = |USL - LSL|/2\sigma$, and then rather than having a goal of 1.33 it would be 4. With these formulas and assumptions, the calculations and the comparisons to goals are easily made.

Kokinos: What happens when the distribution is not normal?

Peismon: Then, we can't calculate certain capability indices like C_p and C_{pk}.

Socrates: That depends.

Peismon: The table of assumptions you and Mathitis produced shows that it is true.

Socrates: What I am suggesting is that we can get around that requirement.

Kokinos: How, if it is a requirement?

Socrates: Rather than stating that C_p is the formula Peismon gave us, we define C_p and all capability indices as a measure of the actual or potential percentage of meeting specifications using different cases of centering and variation.

Kokinos: Then, the formulas don't work.

Socrates: They still work as a shortcut for the normal distribution with two specifications. Let's see for other cases. Let's look at the first table (Table 9.1)—and recall what Mathitis said about the other tables showing the relationship among several of these indices.

Kokinos: Not the one with the assumptions but the first one?

Socrates: That's right. What is listed in the third column?

Kokinos: The percentage within specifications.

Socrates: Why not calculate the percent meeting specifications—whether one-sided specification or two-sided, whether normally distributed or not, whether short-term or long-term variation, and so on—and convert that to any index you want?

Mathitis: You mean I can have just a USL and a nonnormal distribution, typical for cycle time measures, and calculate the percent meeting that specification and then convert to C_p or C_{pk}?

Socrates: Exactly. If you think C_p and C_{pk} are codes for the percent meeting specifications, then any percent can be coded. Just make sure you meet the requirements of centering and variation.

Kokinos: It's just converting from one scale to another. Like Fahrenheit to Centigrade or centimeters to inches.

Socrates: Yes. For example, assume we have an exponential distribution for Mathitis' cycle time case. The specification is USL = 10 and we have estimated the short-term variability to be 2.5.

Mathitis: The exponential cumulative distribution function is $f(X \leq x) = 1 - e^{-x\lambda}$ where $1/\lambda$ is both the standard deviation and the mean. We substitute x = USL = 10 and $1/\lambda$ = 2.5, we get 0.9817.

Peismon: But what is that as C_p?

Socrates: We don't know until we have a table or coding equation that converts it into C_p.

Peismon: Then, we have to do another calculation to understand it and compare to C_p.

Socrates: Consider this. C_p = 1 corresponds to 99.73%. What do C_p = 0.834 and C_p = 1.472 correspond to?

Peismon: I don't know. I just know that the first is less than 1, so it's not capable enough and the second is more than 1.33 so it is capable enough.

Mathitis: I see what Socrates is getting at. We memorize three or four C_p values but we don't know what every C_p value represents as a percent. Like the sigma scale, to explain it to others, we explain it in terms of percentages or portions of the normal distribution curve. Then, we don't need to use C_p. Since C_p = 1 corresponds to 99.73% and C_p = 1.33 corresponds to 99.9934%, then we only need to compare percentages. In the exponential example, we got a probability of 0.9817 or 98.17%, which is less than 99.73%—not good enough.

Socrates: If the best variation was λ = 1, then the percent within the USL = 10 would be 99.996%, which exceeds the equivalent value of C_p = 1.33.

Kokinos: What I get out of this is that we should understand the meaning of the formula and not just how to use it.

Mathitis: I like that. In fact, that explains why some software calculate capability values for nonnormal distributions.

Peismon: But that wouldn't be using the formula.

Mathitis: No, but the formula of tolerance divided by six times the standard deviation is just a shortcut when the distribution is normal.

Socrates: We can use this shortcut because for all normal distributions, regardless of the mean and standard deviation, 99.73% of the middle of the curve occurs between -3σ and $+3\sigma$, or a range of 6σ.

Mathitis: Couldn't we derive a shortcut for the exponential distribution? We said that the cumulative distribution function is $\mathrm{Prob}(X > x) = 1 - e^{-\lambda x}$. We want that to be at least a particular probability, say p, for $x = $ USL, and we know that $\sigma = 1/\lambda$ for the exponential distribution. Substituting, we have $1 - e^{-\mathrm{USL}/\sigma} \geq p$. So, $\mathrm{USL}/\sigma \geq -\ln(1 - p)$.

Kokinos: That's similar to C_p: specifications divided by the standard deviation. But how do we use that?

Peismon: But it won't produce the same scale as C_p.

Mathitis: But we can create a new scale. For $C_p = 1$, $p = 99.73\%$. Then, $-\ln(1 - 0.9973) = 5.9145$. So, if $\mathrm{USL}/\sigma \geq 5.9145$, it is equivalent to a $C_p = 1$. For $C_p = 1.33$, $p = 99.9936\%$. Then, $\mathrm{USL}/\sigma \geq 9.6623$.

Socrates: We could have C_{p_normal} and $C_{p\_exponential}$ and so on. If families of other distributions had similar properties like the normal, then we could create shortcuts for those distribution families.

Peismon: Are there any other distributions like the normal?

Socrates: A simple one is the uniform distribution. A fair die is a discrete uniform distribution—in the sense we spoke about before (Myths 1 and 2). The quantitative, continuous uniform distribution has a range between two values, say $a < b$. The probability of a value in any subset of that range is simply the proportion the subset is of the total range.

Kokinos: Can you give us an example?

Socrates: Suppose the uniform distribution is from 0 to 10. Then, the probability of being in the interval (2.13, 4.76) is simply $(4.76 - 2.13)/(10 - 0) = 0.213$.

Mathitis: I get it. Every interval of the same size has the same probability. This is too simple then. If we want a capability, assuming this uniform distribution of 99.73%, then we just want to know whether $(10 - 0)/(USL - LSL) \geq 0.9973$.

Socrates: Or, generally, $C_{p\_uniform} = (b - a)/(USL - LSL)$.

Peismon: But that's upside-down of the capability indices. They have tolerance divided by the standard deviation.

Mathitis: Then, take the reciprocal: the ratio is $1/0.9973 = 1.0027$. The advantage of the first way is that it states the percent within specifications. It's clear to everyone without additional explanations.

10

Myth 10: p = *Probability of Making an Error*

Scene: *Break room (after lesson on hypothesis testing)*

BACKGROUND

Socrates: Daskalos, would you mind briefly summarizing this morning's lesson on hypothesis testing? I may have misunderstood something.

Daskalos: Certainly, Socrates. There are two competing hypotheses. One hypothesis is called the null hypothesis denoted H_o and the other hypothesis is called the alternative denoted H_a. The test involves deciding which to accept or reject based on a probability called p. We reject H_o when $p <$ alpha. Do you recall this decision table (Table 10.1)?

Socrates: Yes.

Daskalos: In the two cells where I wrote "No error," a correct decision has been made. In the other two, an error occurs. Deciding to "Accept Alternative (H_a)" when "Null (H_o)" is true is a Type I error and deciding to "Accept Null (H_o)" when it's not true is a Type II error.

Socrates: When we use statistics to make decisions, are there probabilities of these errors?

Daskalos: That's right. The probability of making a Type I error is alpha. The probability of making a Type II error is beta. Here, let me redraw the same table and insert this information (Table 10.2). This table has the probabilities for each decision. Confidence

TABLE 10.1

Hypothesis Testing Decisions and Errors

	Decision	
Truth	Accept Null (H$_o$)	Accept Alternative (H$_a$)
Null (H$_o$)	No error	Error (Type I)
Alternative (H$_a$)	Error (Type II)	No error

TABLE 10.2

Hypothesis Testing Probabilities of Decisions

	Decision	
Truth	Accept Null (H$_o$)	Accept Alternative (H$_a$)
Null (H$_o$)	Correct: confidence	Error: alpha
Alternative (H$_a$)	Error: beta	Correct: power

level is the probability of making the right decision when H$_o$ is true; power is the probability of making the right decision when H$_a$ is true. Similarly, alpha and beta are the probabilities of making the wrong decisions when H$_o$ and H$_a$ are true, respectively. Since you can only be in one row, the probabilities sum to 1 or 100% across each row.

Socrates: What is *p*?

Daskalos: It's the probability of making an error.

Socrates: Why is that?

Peismon: As *p* gets smaller and smaller, you are less and less likely to make an error.

Kokinos: That's the way it was explained to me.

ONLY TWO TYPES OF ERRORS

Socrates: That's what I misunderstood. If *p* is the probability of making an error, what error is it?

Daskalos: An error in the decision you make.

Socrates: An error in any decision? For example, if I decide to have a donut for breakfast rather than fruit, is *p* the probability of making an error in that decision?

Daskalos: No, *p* applies to decisions about what is true. The donut versus fruit decision is not about which hypothesis is true. Those are choices you made about what you want, not what is true about the world.

Socrates: How does the error in hypothesis testing occur?

Daskalos: If I accept a hypothesis that is not true or reject one that is true, then an error occurs. If after you make a decision for breakfast, someone asks me which one you chose, I make an error by saying you chose the donut when you in fact chose fruit. That's about reality.

Socrates: To make an error, does one have to make a decision?

Daskalos: Yes. We make a decision based on the comparison of *p* to alpha. Then, we compare the decision to what is true to see if an error is made. If the decision agrees with the truth—the two decisions in the top left and bottom right—then no error is made, and if the decision doesn't match truth—the other two cells (top right and bottom left)—then an error occurs. These two errors are called Type I and Type II.

Socrates: Are these the only two errors possible?

Daskalos: Yes, in this context. We sometimes say there is a third type of error, which is solving the wrong problem or testing the wrong hypothesis. But that's a different issue.

Socrates: Then, *p* is the probability of making one of these two errors. *p* is either the probability of making a Type I or Type II error. That is, is *p* alpha or beta?

Peismon: No—he just said it wasn't. That would be redundant.

Socrates: Then, if there are only two errors that can be made and *p* is not the probability of making either one, then it must be the probability of something other than making an error.

Peismon: I took this course at another company, Socrates. This material and the material we used there have the same thing: *p* is the probability of making an error. Both courses can't be wrong. Everyone uses these materials and they have been used for years.

Socrates: Are you saying that we should accept this simply because others have used it?

Peismon: No, I'm not saying that. But, if others who are teachers have not questioned it, they must understand it to continue teaching it. Are you saying that because you do not understand it, it isn't true or shouldn't be taught?

Socrates: No, not at all, Peismon. But, perhaps you can understand my confusion. Daskalos says there are only two errors that can be made. Their probabilities of occurrence are alpha and beta. Then, he says *p* is the probability of making an error. But that error is neither of the only two errors that can be made. So, what error is it?

DEFINITION OF AN ERROR
ABOUT DECIDING WHAT IS TRUE

Daskalos: I understand your confusion. I should have specified that *p* is the probability of making an error, in general.
Socrates: You described hypothesis testing as making a decision about what is true and comparing it to what is true. If it matches what is true, no error is made, and if it doesn't match what is true, an error is made.
Daskalos: That's right. But the decision for each is different and what is true for each is different. A Type I error can only occur when H_o is true and you decide to reject H_o. A Type II error can only occur when H_a is true and you decide to reject H_a.
Socrates: Then, when you say that *p* is the probability of making an error, even in general, it is incomplete. It is missing both the decision that is made and what is true.
Daskalos: Yes, I guess you are right. It is missing this information.
Socrates: For *p* to be the probability of making an error, then a table similar to the one we started with must exist like in Table 10.3. We could have the rows represent truth and the columns represent

TABLE 10.3

Hypothesis Testing Decision Table Assuming *p* = Probability of Making an Error

Truth	Decision	
	Accept Null	**Accept Alternative**
Null	No error	Error (*p*?)
Alternative	Error (*p*?)	No error

the decisions. Then we have to state the hypotheses and which error is associated with p. Is there such table even though we don't know in which row-column combination to place p?

Daskalos: No, there isn't and for that reason.

Socrates: Then, what is the decision and what is true?

Peismon: p is the probability of an error in whichever decision you make in hypothesis testing.

Socrates: So, p is the probability of making either error regardless of the decision?

Peismon: Yes.

Socrates: Then, when $p = 0$, regardless of the decision, no error is made. I accept H_o and you reject H_o and neither one made an error.

Mathitis: That is not right. Both can't be true.

Socrates: When $p = 1$, regardless of the decision, an error is made. I accept H_o and you reject H_o and we are both wrong.

Mathitis: That isn't true either. I understand your confusion as I am now confused.

Socrates: Help me make it clearer. First, tell me what is true and what decision is made. Then, we can compare the decision to what is true to see if an error is made. The concept of an error in general does not contain either component of the definition. We agreed that there can only be two kinds of errors. Daskalos, you said in class and now that only one of these can be true. In other words, we can only be in one row or the other but not both. The error in general is neither.

Mathitis: I see that now. We are never in either state—H_o is true or H_o is false—or in both states—H_o is true and false. We are always in just one of those states. We just don't know which. To make an error, we have to specify which state we are in—determine what is true—and then compare our decision to that state or what is true.

Socrates: Then, p cannot be the probability of making an error of either kind or an error in general? Do you see the contradiction?

Mathitis: I think I do.

CALCULATION OF *p* AND EVIDENCE FOR A HYPOTHESIS

Socrates: Let's consider how p is calculated. Daskalos, would you remind us?

Daskalos: Sure. In general, H_o is a mathematical expression of "no difference." For example, H_o could be the hypothesis that a population mean is equal to 15. The alternative hypothesis is that the mean is not 15. When a sample is taken, we get a sample mean that typically differs from the hypothesized value. Then, assuming H_o, p is calculated as the probability of getting the sample result or a result more contrary than what H_o hypothesizes.

Socrates: Can you give us more details of the example?

Daskalos: Suppose H_o is the mean = 15 and the sample mean was 17, then p is the probability of getting a difference of $17 - 15 = 2$ or greater assuming H_o is true.

Socrates: For a coin example, if we tossed the coin five times and got four heads, would p be the probability of getting four or more heads?

Daskalos: Exactly.

Socrates: If the probability of heads is 0.5, what is the probability of getting four or more heads?

Daskalos: It would be the probability of getting exactly four heads in five tosses plus the probability of getting exactly five heads in five tosses. Those probabilities are 5/32 + 1/32, so the sum is 6/32 or 0.1875.

Socrates: That is p? $p = 0.1875$?

Daskalos: Yes.

Socrates: If we assumed a different probability for a single head, such as 0.6, then the calculation of p for four or more heads would be different.

Daskalos: That's right.

Socrates: Then, alpha and beta are the probabilities of errors before collecting the data and making a decision. They are the probabilities of Type I and Type II errors because they determine when to accept or reject H_o. And p is nothing more than a calculation of a possible result assuming a certain hypothesis. p can and is calculated no matter which hypothesis is true.

Daskalos: Yes, any hypothesis can be true. However, p is calculated assuming a particular hypothesis. We calculate p assuming H_o is true.

Socrates: Then, are we not in the H_o row?

Daskalos: I guess you are right. So, p must be the probability of making an error when H_o is true.

Mathitis: Doesn't that mean that p is alpha?

Daskalos: In general, repeating the test many times, alpha is the probability of a Type I error when H_o is true. But for a particular case, if H_o is true, the probability is p.

Kokinos: Now, you are in only one of the rows.

PROBABILITY OF MAKING AN ERROR FOR A PARTICULAR CASE

Peismon: That makes sense. Alpha is the probability of making a wrong decision when H_o is true over many applications of this rule. By using alpha as the cutoff value, then $p \geq$ alpha will occur 1 – alpha times, and $p <$ alpha will occur alpha times when H_o is true. Alpha applies to all possible cases. However, for a particular case, p is the probability of making an error in that case.

Socrates: But Daskalos said—and you agreed—that p is the probability of making an error in general. Now, you are saying it is making an error in particular.

Peismon: Yes, for a particular case, p is the probability of making an error.

Agrafos: But for a Type I error, right?

Socrates: It must be for one or the other. Otherwise, when $p = 0$, then whether I accept or reject H_o, the probability of making an error is zero, and when $p = 1$, both accepting or rejecting is certainly wrong with probability 1.

Mathitis: That doesn't make sense.

Peismon: Then, p is the probability of making a Type I error for a particular case.

Daskalos: That's what I was saying. Since p is calculated assuming H_o, then it is the probability of a Type I error for this particular case. We are in the H_o truth row, so we know what is true, H_o, and we know what decision we made, reject H_o.

Socrates: How do we determine that probability in the long-run?

Peismon: We could repeat the analysis thousands of times, taking a different sample each time assuming H_o is true. For each case, we would see whether $p \geq$ alpha or $p <$ alpha. Then, we would calculate the number of times $p <$ alpha and divide by the total number of repetitions. That proportion is approximately the probability of being wrong. It should be alpha.

Socrates: In each, you check whether you were right or wrong?

Peismon: Yes.

Socrates: Then in each case, what is the probability?

Daskalos: There is no probability. Well, it's either 0 or 1.

Socrates: Each case is one of those particular cases you were talking about.

Mathitis: I see what you mean. In a particular case, H_o is either true or not.

Peismon: That doesn't make sense to me.

Socrates: Consider this. If we have a deck of 52 cards, four suits and 13 cards in each suit, what is the probability of randomly getting the three of hearts?

Peismon: It's 1/52.

Socrates: Suppose you randomly select a card and look at it. Now what is the probability?

Peismon: If I look at it, I know what it is. There is no probability—well, it's either 0 or 1. It either is or it isn't the three of hearts.

Socrates: Is that not the same situation for a particular case? Before we collect the data, calculate p, and compare to alpha, alpha and beta are the probabilities, not p. After we collect the data, calculate p, and compare to alpha, then the probability of being right or wrong is either 0 or 1.

Mathitis: I think I understand. After you compare to alpha, you make a decision. Then either that decision is right or it is wrong. After you select the card, it is either the three of hearts or it isn't.

Socrates: Another way to look at this is to consider confidence intervals. Why are they not called probability intervals? Remember, Daskalos, you told us why in class?

Daskalos: Because once you collect the sample, the interval either contains the population parameter or it doesn't. Just as you have said about the decision once it is made. I see what you mean but it isn't what we've been teaching.

Socrates: When $p = 1$, is it always true that I make a Type I error when I reject H_o?

Kokinos: I don't know.

Daskalos: Based on this interpretation of p, it seems so.

Socrates: When does $p = 1$?

Daskalos: It has to be only when H_o is true. But that doesn't make sense.

Socrates: Why not?

Daskalos: There may not be sufficient evidence that H_o is false.

Socrates: Let's use the coin example where we are testing the fairness of the coin tossing. Then

H_0: probability of heads = 0.5 versus
H_a: probability of heads ≠ 0.5.

When will $p = 1$?

Daskalos: I can only get $p = 1$ when I have included all possibilities because p is based on the actual results plus all the more extreme possibilities.

Socrates: How can that occur?

Daskalos: Since H_0 is the probability of heads = 0.5, then I need 50% heads in my sample. Then, p would be the probability of getting 50% heads plus the probabilities of all other extreme results from 0% heads to 100% heads, which would be everything.

Socrates: If I get exactly 50% heads, does that guarantee that the probability of heads is 0.5?

Daskalos: No. The coin could be slightly favoring tails, and I could get equal number of heads and tails.

Socrates: Then, p can't be the probability of a Type I error even for a specific case. When $p = 1$, what would be the probability of making a Type I error if I reject H_0?

Daskalos: It would be 1—guaranteed to be wrong. But we see that p can equal 1 even when H_a is true. The probability is not 1 when rejecting H_0 when $p = 1$.

Mathitis: Then, p is merely the conditional probability of getting certain results assuming a particular hypothesis.

Agrafos: Then, as we said initially, there are only two types of errors, their probabilities are alpha and beta, and p is simply a conditional probability.

PROBABILITY OF DATA GIVEN H_0 VERSUS PROBABILITY OF H_0 GIVEN DATA

Agrafos: I'm confused about alpha and beta. We said that they are the probabilities of Type I and Type II errors. But we also said that once the sample is drawn and we make a decision, the probabilities are either 0 or 1.

Daskalos: That's because we are taking a frequentist approach to statistics. Think of the deck of cards. There are 52 unique cards. Each card occurs once. However, there are four cards that have the value six. That frequency is 4 out of 52. There are 13 cards that are spades, so the frequency of spades is 13 out of 52. A frequentist view of statistics uses the frequency as the probability.

Socrates: Then, when we draw a card at random, do we assign its frequency as its probability?

Daskalos: Yes. With the exception you noted. Once the card is drawn, then it either is or isn't a particular card. The frequentist approach assumes a single probability distribution. If the probability distribution is a member of a parametric distribution family, then the frequentist approach assumes the distribution is one of the members. Therefore, the parameters are fixed.

Mathitis: You mean, for example, if I assume a normal distribution, then it is one specific normal distribution with a given or fixed mean and variance.

Daskalos: That's right. Each member of the normal distribution family is defined uniquely by those two parameters, which makes the normal distribution a two-parameter family.

Mathitis: That explains why we have hypotheses tests on the mean, for example, to identify which normal distribution it is.

Daskalos: That's right.

Agrafos: What if I want a probability rather than confidence?

Socrates: That would require that the probability varies so it can be treated as a random variable.

Daskalos: That's right.

Agrafos: Can we do that?

Socrates: Why not? It's just a model. Do you think that every time you shuffle the cards or randomly select a card the randomization is the same?

Agrafos: Probably not.

Kokinos: Then, it would be the same with tossing a coin. If we toss a coin 10 times, each toss is slightly different—the height, the direction, which side is up when we start, and so on. Yet, we assume that the probability of heads is constant.

Agrafos: But without those slight differences or variations, shouldn't we get the same result?

Kokinos: I guess we assume that the probability is the same even given those variations but they are what explain why the results are not the same every time.

Mathitis: Or, we could assume that with each variation of those factors the probability changes. I guess I hadn't thought of the probability of heads being a random variable.

Kokinos: If we assume that the mean is a random variable, then we can calculate the probability that the mean is a certain value or in a certain range.

Socrates: We started this discussion by thinking that p was the probability of making an error. We discovered that it is merely the conditional probability of getting the actual or more extreme results assuming the null hypothesis was true. Is that even the question we want to answer?

Daskalos: Why not? It's how hypothesis testing is done.

Socrates: Think of what we were saying about p. We wanted to say it is the probability of being right or wrong, or making an error, as Peismon put it.

Mathitis: You're right. We want to know the probability that H_o is true given the data.

Socrates: But what does p give us?

Daskalos: The probability of the data given H_o.

Agrafos: If that's the reverse of what we want, why do we do it that way?

Socrates: Consider the null hypothesis that the mean equals 10. Using the frequentist approach, we can only test whether if the null hypothesis is true what results do we expect. Do we think the mean is exactly 10?

Kokinos: No, not exactly 10.

Mathitis: Deming has said we already know it is not, so why test?

Kokinos: It could be 10.000001, which is not 10. I think that's what Deming meant. Test whether it is within a range.

Socrates: If we assume that the mean is a random variable, then we can estimate the probability it is within that range using the data.

Daskalos: That's the Bayesian approach.

Socrates: Consider this. Kokinos may want to test whether a treatment is better than a placebo or control. The frequentist approach answers what question?

Kokinos: We assume there is no difference, that the treatment has no effect. The question might be "Does the treatment have an effect?"

Socrates: When you are done with the data collection and analysis, what answer do you have?

Kokinos: There is evidence of an effect because we reject the assumption of no difference. Or, we fail to reject the assumption of no effect or we accept that there is no difference.

Agrafos: Isn't that what we want to answer?

Socrates: Or, is it this: "What is the probability that the effect is at least delta?"

Mathitis: That question makes more sense to me.

Socrates: To apply a probability distribution, the second question requires the effect to be variable so we can model the variability with a probability distribution.

Kokinos: The effect certainly is not the same for every application. I would say that a variable effect or variable mean reflects reality more than a fixed or constant effect or mean.

NONPROBABILISTIC DECISIONS

Socrates: While we have talked about statistical hypothesis testing, the same framework applies to all decisions whether statistical or not.

Mathitis: How?

Socrates: Whenever you have one theory or hypothesis or assumption or guess, we can consider it as the null hypothesis and its negation as the alternative. Let's create that generic table.

Daskalos: Like in Table 10.4.

TABLE 10.4

Hypothesis Testing for Nonstatistical Hypotheses as Statements about What Is True

	Decision	
Truth	**Accept Null**	**Accept Alternative**
Null (statement true)	No error	Error (Type I)
Alternative (statement false)	Error (Type II)	No error

Socrates: Yes.

Mathitis: Every decision about what is true is simply implies that a statement is either true or not.

Socrates: Do we make such decisions in solving problems?

Mathitis: Sure. We often use matrices to prioritize problems or solutions. Sometimes, we make decisions about possible causes. We assign weights and then rank or vote on the options.

Socrates: Let's look at the prioritization decision. We could hypothesize that solution A is best, by some definition of best, for example, most effective, fastest, cheapest. It doesn't matter the reason. What is the null hypothesis?

Daskalos: The statement "Solution A is the best." The alternative is that it is not the best.

Socrates: Then, we decide by whatever way you mentioned—ranking, voting, throwing darts—whether to accept that solution A is the best. Can we make errors or be wrong in our decision?

Mathitis: Yes, of course. If it is the best and we reject it, we made one type of error, and if it is not the best but we decide it is, we make a different type of error.

Daskalos: Just as it is shown in the table.

Socrates: Then, do the concepts of confidence, power, and the two types of errors apply to all these decisions?

Daskalos: It appears so.

Peismon: But we don't know what the probabilities are.

Socrates: I think that is why we don't view these decisions as having errors to them. But also remember that probabilities are just mathematical models. We could apply them.

Mathitis: In fact, in some decision-making, we use one factor for decision, "probability of occurrence." We don't use formal probabilities as we do when calculating p, but we still estimate the relative probabilities, for example, low, medium, high, or on a scale of 1 to 10.

Socrates: Then, even when we do not use statistics to make decisions but use opinions, we still have Type I and II errors and may not have enough power to reject the null hypothesis.

Daskalos: I hadn't thought about that. So, when we use subjective decision-making tools, we believe that these types of errors don't occur but they really do.

Socrates: And, there are only two types of errors associated with these two-by-two tables.

Mathitis: Then, for every decision, we have these two types of errors. That means that the chances of making an error increases the more decisions we make.

Kokinos: Sometimes, we are comparing three or more solutions using five or more criteria. Are you saying that, for each combination of solution and criterion, we have the two types of errors?

Mathitis: It certainly appears so.

11

Myth 11: Need More Data for Discrete Data than Continuous Data Analysis

Scene: *Break room (instructor introduced sampling in the morning session)*

BACKGROUND

Mathitis: How was class this morning?

Socrates: Interesting. It covered some topics we discussed the other day about data types. Daskalos repeated what Peismon had said that we need to know the type of data because statistical analyses depend on the type of data.

Mathitis: That's right. There are discrete data statistical tests and continuous data statistical tests.

Socrates: He also said that if you have a choice, use continuous data because you need a larger sample size for discrete data analysis. Why is that?

Mathitis: For example, I could say something is on time or late, those being the two choices. Those data are discrete. But it is more informative to know how much late or how much on time. Knowing the amount of time something took is more informative as it tells us the degree of lateness. We can always convert the times to on-time or late but not the reverse.

Socrates: That makes sense. But what about when continuous data do not exist? For example, colors of the car are discrete data and can't be made continuous by simply numbering the colors.

Peismon: That's true.

Socrates: Is that why the advice is to use continuous data when possible but not always possible?

Peismon: Yes. The advantage is that you need less data when doing statistical analysis with continuous data than with discrete data.

Kokinos: That's what I was taught. There wasn't an explanation, but some examples were shown.

Socrates: This refers to sampling and making inferences based on the sample.

Peismon: Yes.

Socrates: Remember that we had determined that discrete applies to both qualitative and quantitative scales (Myth 1). Counts are discrete but quantitative, but distances are continuous and quantitative. Are you saying that quantitative discrete data requires a larger sample size than quantitative continuous data for making inferences?

Peismon: Yes. Discrete data tests require a larger sample than continuous data tests.

Kokinos: Testing proportions requires more data than testing means, for example.

DISCRETE EXAMPLES WHEN $n = 1$

Socrates: Let's look at some examples so I can understand. Suppose you want to test whether replacing a lightbulb will result in the light going on when you flip the switch. Is this a discrete data test or continuous?

Peismon: Discrete because there are only two choices, it's binary. It either works or does not work, light versus no light, or pass versus fail.

Socrates: How many times would you need to flip the switch to see if it is good or bad?

Agrafos: (laughing) Actually, I do flip the switch a couple of times.

Mathitis: But, you're right Socrates, you just need to flip it once.

Socrates: So the sample size is one?

Mathitis: Yes.

Socrates: Can a continuous test need less than one as a sample size?

Mathitis: No.

Socrates: So, why do you think discrete data tests require more data?

Peismon: Well, this isn't really a probability test.

Socrates: Don't probabilities range from 0 to 1? Isn't this a case of the probability being either 0 if it doesn't work or 1 if it does?

Peismon: Probabilities do range from 0 to 1, but what is the random variable? You need a random variable for probabilities to apply.

Socrates: It is a random variable as defined by the two choices and the frequency for each. In this case, the frequencies are 0% and 100%.

Peismon: I'm not convinced. This case is deterministic not probabilistic.

Socrates: Okay, let's consider another case then. I have two jars each with 100 marbles. Jar A has 99 red marbles and 1 blue marble and jar B has 99 blue marbles and 1 red marble. If I randomly select from jar A, what is the probability of getting a red marble?

Peismon: Ninety-nine percent—the same probability of getting a blue marble from jar B.

Socrates: So, do we now have probabilities?

Peismon: Now we do.

Socrates: If I randomly select a jar, how many marbles must I take to be 95% confident I selected jar A? If my two hypotheses are H_o: jar A and H_a: jar B, how large of a sample size must I take?

Peismon: I don't know if the percent of marbles in the jars gives me the confidence level.

Agrafos: Why wouldn't it?

Kokinos: If you randomly select from jar A, then the probability of getting a red marble is 99%. But we want to know if you get a red marble, what is the probability that it came from jar A? That's a conditional probability.

Socrates: What is that probability?

Mathitis: If we let "|" represent "given that" or "conditional on," then the conditional probability is: Prob(jar A|red marble) = Prob(red marble|jar A) × Prob(jar A)/[Prob(red marble|jar A) × Prob(jar A) + Prob(red marble|jar B) × Prob(jar B)] = 0.99 × 0.5/[0.99 × 0.5 + 0.01 × 0.5] = 0.99 × 0.5/0.5 = 0.99.

Socrates: This discrete test only needs a sample of $n = 1$. If we change the number of red or blue marbles in each jar to 98, 97, 96, or 95, will my sample size change to be 95% confident?

Peismon: I don't know. I'd have to check.

Agrafos: By what Mathitis said, the probability of a red marble coming from jar A given that we have a red marble is simply the probability of getting a red marble from jar A.

Peismon: I'm still not convinced.

Socrates: How many red marbles are there altogether?

Peismon: One hundred.

Socrates: Suppose we line only the red marbles so that the first 95 are from jar A and the next 5 are from jar B. What is the probability of selecting a red marble from jar A?

Peismon: However, these aren't random. You've ordered them.

Mathitis: But if we randomly select a number from 1 to 100 and let that tell us which marble we choose, then it is random. It doesn't matter whether they are lined up randomly. Just as in the jar, it didn't matter whether they were randomly placed in there or not.

Socrates: Once we only have red marbles, and select one knowing whether it came from jar A or from jar B, we can now know the probability of having selected jar A if the marble is red.

Mathitis: As I said before, it is simply the proportion of red marbles from jar A.

Agrafos: Then, we do have at least 95% confidence if the percent of red marbles is at least 95%.

Socrates: How large of a sample size did we need?

Agrafos: Only one.

Socrates: Then we have several cases—and we can create an infinite number of similar cases—where we would only need an $n = 1$ using discrete data. We know that many instances of continuous data analyses require more than a sample size of 1. Do you agree, then, that not every discrete data analysis will require more data than every continuous data analysis?

Mathitis: Yes, now it's obvious.

Peismon: I didn't mean every single analysis. I meant generally you need more data when the data type is discrete rather than continuous.

Socrates: When you say generally, what do you mean by that?

Peismon: If we have continuous data to analyze and convert that to discrete data and analyze it, we will need more data generally for the discrete data analysis.

Socrates: And when you say generally but not always, do you mean the majority of the time?

Peismon: Yes, I would say so.

FACTORS THAT DETERMINE SAMPLE SIZE

Socrates: Let's see if that's the case. Returning to the general concept of hypothesis testing, Peismon, what are the factors that determine sample size?

Peismon: They were the probabilities of Type I and II errors, alpha and beta; the variability or sigma of the population; and the difference delta we want to detect.

Socrates: Did we choose the values for these factors or are they provided to us?

Peismon: We choose alpha and beta, and we also choose the difference delta we want to detect. The data provide us the variation, sigma.

Socrates: Do we choose the values on the basis of whether the data are continuous or discrete?

Peismon: No, we don't. We typically choose alpha = 0.05 and beta = 0.1 or 0.2. The difference to be detected, delta, is based on how much of a change we want to detect.

Socrates: The only factor we don't choose is sigma, the variation.

Kokinos: You can't control the variation—it is what it is.

Socrates: What is the relationship between population variation and sample size?

Kokinos: The larger the variation, the larger the sample size.

Socrates: For discrete data analysis to require more data than continuous data analysis, assuming that alpha, beta, and delta are the same, would require that the variation for discrete data be generally larger than that for continuous data. Is that the case?

Mathitis: I don't know but I see your point. If three of the four factors that determine sample size are the same, then only the fourth must be different.

Agrafos: How would we determine whether it is?

Socrates: What are some data analyses?

Peismon: You can do hypothesis testing, confidence interval estimates, control charting.

Socrates: Let's start with confidence intervals since we only need three of the four factors: alpha, delta, and sigma. What is a formula for calculating a confidence interval for the mean?

Peismon: That would be for continuous data.

Socrates: Let's assume for a normal distribution.

Mathitis: A confidence interval for the mean has the form sample mean $\pm\Delta$, where Δ is the margin of error or sampling error. If we assume a normal distribution, then we can use the standard normal distribution or Z score, where $1 - \alpha$ is the confidence level and s is the sample standard deviation, then $\Delta = z_{1-\alpha} \times s/\sqrt{n}$.

Socrates: Rearrange the formula for n. Let's denote this sample size as n_c for continuous data.

Mathitis: Solving for n_c, I get

$$n_c = \left(\frac{z_{1-\alpha} \times s}{\Delta}\right)^2 = s^2 \times \left(\frac{z_{1-\alpha}}{\Delta}\right)^2 \tag{11.1}$$

Socrates: Now, what is a confidence interval formula for a proportion, discrete data?

Peismon: Well, we can use a normal distribution approximation so the notation would be the same. The discrete data sample size would be

$$n_d = p \times (1-p) \times \left(\frac{z_{1-\alpha}}{\Delta}\right)^2 \tag{11.2}$$

Socrates: The claim is that $n_c < n_d$, not always but the majority of the time. Are there cases when the reverse is true, i.e., when $n_c > n_d$? Solve this inequality and see if it ever occurs.

Peismon: Then, $n_c > n_d$ whenever $s^2 \times (z_{1-\alpha}/\Delta)^2 > p \times (1-p) \times (z_{1-\alpha}/\Delta)^2$. Canceling the last terms, $n_c > n_d$ whenever $s^2 > p \times (1-p)$. In addition, we know that since p is from 0 to 1, $p \times (1-p)$ is its maximum when $p = 0.5$. Then, $n_c > n_d$ when $s^2 > 0.5 \times (1-0.5)$ or when $s > 0.5$.

Socrates: Is s ever greater than 0.5?

Kokinos: Of course.

Socrates: Is $s > 0.5$ less than half the time?

Kokinos: I don't know. It seems like it wouldn't be.

Peismon: But wait! We could change the scale and make $s < 0.5$. Rather than using millimeters we use centimeters, for example. If $s = 2$ mm, then it would be 0.2 cm.

Socrates: That's clever. But does that make sense that if we change the scale, such as using millimeters instead of centimeters, we would need a smaller sample size?

Peismon: No, it doesn't. So what's wrong with the idea of changing scales?

Socrates: Consider the terms in the equation. What units are used for s and p?

Peismon: s will always have units but p is unitless.

Agrafos: Isn't that comparing apples to oranges?

Socrates: It appears so. That's one reason it doesn't make sense to say you need more data for discrete measures, unitless in this case, than for continuous measures, which have units.

Agrafos: But the sample sizes are unitless.

Socrates: The formulas have a ratio of the variation to the difference: sigma to delta.

Agrafos: Oh, I see. The margin of error Δ has the same units as s in the formula for n_c. So, they cancel when determining the sample sizes. But in the formula for n_d, Δ and p are both unitless.

Socrates: What about hypothesis testing? Let's compare sample sizes for testing two means versus two proportions for the same delta (excluding the units), confidence level, and power. Do you think that the sample size for continuous data will always be smaller than that for discrete?

Kokinos: Not anymore, but let's do the math to confirm.

Socrates: What formulas should we use?

Mathitis: For comparing two means and assuming a normal distribution with a known standard deviation σ, we can use this formula:

$$n_c = 2 \times \left(\frac{\sigma}{\Delta}\right)^2 \times (z_{\alpha/2} + z_\beta)^2 \qquad (11.3)$$

where Δ is the difference between the two means we want to detect, $z_{\alpha/2}$ is the factor for $100 \times (1 - \alpha)\%$ confidence level, and z_β is the factor for $100 \times (1 - \beta)\%$ power.

Socrates: For discrete data, proportions say, what would we replace in this formula?

Mathitis: The standard deviation σ is replaced by the square root of $p \times (1 - p)$ and the difference Δ is replaced by whatever difference we wanted to detect between the two proportions p_1 and p_2:

$$n_d = 2 \left(\frac{\sqrt{[p(1-p)]}}{\Delta}\right)^2 (z_{\alpha/2} + z_\beta)^2 \qquad (11.4)$$

where $p = (p_1 \times n_1 + p_2 \times n_1)/(n_1 + n_1)$ and simplifying by letting $n = n_1 = n_2$.

Socrates: Then, the ratio $n_c/n_d = \sigma/[p \times (1 - p)]$. If in general we need more data for discrete data analysis of this type, then this ratio must be less than 1. Again, the maximum $p \times (1 - p)$ can be is 0.5. Thus, n_c/n_d will be less than 1 when $\sigma < 0.5$.

Kokinos: But we know that standard deviations can be and are greater than 0.5.

Socrates: And for other tests, the formulas just become more complicated but the same analyses can be done with similar conclusions.

Mathitis: Sample size depends on several factors—none of which include whether the data are discrete or continuous.

Agrafos: So we can't say that the analysis for one type of data requires more or less data than for another type of data.

Kokinos: But is that a fair comparison? You're comparing one analysis on one thing with other analyses on other things.

Agrafos: What do you mean?

Peismon: I agree with Kokinos. This is comparing tests on proportions with tests on means.

Socrates: What would be a fair comparison?

Peismon: Comparing the same thing: means versus medians, for example. Both are measures of central tendency.

Socrates: But means and medians are not the same unless the distribution is symmetrical. Otherwise, the means could be different and the medians not, or vice versa.

Kokinos: Wouldn't both means and medians be for continuous data anyway, so we wouldn't really be comparing continuous versus discrete?

Socrates: That's often true but not always. Let's finish this discussion first.

RELEVANCY OF DATA

Agrafos: But I'm still confused why we don't need less data for continuous measures when it has more information. That's still true.

Socrates: Yes it is.

Kokinos: Yet, the calculations we did comparing different situations showed that it's not always true and maybe not even the majority of the time.

Peismon: But it also follows naturally from the comment that continuous data are more informative. Having more information allows you to collect less data.

Socrates: Would that depend on whether the additional information is relevant?

Mathitis: I'm not sure I follow.

Kokinos: I don't understand either.

Socrates: Suppose I want to compare the amount of space in two houses. Knowing the square footage of both houses is sufficient to tell me that. Knowing when they were built is additional information. However, that information doesn't help me in analyzing which house has more space. In this case, more information is not necessarily better.

Agrafos: I agree with Socrates. The information has to be relevant.

Mathitis: Yes, you're right.

Socrates: When there is no more relevant information—as in the case of discrete data—then why would more data be required?

Peismon: I would rather know not just whether something is good or bad but the extent it is good or bad.

Socrates: But we agreed that more information may or may not always be relevant.

Peismon: That I understand. But in the case of good/bad versus degree good/bad, it is relevant.

Socrates: Maybe. It depends on why you are collecting the data. Knowing that one car has a 15-gallon gas tank and gets 30 miles to the gallon is more information than just knowing another car can take you 480 miles on a full tank of gas. However, the additional information does not help you anymore in determining which car will go farther on a tank of gas.

Peismon: But how far away a result is from a specification limit is relevant information. I want to know whether I am really good or barely good or barely bad, not just good or bad.

Socrates: My question isn't whether that information is relevant or not at all in a vacuum or by itself or generally, but whether it is relevant with respect to determining a sample size.

Peismon: How could it not be?

Socrates: Why are you collecting data? What are you going to do with it?

Peismon: I would analyze it.

Socrates: Specifically, what analysis?

Peismon: I could estimate my average performance or variation to see how well I am doing. I could do some hypothesis testing.

Socrates: Look again at the formula for determining sample size for estimating the mean. Does it ask for the specifications?

Peismon: No.

Mathitis: And, neither does the formula for estimating proportions or any
other parameter.

Kokinos: But the proportion test requires you apply the specifications to
get the proportion.

Mathitis: Yes, for proportion good or defective.

Socrates: Let's look at it another way. Assume you have lower and upper
specifications and the target is exactly in the middle. Suppose
you want to know whether the average is statistically the same as
the target. Does the sample size needed to test that hypothesis of
no difference between the mean and target change whether you
have wide or narrow specifications?

Mathitis: No, it's independent of the specifications.

Socrates: What about testing whether two or more means are the same?
Does the amount of data needed depend on where or how wide
the specifications are?

Mathitis: No.

Kokinos: I see now why you asked whether the formulas for sample size
include specifications. Since they do not, then how close a result
is to a specification limit is irrelevant information for these
types of analyses.

Socrates: For determining sample sizes for these types of analyses. The formulas tell us that the variation, the confidence level, the power,
and the difference to detect are the only factors that determine
sample size. Consider this: which is larger, 3% or 2.5 inches?

Kokinos: You can't answer the question as they are in different units.

Mathitis: But you don't need to. Whether they refer to the difference Δ or
the standard deviation σ, it doesn't matter. What does matter
is the relative size of the standard deviation to the difference in
whatever unit each is.

Socrates: So, we can conclude that there is no general rule that more data
are needed for discrete data analyses versus continuous data
analyses.

Agrafos: But what was the point of saying that you need more data for
discrete data analysis?

Socrates: Excellent question, Agrafos. What do you think was the point?

Kokinos: I think it goes back to why Peismon was arguing it was true: continuous data have more information. It was to try to get people to use continuous measures rather than discrete measures.

Mathitis: Then just say that.

Agrafos: Also, as Socrates mentioned, you can't always have continuous data so that might lead people to get the wrong measure—as I was told to do by converting from measuring whether a call was transferred to what proportion of calls were transferred. It was the wrong measure.

Mathitis: Or, at least, a measure on a different thing.

12

Myth 12: Nonparametric Tests Are Less Powerful than Parametric Tests

Scene: *Break room (Instructor introduced sampling in the morning session)*

BACKGROUND

Agrafos: Daskalos also said that parametric tests are more powerful than nonparametric tests.

Peismon: Recall that hypothesis testing can have two types of errors because there are two competing hypotheses. We collect data and decide whether to accept or reject one of these hypotheses. If we reject, we could be wrong; if we accept, we could be wrong. Power is the ability to make the right decision when the alternative hypothesis is true.

Agrafos: Okay, so when is one test less powerful than another?

Peismon: It simply means that a less powerful test needs more data or a larger sample size.

Agrafos: Isn't that the same as discrete versus continuous? Aren't nonparametric tests discrete tests?

Mathitis: No, not all parametric tests are applied to discrete data.

Socrates: What is the difference between a parametric test and a nonparametric one?

Mathitis: In general, a parametric test assumes a probability distribution while a nonparametric does not. Nonparametric tests are also called distribution-free tests, for that reason. Typically, the assumed distribution is a family of distributions with each member distinguished by one or more parameters. A nonparametric test does not assume a parametric distribution, so we don't have to estimate any parameters.

Kokinos: In my class, nonparametric meant it did not assume a normal distribution. It could assume some other distribution.

Mathitis: I think there are different versions of what nonparametric and parametric mean. Parametric certainly means that the distribution you assume requires the value of one or more parameters to be either assumed or are estimated. That's why assuming a normal distribution to do a test would be a parametric test. It has two parameters, the mean and standard deviation, which must either be assumed to have specific values or must be estimated or tested to have specific values.

Agrafos: So, a test on the mean of a normal distribution is a parametric test.

Mathitis: That's right. Like the z test or t test.

Peismon: But note that the z test would assume you knew the standard deviation while the t test assumes you do not know either the mean or the standard deviation.

Kokinos: But what about testing whether a sample comes from a normal distribution? We aren't assuming any values for the mean or standard deviation nor test on those parameters.

Mathitis: My understanding of the literature is that tests like that can be classified as nonparametric. Testing the shape of a distribution, not just whether it is normal or exponential or Weibull, but also whether it is symmetrical, would be nonparametric tests.

Agrafos: Would that include whether two or more distributions are the same?

Mathitis: Yes, I would say so.

Socrates: Is a test on proportions be a nonparametric test?

Mathitis: Yes, because the proportion could come from any distribution. The test does not require the assumption of a particular distribution. It could be symmetrical or not, continuous or not, one mode or more.

Socrates: Recall our discussion of two jars with different amounts of red and blue marbles. Would a test on the proportion of red marbles be a nonparametric test or a parametric test?

Mathitis: That would be nonparametric.

Socrates: What kind of test would it be on the mean diameter of the marbles?

Peismon: If you assumed a normal distribution, you would be using a parametric test.

Mathitis: But you can also assume other parametric distributions besides the normal distribution and the tests would be parametric.

Socrates: If a test on the mean diameter of the red marbles required a larger sample compared to a test on the proportion of red marbles, would the nonparametric test be more powerful?

Mathitis: If that were the case, then it would be, but I doubt that it is.

Socrates: What if I tested using a nonparametric test and higher error rates—for example, alpha = 0.1 and beta = 0.2—versus a parametric test using lower error rates—for example, alpha = 0.01 and beta = 0.05? Would I still require more data for the nonparametric test?

Kokinos: Of course not, Socrates. You have to be comparing under the same conditions.

Socrates: Then, the comment the instructor made does not mean that every nonparametric test requires more data than every parametric test.

Peismon: No, not at all. In my experience you typically need more data to do nonparametric tests than parametric tests because they are less powerful.

Kokinos: Isn't that redundant if less powerful means larger sample size?

DISTRIBUTION FREE VERSUS NONPARAMETRIC

Socrates: Mathitis, you said that distribution-free tests are the same as nonparametric tests.

Peismon: I think the terms are used interchangeably.

Socrates: By distribution free, we mean there is no distribution assumed?

Mathitis: I think it means no particular distribution.

Peismon: No, it means no distribution at all.

Agrafos: Isn't that the same?

Mathitis: No particular distribution means there is one but we don't know which one or it could be anyone. For example, it's the difference between saying "a car" and "the car." You forget where you parked your car and ask "Where's the car?" referring to your car in particular. You wouldn't ask "Where's a car?"

Agrafos: I understand. So Peismon is saying there isn't any car—or distribution.

Socrates: By distribution, we mean a probability distribution?

Peismon: That's right, like a normal distribution or exponential.

Socrates: But when we do hypothesis testing, we always calculate p (Myth 10)?

Peismon: Yes, and compare to alpha.

Socrates: p is a probability.

Peismon: Yes, that's right. We discussed this the other day.

Socrates: How can we calculate a probability if there is no probability distribution?

Mathitis: We don't when we do descriptive or enumerative statistics.

Agrafos: You mean like sports statistics: batting average, shooting percentage, and so on?

Mathitis: That's right. Those statistical analyses do not need assumptions about the distribution.

Kokinos: Those are mathematical calculations that we call statistics. It is one type of statistical analysis. We have the entire population.

Socrates: But even when we have the entire population and we want to describe some probabilistic characteristics, we must assume a probability distribution.

Kokinos: Why? Give me an example.

Socrates: What is the probability of getting 3 heads in a row when flipping a coin?

Kokinos: One-half to the third power—1/2 times 1/2 times 1/2—that's one-eighth.

Socrates: Why did you use 1/2?

Kokinos: That's the probability of getting one head.

Socrates: Why?

Mathitis: Kokinos, you assumed that it was a fair coin. That's a probability distribution.

Socrates: I didn't say it was a fair coin, I just said flip a coin, an arbitrary coin. What if the probability of one head is 0.1 or 0.783? Does that change your calculation?

Kokinos: Yes, but if I don't know that how could I use the right probability?

Mathitis: You can't calculate p if you don't know what probability to use. That's Socrates' point.

Socrates: Knowing or assuming it. What about inferences from a sample to a population?

Mathitis: When we do statistical testing and calculate p, it is always based on the assumption that the null hypothesis is true.

Peismon: Yes, that's true, as we discussed earlier.

Socrates: Therefore, we do make an assumption: we assume the null hypothesis is true.

Peismon: That's right. But it isn't always an assumption about a distribution.

Socrates: Can you give me an example?

Peismon: Testing whether two means are equal.

Socrates: Two means from a normal distribution?

Peismon: We do sometimes assume a normal distribution but it isn't required. They could be skewed distributions, but the central limit theorem states that the sample means have approximately a normal distribution.

Socrates: It's the distribution of the statistic, the sample mean in this example that is approximately normally distributed?

Peismon: That's right. We made no assumption about the original population from which the samples came.

Mathitis: But I see what Socrates is getting at. When we do the test on the statistic, we are assuming that the distribution of the statistic is approximately normal.

Socrates: If you did not know the standard deviation, you would use the *t* test.

Mathitis: That's right.

Socrates: Are you not assuming then that the sampling distribution has a *t* distribution?

Mathitis: Yes, of course. I see. Then, every test assumes that the sampling statistic has a particular distribution. Using the *F* test, you assume that the sampling distribution is the *F* distribution, the chi-square test assumes chi-square distribution, and so on.

Peismon: But that is not an assumption about the population distribution.

Socrates: Can we assume a sampling distribution without an assumption about the population distribution? In either case, we are making assumptions about probability distributions.

Kokinos: Then, there really isn't any distribution-free test?

Peismon: I'm not convinced.

COMPARING POWER FOR THE SAME CONDITIONS

Socrates: Let's return to the issue. You said you should be comparing under the same conditions. Should you also be comparing the same things?

Mathitis: I'm not sure I understand. Aren't the same things part of the same conditions?

Socrates: The examples the instructor gave were not on the same thing. He gave examples of nonparametric tests on medians compared to parametric tests on means. Means are not medians.

Agrafos: What Socrates says is true.

Mathitis: The mean equals the median only when the distribution is symmetrical.

Kokinos: Did he actually compare a median test to a mean test?

Agrafos: No, these were just examples of different kinds of tests.

Peismon: But if you tested the same thing under the same conditions except for the type of test, a parametric test is more powerful than a nonparametric test, in general.

Socrates: What are all those same conditions?

Peismon: The four we spoke of before: the probability of a Type I error or alpha, the probability of a Type II error or beta, the difference or delta that the test is to recognize, and the variation or sigma of the population from which the samples come.

Mathitis: Assuming that alpha, beta, delta, and sigma are the same for two tests, the test that needs a smaller sample size to comply with these values is the more powerful one.

Socrates: Are we saying then, that every nonparametric test, regardless of the values of alpha, beta, delta, and sigma, always requires more data than every parametric test?

Peismon: No, only for the same values of alpha, beta, delta, and sigma.

Socrates: If these four values are the same, won't the sample size be the same for different tests?

DIFFERENT FORMULAS FOR TESTING THE SAME HYPOTHESES

Mathitis: No, because the formulas are different as they depend on different assumptions.

Socrates: Assumptions about probability distributions? Can we go through an example? You said that a test on means is parametric and a test on proportions is nonparametric. Correct?

Mathitis: Yes, that's correct. However, the formulas for power often don't exist or require software because the equations are intractable and so are approximated or solved iteratively.

Socrates: The instructor mentioned several statistical software programs. Could we use them to check the claim that nonparametric tests are less powerful than parametric tests?

Mathitis: Why not? I have some software packages on my laptop we could use.

Socrates: Because proportions can only be in the range of 0 to 1 and the maximum sample size occurs when $p = 0.5$, let's use these numbers.

Mathitis: The nonparametric proportion test is testing whether a proportion $p = 0.5$ versus $p \neq 0.5$, and for the parametric mean test, we are testing whether a mean $\mu = 0.5$ versus $\mu \neq 0.5$?

Socrates: Yes. The standard deviation for the proportion test is the square root of $p \times (1 - p)$, which is 0.5 when $p = 0.5$. For the means test, we can use sigma = 0.5. Let's test at 95% confidence or alpha = 0.05 and 80% power so beta = 0.2.

Mathitis: What difference, delta, should we use?

Socrates: Proportions range from 0 to 1. Select something in that range.

Kokinos: Let's start with 0.1.

Mathitis: With one program, using delta = 0.1, I get $n = 199$ for the one-sample t test on the mean, $n = 197$ for the z test on the mean, and $n = 194$ for the proportion test. Using another, I get $n = 198.15$ for a one-sample t test and $n = 196.22$ for the proportion test. With a third, I get $n = 199$ for the mean test and $n = 187$ for the proportion test. I don't know what formulas are used in the third program. But the results surprise me.

Socrates: It doesn't appear that for the same conditions of all four factors, nonparametric tests require more data than parametric tests.

Kokinos: I don't understand why. Are you sure you entered the right information?

Mathitis: Here, look.

Kokinos: I still don't believe it.

Socrates: Also, there was a larger difference with the third program.

Mathitis: Software programs use different formulas. The only thing I can think of is that for the test on proportions, programs often use a large-sample approximation rather than an exact test.

Socrates: How would that affect the results?

Mathitis: The large-sample approximation is based on normal distribution theory. In that case, the formulas would be almost the same. Using the same numbers in the nearly identical equations will produce nearly identical answers—which is what occurred.

Socrates: Is the nonparametric test more powerful since it always had a smaller sample size?

Mathitis: Yes, more powerful means needing less data.

Peismon: These must be special cases. Typically, the differences we want to compare with proportions are smaller than the difference with means, so a larger sample size is needed.

Kokinos: It certainly seems that way to me.

Mathitis: But that could be just coincidence with your applications. A finite number of cases doesn't mean that it's true for an infinite number of possibilities. As we have shown, for the same four factors that determine sample size, we get very similar results.

ASSUMPTIONS OF TESTS

Socrates: Let's look more carefully at those assumptions. You said the normal distribution is a parametric distribution with two parameters.

Mathitis: Yes, the mean and the standard deviation.

Socrates: Then, depending on which we don't know, we use different tests: z or t test.

Mathitis: Yes. That's what a parametric test would determine.

Socrates: How does the assumption of normality being wrong affect the test? What if the value you assumed for the standard deviation is wrong?

Mathitis: We would be more likely to make an error in our decision. If we thought we were testing at 95% confidence, it could be 75%. Or, it could be much higher.

Socrates: And power?

Mathitis: The same. If we thought we had 90% power, maybe it's only 50% or 5%.

Socrates: And sample size is a function of both confidence and power?

Mathitis: Yes. If those were actually a lot less, we would need more data, a larger sample size.

Socrates: It appears that if a parametric test is more powerful, it could be because the assumptions are correct.

Mathitis: Or nearly so.

Socrates: Do you know how "nearly so" it has to be to be more powerful?

Mathitis: I don't. But the large-number theorems identify conditions under which normal distribution approximations can be used. These theorems give us confidence that parametric tests are more robust against small deviations from the assumptions.

Socrates: How does that apply?

Mathitis: If we have large sample sizes, like we did for the proportion versus mean example, the assumption of normality is not critical. However, for small sample sizes, not only can we not confidently check our parametric assumptions, but we can't rely on these large-number theorems.

Socrates: Are you saying that one application of nonparametric tests is when sample sizes are small because of the lack of robustness?

Mathitis: That's what I'm saying. That might be enough to warrant using a less powerful test.

Socrates: If the assumptions are true, why use the less powerful test when you have less data?

Mathitis: It's because we don't know whether the assumptions are true and can't confidently check them.

Socrates: If fewer assumptions are needed, then it is less likely to be invalid, right?

Mathitis: I suppose so. So, I guess the question is whether the loss of information by converting to a nonparametric test has a greater impact than violating an assumption about the distribution.

COMPARING POWER FOR THE SAME CHARACTERISTIC

Kokinos: But the more I think about the proportion versus mean example we did, the more I think it's not a good case for determining whether parametric tests are more powerful. When we want to test the mean and don't have a normal distribution, then rather than use a one-sample *t* test, we would switch to a nonparametric test on the median, not on a proportion. Your comment about not testing the same thing is why the proportion versus mean comparison didn't work.

Mathitis: That's true. While the mean and the median are not the same, they are both measures of central tendency.

Socrates: Why would you test the medians rather than the means?

Mathitis: When the distribution is not normal because of skewness or asymmetry, then the mean does not always represent the middle of the distribution. But for skewed distributions, the median will be closer to where the bulk of the data are.

Socrates: As Kokinos said, proportions or medians versus means are not comparing the same things.

Mathitis: I guess you are right about that. However, there are tests that would compare the same thing. For example, we might use a sign test that is distribution-free or nonparametric.

Socrates: How does it work?

Mathitis: I have some data from a recent laboratory study we did. Let's use that (Table 12.1). There were 11 subjects and we applied a control and a treatment to each. The Difference column is Treatment—Control. The null hypothesis is that, on average, the Treatment and Control have the same effect. Because each subject had the same two conditions, we used a paired *t* test.

Socrates: You look at the difference between Treatment and Control for each subject.

TABLE 12.1

Control and Treatment Data to Compare Paired *t* Test versus Sign Test

Subject	Control	Treatment	Difference	Sign
1	11.8	10.6	−1.2	0
2	23.35	34.6	11.25	1
3	42.25	112.6	70.35	1
4	14.2	40.6	26.4	1
5	14.5	7.6	−6.9	0
6	37.6	118.6	81	1
7	19.15	91.6	72.45	1
8	16.75	22.6	5.85	1
9	32.05	57.1	25.05	1
10	42.55	72.1	29.55	1
11	29.35	16.6	−12.75	0

Kokinos: That's right. You need two results from the same subject or testing unit. Another common use of this test is to compare knowledge before and after an event, for example, a lesson.

Socrates: Now, the null hypothesis H_o is that the average difference is 0 versus the alternative H_a that it is not zero. Is that correct?

Kokinos: Yes, that is one possibility. But there are two others. We could be testing whether the Treatment is more than the Control or is less than the Control, on average.

Mathitis: In our case, we expected the Treatment to result in higher values.

Socrates: Then, we have this: H_o: mean difference ≤ 0 versus H_a: mean difference > 0.

Mathitis: Right. The sign test merely identifies each difference as consistent with H_o by assigning the difference 0 or consistent with H_a by assigning the difference 1.

Kokinos: It's like tossing a coin. Assign 0 if it is tails and assign 1 if it is heads. Then, we are checking whether we had a greater number of heads than expected.

Mathitis: Kokinos is right. In this case, we got the equivalent of 8 heads out of 11 coin tosses.

Socrates: Is that significant?

Mathitis: The mean difference is 27.37 and the standard deviation is 33.30. Using the paired *t* test and the assumption of a normal distribution for the differences, the probability of getting at least this large of an average difference is $p = 0.011$, which is less than 0.05.

Socrates: If we are testing at a 95% confidence level, or alpha = 0.05, then there is a significant difference. Since the average difference is positive, it says that, statistically, the average for Treatment is greater than the average for Control.

Kokinos: The parametric test had enough power to detect the average difference of 27.37.

Agrafos: I see that, since $p <$ alpha. But what about the sign test?

Mathitis: The probability of getting at least 8 subjects with Treatment > Control is the same as getting at least 8 heads in 11 coin tosses. The probability of *h* heads assuming $p = 0.5$ is the combination of 11 things taken *h* at a time multiplied by $0.5^h(1 - 0.5)^{11-h}$. We do this calculation for every value of *h* from 8 to 11 and sum the probabilities. The sum is 0.113, which is greater than 0.05.

Socrates: Now, the nonparametric test is less powerful than the parametric test.

Peismon: Yes. And, we can see why. The parametric test looks at the degree of difference while the nonparametric test looks at just whether Treatment is greater than Control.

Agrafos: That's what you said before: continuous data are more informative than discrete data.

Peismon: Thank you. This proves my point. Without that additional information about the degree of difference, the test was not as powerful. That's why parametric tests are more powerful.

Socrates: The sign test is the same as a proportion test as it considers combinations of 11 things taken 8 or more at a time and their probabilities. It assumes a binomial distribution, doesn't it?

Mathitis: That's true. We calculated the probabilities using $p = 0.5$ as we did in the previous example where we found that the sample sizes were less than for the one-sample t test.

Socrates: Then, this again doesn't really compare the same things.

Peismon: Why not?

Socrates: The sign test is evaluating the difference using a proportion while the paired t test is evaluating the difference using a mean. The sign test checks whether the proportion $8/11 = 0.727$ is different from 0.5, while the t test checks whether the mean difference 27.37 is different from 0.

Kokinos: But the hypotheses are the same. We reject or accept H_0 that the Treatment is no better than the Control.

Socrates: Those aren't the hypotheses. Those statements are general descriptions of the issue.

Agrafos: My mentor said we start with a practical problem and convert it to a statistical problem. You solve the statistical problem and then convert the statistical solution into a practical solution. I think Kokinos stated the practical problem.

Mathitis: We often use that structure. The statistical problem uses precise hypotheses. I agree with Socrates. Note how we stated the precise hypotheses. The parametric test tested whether the mean difference was greater than the zero, which occurs when the Control and Treatment are the same. The nonparametric test tested whether the proportion of differences that are positive is greater than 50%.

Agrafos: I think that when Kokinos said reject or accept that the Treatment was no better, he was stating the practical solution. The statistical one is whether there is a statistical difference in two proportions or in a mean and zero.

Peismon: But in this case, I think those two tests are equivalent.

Mathitis: They are not equivalent if they produce different results or conclusions.

Socrates: I agree. Look at it this way. Is a paired *t* test more powerful than a one-sample *t* test?

Mathitis: Actually, we could compare the paired *t* test to a two-sample *t* test because there are two samples, Control and Treatment.

Socrates: Can we check that?

Mathitis: Sure. Using one program to test H_o: Control Mean ≥ Treatment Mean versus H_a: Control Mean < Treatment Mean, I get $p = 0.026$, which is less than alpha but less than the p for the paired *t* test. This indicates that it is less powerful but still showed a statistical difference.

Socrates: Does that mean we should always use a paired *t* test rather than a two-sample *t* test?

Kokinos: If we could design the study that way, then yes. The paired *t* test removes the variability of the test units and we know that with less variability a smaller sample is needed.

Socrates: Then, doesn't the same apply to nonparametric versus parametric tests?

Mathitis: I don't understand what you mean.

Socrates: Similarly, if we could design the study to use a parametric test rather than a nonparametric test, do so, because it is more powerful.

Mathitis: Exactly!

Socrates: Can we always use a parametric test?

Mathitis: Unfortunately, no.

Socrates: When can't we?

Mathitis: Parametric tests apply to quantitative data. If we have qualitative data, either nominal or ordinal, then we can't use parametric tests.

CONVERTING QUANTITATIVE DATA TO QUALITATIVE DATA

Socrates: The parametric test is always more powerful than the nonparametric test for quantitative data?

Mathitis: In my experience it is.

Peismon: Mine too for the reason I stated before. It provides more information.

Socrates: Even comparing the same characteristic?

Peismon: Yes.

Socrates: And you both think it is because the paired *t* test takes into account the degree of difference. Doesn't that suggest that if the degree of difference was consistently small compared to the variation, it might not be more powerful than the nonparametric test?

Peismon: Why?

Socrates: Let's adjust some numbers and see (Table 12.2).

Kokinos: The mean difference is only 7.06 with a standard deviation of 16.94. But the number of cases where Treatment > Control *increased to 9.*

Mathitis: Now, the paired *t* test yields $p = 0.098$ and the two-sample *t* test has $p = 0.165$; neither are significant. But the sign test has $p = 0.033$, which is significant.

Socrates: Suggesting it is more powerful. It seems that even for cases that compare the same hypotheses, there are more factors to consider than whether the test is parametric or nonparametric.

Mathitis: That's for this example.

Kokinos: In this case, the ratio of the mean difference to the standard deviation is $7.06/16.93 = 0.42$, while in the first case, it was $30.23/33.65 = 0.90$. That's what you suggested, Socrates: a smaller but consistent difference. Is that why?

TABLE 12.2

Modified Control and Treatment Data to Compare Paired *t* Test versus Sign Test

Subject	Control	Treatment	Difference	Sign
1	11.8	4.5	−7.3	0
2	23.35	34.6	11.25	1
3	42.25	51.4	9.15	1
4	14.2	33.6	19.4	1
5	14.5	16.7	2.2	1
6	37.6	67.9	30.3	1
7	19.15	43.7	24.55	1
8	16.75	22.6	5.85	1
9	32.05	37.8	5.75	1
10	42.55	52.1	9.55	1
11	39.35	6.3	−33.05	0

Mathitis: It could be. It could also be that the distribution for the differences is skewed so the assumption of normality or symmetry is not met. There appear to be several possible reasons.

Peismon: The issue for me is that Socrates manipulated the data to show what he wanted.

Socrates: Yes, I did change the data. Since we don't know what the true parameter values are, we don't know which test is correct. Those are issues that help us understand whether the claim about samples sizes for parametric and nonparametric analyses is universally or generally true. And if not, then when is it not and what can we do about it?

Mathitis: Looking at the Difference column, I see some skewness. That would be an issue of the assumptions of the test. For the *t* tests on means, we typically think that normality is reasonable because of the central limit theorem—especially for large sample sizes. But normal distributions are not skewed, as they are symmetrical.

Kokinos: Maybe that is more of a critical issue for the paired *t* test. We can plot the data.

Mathitis: Or, compare the median to the mean. If they are not equal, then it's skewed. But I suspect we might have too few data to really tell about the population these came from.

Peismon: In the first case, each group's median is greater than its mean: 25.78 versus 23.35 and 53.15 versus 40.6. The Treatment group is much more positively skewed. That could be enough to violate the assumption of normality. In the second case, the control's median is 26.69, which is greater than its mean of 23.35, but the Treatment median is 33.75, less than its mean of 34.6.

Mathitis: For the paired *t* test, we need to look at the differences. In the first case, the median is 25.05 versus an average of 27.37, while in the second case, the numbers are 9.15 versus 7.06.

Kokinos: But that is a critical point we talked about earlier. Parametric tests make assumptions about the distributions. If those are violated, then they may not be as powerful as nonparametric tests because their assumptions are less stringent.

Kokinos: Which gets back to the point that we can't just say parametric tests are more powerful than nonparametric tests.

Peismon: Okay, but we can add the condition "as long as the parametric assumptions are not violated or not violated too much."

Mathitis: But we don't know when it is too much.

Socrates: Consider this situation (Figure 12.1). Because of their skewness in opposite directions, the means can be equal but the medians are not. Is the probability of randomly selecting a value from A that is greater than the common mean $\mu_a = \mu_b$ from B the same as it being less than the common mean $\mu_a = \mu_b$ from B? In other words, would we expect the same test result using a paired *t* test as the sign test?

Mathitis: No. Clearly, B has greater values although the means are the same.

Peismon: That's an example of the assumption being violated too much.

Kokinos: But we can't tell from the table of data. I like the graph to help us understand and perhaps check the assumptions. Besides, the graph is general not specific, like the example.

Socrates: You also have to know the assumptions.

Mathitis: Then, we should change what we say. We now know that "parametric tests are more powerful than nonparametric tests" is false in a universal or general sense when applied to different things under different conditions, to different things under the same conditions, and to the same thing under the same or different conditions. The conditions include the four factors of two types of errors, the variation, and the difference to be detected.

Kokinos: And, the assumptions.

Agrafos: What should we say?

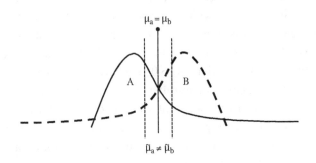

FIGURE 12.1

Two skewed distributions A and B compare the power of the paired *t* test versus sign test. Dotted lines are the unequal medians ($\tilde{\mu}$) while the solid line with the round top is the common mean (μ).

Socrates: As this last example shows, perhaps we should say "Depending on your specific case, parametric or nonparametric tests may be more powerful. Design your test to take advantage of the more powerful test. But also use both to make your decision."

Kokinos: I like that. Doing the analysis is the easy part. Collecting data is hard.

13

Myth 13: Sample Size of 30 Is Acceptable (for Statistical Significance)

Scene: *Break room (continuation of sampling discussion from the morning session)*

BACKGROUND

Agrafos: It just occurred to me that while we have discussed which requires a larger sample size—discrete or continuous data analysis—or whether parametric tests are more powerful than nonparametric tests, neither discussion told us how much data to collect.

Mathitis: Great point, Agrafos. Probably the most frequently asked statistical question is "What sample size do I need?"

Peismon: But, we know that, as a general rule, we should collect sample sizes of 30.

Kokinos: That's what I learned in class also.

Socrates: What is the basis for that?

Peismon: I'm sure others have developed the basis for that guidance. It works for me.

Socrates: I would like to understand that basis. If the rule is not always applicable, then shouldn't we know when it isn't and why? Knowing that would allow us to find more appropriate sample sizes or create ways that are more efficient. Should we oppose such efforts to understand?

Mathitis: I don't think so. We should seek to have greater understanding, especially if we are going to mentor or teach others.

A RATIONALE FOR $n = 30$

Socrates: So, what is a rationale for $n = 30$?

Kokinos: Someone in my class researched the reason and found one explanation relating the t distribution to the normal distribution and another about estimating the standard deviation.

Mathitis: I've seen discussions on the topic related to testing for normality.

Socrates: How does the t distribution and normal distribution affect the sample size?

Mathitis: She may have referred to the central limit theorem. For large samples, the sample mean approximately follows a normal distribution so you can switch from using the t distribution to the normal distribution to approximate the confidence interval.

Peismon: Remember confidence intervals? Confidence intervals for means have the form: sample mean ± margin of error. Using a normal distribution, the margin of error is this:

$$\text{margin of error} = Z_{1-\alpha/2} \times \sigma/\sqrt{n} \qquad (13.1)$$

where $Z_{1-\alpha/2}$ was the value for the normal distribution corresponding to $100(1 - \alpha)\%$ confidence level, σ is the population standard deviation, and n is the sample size.

Mathitis: That's correct. But if you don't assume a normal distribution for the individual values, you can use the t distribution to approximate the distribution because of the central limit theorem. Then, the margin of error becomes this:

$$\text{margin of error} = t_{1-\alpha/2,\,n-1} \times s/\sqrt{n} \qquad (13.2)$$

Kokinos: Then, $t_{1-\alpha/2,\,n-1}$ replaces $Z_{1-\alpha/2}$ and is the value corresponding to $100(1 - \alpha)\%$ confidence level using the t distribution with $n - 1$ degrees of freedom, s is the estimate of the population standard deviation, and n is the sample size.

Agrafos: While you were doing that, I checked the tables of the normal and t distributions. For 95% confidence, $Z_{1-\alpha/2} = 1.96$. The t distribution values range from 2.093 when $n = 20$ to 1.962 when $n = 1000$. For $n = 30$, $t_{1-\alpha/2,\,n-1} = 2.045$. That's slightly more than 4% more than $Z_{1-\alpha/2} = 1.96$.

Kokinos: What are the differences when $n = 20$ or 40?

Agrafos: For $n = 20$, it's 6.8%, and for $n = 40$, it's approximately 3.1%. For $n = 60$, it drops to 2%.

Kokinos: I suppose you could argue that 4% is not too much of a difference and that more data doesn't get you much of a gain unless it's much larger than 30.

Agrafos: It seems to decrease by half a percent for each 20 additional data points. At $n = 80$, the difference is 1.6%, and at 100, it's 1.2%.

Peismon: That's not very much. That seems to prove the point of $n = 30$.

Agrafos: But when we talked about the sigma scale, you wanted to distinguish between 99% and 99.9% and 99.99%, which are differences less than 1%. Now you are saying that a 2%, 3%, or 4% difference is inconsequential.

Peismon: That's because it is for different applications.

Socrates: Are you saying to use $n = 30$ regardless of the purpose or statistical analysis?

Peismon: Yes, for sample size determination.

CONTRADICTORY RULES OF THUMB

Socrates: But that contradicts the other rules of thumb.

Peismon: What do you mean?

Socrates: You thought discrete data analyses required more data than continuous data analysis. Then, why did you think that $n = 30$ was appropriate for both types?

Peismon: I hadn't thought of those views as being contradictory.

Socrates: You thought that parametric tests were more powerful than nonparametric tests—requiring less data. Then, why did you think that $n = 30$ was appropriate for both types of tests?

Kokinos: Because that's what we were taught.

Socrates: I acknowledge that we seem to be taught these inconsistent rules. Now, we are discussing the basis for those teachings. We are told $n = 30$ is always appropriate.

Peismon: Not always appropriate but in general you can start with $n = 30$.

Socrates: But that's the contradiction or inconsistency. If we need a larger sample for discrete data and nonparametric testing, does that mean we need more than 30? Or, is it that we can have less than

30 for continuous and parametric analyses? Do you see the contradiction?

Mathitis: Don't forget that we showed that the rules for discrete versus continuous and parametric versus nonparametric were false. I gather that you think this rule of thumb might also be wrong.

Peismon: But $n = 30$ could be a good rule of thumb if the variation is small. If it's 27 or 32 rather than 30, it's still a good rule of thumb.

Mathitis: But if the difference is small and negligible, then why make a point of saying that we need more data for discrete than for continuous data analysis and for nonparametric than for parametric data testing?

USES OF DATA

Socrates: We can also consider the purpose for collecting data. Assume you told someone to use $n = 30$. They collect a sample of 30 and then ask, "Ok. I have my 30 results. Now what do I do with the data?" What would you say?

Kokinos: That doesn't make sense. We should know the purpose before collecting data.

Mathitis: It's possible you don't even need to collect data. We saw several examples with the jars of marbles where $n = 1$ was enough.

Socrates: What are the different uses of samples?

Mathitis: They could be estimating or hypothesis testing. They may want to check process stability, so they would be collecting data for control charts, for example. They may be checking whether the data come from a particular distribution.

Socrates: Is the rule of thumb of $n = 30$ applicable to all these uses? That is, would you still recommend 30 regardless of which use?

Mathitis: No, I guess not.

Socrates: Would you not want to know the use of the data before you suggest $n = 30$?

Mathitis: Yes, I would—I should.

Socrates: Let's consider two examples. Sample sizes for estimating a mean using confidence intervals and for testing a mean. Let's suppose the distribution is normal. What are the formulas?

Kokinos: For a two-sided $100(1 - \alpha)\%$ confidence interval estimate assuming we know the standard deviation σ, the sample size is

$$n_e = \left[Z_{1-\frac{\alpha}{2}} \times \frac{\sigma}{\Delta} \right]^2 \tag{13.3}$$

where Δ is half the confidence interval width, σ is the population standard deviation, and $Z_{1-\alpha/2}$ is the standard normal distribution's $1 - \alpha/2$ percentile.

Mathitis: For hypothesis testing at $100(1 - \alpha)\%$ confidence and $100(1 - \beta)\%$ power, the sample size is

$$n_h = \left[\left(Z_{1-\frac{\alpha}{2}} + Z_{1-\beta} \right) \times \sigma/\Delta \right]^2 \tag{13.4}$$

where $Z_{1-\beta}$ is the standard normal distribution's $1 - \beta$ percentile and Δ is the difference to be detected at $1 - \beta$ power.

Socrates: Everything is the same for these two formulas except that there is the extra factor of $Z_{1-\beta}$ in the hypothesis testing formula. How much larger does this make the required sample size for hypothesis testing?

Mathitis: If we divide n_h by n_e, we get

$$\frac{\left(Z_{1-\frac{\alpha}{2}} + Z_{1-\beta} \right)^2}{\left(Z_{1-\frac{\alpha}{2}} \right)^2} = \left(1 + \frac{Z_{1-\beta}}{Z_{1-\frac{\alpha}{2}}} \right)^2 \tag{13.5}$$

Kokinos: Typically, testing is done with power less than confidence—80% and 95%, respectively—so $Z_{1-\beta} < Z_{1-\alpha/2}$. In either case, the ratio $n_h/n_e > 1$.

Mathitis: For that power and confidence level, the Z distribution values will be 0.84 and 1.96. The ratio is 1.43, meaning that we need $(1.43)^2 = 2.04$ or twice the data for hypothesis testing than confidence interval estimation.

Kokinos: If $n = 30$ is appropriate for estimation, then we need $n = 61$ for hypothesis testing.

Socrates: For *this* application and *these* confidence levels and power. It will vary for other applications and other values.

Kokinos: For 90% power, the hypothesis test requires $(1.65)^2 = 2.36$ times more data, or $n = 82$ versus $n = 30$ for estimation.

Mathitis: You are right, Socrates. Sample size certainly depends for what the data will be used.

SAMPLE SIZE AS A FUNCTION OF ALPHA, BETA, DELTA, AND SIGMA

Socrates: Let's just consider hypothesis testing. In this example, we considered alpha and beta, the probabilities of the two types of errors. Let's consider the other two factors.

Mathitis: You mean the difference you want to detect delta (Δ) and the amount of variation in the population denoted as sigma (σ).

Socrates: Yes. Suppose $n = 30$ is appropriate for a particular difference delta. If you increase delta, is $n = 30$ still appropriate?

Kokinos: Of course not. It would be too much. If delta was smaller, it wouldn't be enough.

Socrates: If $n = 30$ is enough for one alpha, is it enough if alpha is smaller or larger?

Kokinos: Again, no. It would be insufficient for smaller alpha and too much for larger alpha.

Mathitis: The same is true for beta. If, for a given beta, $n = 30$ is appropriate, then 30 would not be enough for smaller beta and too much for larger beta.

Kokinos: And also for sigma. If the variation σ is smaller, then a smaller sample size is needed, and if σ is larger, a larger sample size is needed.

Mathitis: As alpha, beta, and delta decrease, n increases; as sigma increases, n increases.

Socrates: Yet, knowing all these facts, we thought that $n = 30$ was always—not always, but in general—appropriate no matter the conditions, the uses, and the types of analyses.

Mathitis: I thought all those things. I guess I hadn't viewed them together but as separate, independent ideas. I see what you mean about contradictory beliefs.

Kokinos: But I asked my instructor if I needed to use the formulas we learned in class, and he said that we didn't want to spend all day calculating sample sizes. He recommended $n = 30$.

Agrafos: We did quite a few in class using software without taking all day.

Peismon: I agree with the instructor. The reason you didn't take "all day" is because you had the information you needed. You can quickly plug in those numbers and click a button.

Socrates: Then, are you saying when you don't have all the information it will take longer?

Peismon: You can't use the software or calculate the sample size manually.

Socrates: So when we don't know why we are collecting data, we should just use $n = 30$?

Mathitis: It seems just the opposite. When you don't have all the information to determine an appropriate sample size, then don't collect any until you do get that information.

Peismon: Yes, and that takes all day or longer.

Kokinos: Or, you can try different combinations and see what it gives you.

Socrates: Also, we spent more than half a day discussing sample size with several examples of how different factors affect sample size. We applied the formulas we learned and then used software to calculate power or sample size curves. Why spend so much time learning about the factors that determine sample size, if $n = 30$ is "good enough" for all or most cases?

Mathitis: Good question, Socrates.

SAMPLE SIZE FOR PRACTICAL USE

Agrafos: Then why that advice? Where did $n = 30$ come from?

Kokinos: What about the comparison of the normal distribution to the *t* distribution?

Peismon: That showed $n = 30$ made them sufficiently close.

Socrates: That showed that we can substitute the normal distribution for the *t* distribution once the sample size is large enough. It doesn't tell us whether the sample size of 30 is good enough for specific uses.

Peismon: I don't see why not.

Socrates: We're doing it in reverse.

Peismon: What do you mean?

Socrates: We can always say the margin of error is ±infinity without collecting any data.

Peismon: Yes, and that's useless.

Socrates: That's the point. Don't you want the margin of error to be small enough to be useful?

Mathitis: Yes. First, decide on the maximum margin of error and then determine the sample size.

Kokinos: That's where delta, the difference we want to detect, comes into play.

Socrates: Suppose we have delta = 5. Then, the margin of error must be no greater than ±5. If it is exactly 5, a 4% difference amounts to 0.2. That might be fine if we needed a sample of 30. What if all we needed was a sample of 15?

Agrafos: What difference does that make?

Socrates: If $n = 30$ gets us very close but not greater than delta, then it's okay. But if $n = 30$ gives us a margin of error much smaller than delta, we collected too much data. If $n = 30$ gave us a margin of error that's half of what we needed, we didn't need that much data.

Agrafos: How do we determine how much we need?

Mathitis: We can work backward. Suppose the margin of error = 2.5 when $n = 30$. To make it simple, let's use the normal distribution formula where $Z_{1-\alpha/2} = 1.96$ for 95% confidence. Then from (Equation 13.1), $2.5 = 1.96 \times s/\sqrt{30}$. That makes $s = 2.5\sqrt{30}/1.96 = 6.99$. Then from Equation 13.3, the maximum sample size needed is $n = (1.96 \times s/5)^2 = 7.5$ or 8, rounding up.

Socrates: We are oversampling by more than 70. If we had used the t distribution, what would n be?

Mathitis: That's an iterative process since the t value is a function of n. Agrafos, what's the $t_{1-\alpha/2, n-1}$ value for $n = 8$ and $\alpha = 0.05$?

Agrafos: It's 2.365, which is approximately 21% greater than $Z_{1-\alpha/2}$. Should we try 21% greater than 8?

Socrates: Yes.

Agrafos: That's approximately 10.

Socrates: If we rearrange the formula, we want $t_{1-\alpha/2, n-1} \times s/\sqrt{n} = 5$ or $t_{1-\alpha/2, n-1}/\sqrt{n} = 5/6.991 = 0.72$. Let's use your table to see if the t value for $n = 10$ divided by the square root of 10 is approximately 0.72.

Agrafos: That's $2.262/\sqrt{10} = 0.72$. Then, for the margin of error of 2.5, only an n of 8 or 10 using either the Z or t distributions would be enough.

Socrates: Now, there may be little difference between $n = 8$ and $n = 10$, But, we can't say that $n = 30$ is good enough. Even using the t distribution, we would have collected three times as much data as we needed.

Peismon: But another reason for $n = 30$ Kokinos mentioned was to estimate the standard deviation.

Kokinos: If I remember correctly, that showed that the confidence interval for the standard deviation stabilized around a sample size of 30.

Socrates: How was that shown?

Kokinos: The person created a simulation that drew samples of sizes 1 through 100 from a normal distribution with a given σ and μ. For each sample, the program calculated the mean, the sample standard deviation, and a 95% two-sided confidence interval for σ. Then, a graph of s for each sample size was produced with the lower and upper confidence limits as three lines.

Peismon: What did that show?

Kokinos: The author claimed that it showed that, around a sample size of 30, the estimate s stabilized and so did the confidence bands around that value.

Socrates: What definition did he use for "stabilize?"

Kokinos: I don't remember anything specific other than looking at a graph and seeing the line for the limits flatten on the graph.

Socrates: Think about what we did with the t and normal distributions. We said that when $n = 30$, the difference between them was approximately 4% and decreased slowly after that. If someone says that's still too large of a difference, then what do we use to make an objective decision?

Peismon: But 4% doesn't seem too large to me. Rather than $n = 30$, it might be 31. So, what?

Socrates: If your paychecks were 4% less, would you think that was OK?

Peismon: That's different.

Kokinos: It does depend on the situation.

Mathitis: I agree with that.

Socrates: Let's see what we can learn. What is the formula for confidence intervals on the standard deviation?

Mathitis: It uses the chi-square distribution. These are the limits:

$$\sqrt{\frac{(n-1)S^2}{\chi^2_{(\alpha/2;n-1)}}}, \sqrt{\frac{(n-1)S^2}{\chi^2_{(1-\alpha/2;n-1)}}} \tag{13.6}$$

where $\chi^2_{(1-\alpha/2;n-1)}$ is the $1 - \alpha/2$ upper percentile, $\chi^2_{(\alpha/2;n-1)}$ is the chi-square distribution lower $\alpha/2$ percentile, S is the standard deviation, and n is the sample size.

Agrafos: The left quantity is the lower confidence interval limit and the right quantity is the upper limit?

Mathitis: That's right.

Agrafos: Then, we can replace the denominators by the chi-square values for various values of *n* as we did with the confidence intervals using the *t* and normal distributions.

Socrates: Let's simplify this. We want to know how wide the interval is, so just divide the lower confidence limit by the upper confidence limit. The numerators cancel, after squaring, leaving

$$\frac{\chi^2_{(\alpha/2;n-1)}}{\chi^2_{(1-\alpha/2;n-1)}} \tag{13.7}$$

Kokinos: I get it. Now, we can see for different sample sizes how the ratio varies.

Agrafos: Do we need to define *stabilize* before we do so?

TABLE 13.1

Percent Change in Ratio of Upper 2.5% Chi-Square Value to Lower 2.5% Chi-Square Value as a Function of *n*

n	$\chi^2_{(\alpha/2;n-1)}$	$\chi^2_{(1-\alpha/2;n-1)}$	Ratio	% Change
2	0.001	5.024	5115.61	NA
3	0.051	7.378	145.70	97.2
4	0.216	9.348	43.32	70.3
5	0.484	11.143	23.00	46.9
6	0.831	12.833	15.44	32.9
7	1.237	14.449	11.68	24.4
8	1.690	16.013	9.48	18.9
9	2.180	17.535	8.04	15.1
10	2.700	19.023	7.04	12.4
11	3.247	20.483	6.31	10.4
12	3.816	21.920	5.74	8.9
13	4.404	23.337	5.30	7.8
14	5.009	24.736	4.94	6.8
15	5.629	26.119	4.64	6.0
16	6.262	27.488	4.39	5.4
17	6.908	28.845	4.18	4.9
18	7.564	30.191	3.99	4.4
19	8.231	31.526	3.83	4.0
20	8.907	32.852	3.69	3.7

Socrates: That would be a good idea. We could see when the change is no more than 4% to be consistent with what we did before. Can you do that Agrafos?

Agrafos: Let me create the columns in this spreadsheet. Okay, here are some results (Table 13.1).

Kokinos: Well, this is different!

Agrafos: It seems that we need $n = 19$ to get a change in the confidence interval width to be no more than 4%.

Peismon: Is this really the same comparison as before?

Kokinos: Why isn't it?

Mathitis: Even if it isn't, if we thought that 4% difference is small enough to ignore, why isn't it small enough to ignore here?

Socrates: Kokinos is right. Before we looked at the interval width or delta. This is the ratio of the upper control limit to the lower—and we squared the result. Agrafos, can you do the analysis but now look at the change in delta, half the difference between the two limits?

Agrafos: What should we use for *s*?

TABLE 13.2

Percent Change in 95% Confidence Interval Width (Delta) as a Function of *n*

n	LCL	UCL	Delta	% Change
2	0.446	31.910	15.73	NA
3	0.521	6.285	2.88	81.7
4	0.566	3.729	1.58	45.1
5	0.599	2.874	1.14	28.1
6	0.624	2.453	0.91	19.6
7	0.644	2.202	0.78	14.8
8	0.661	2.035	0.69	11.8
9	0.675	1.916	0.62	9.7
10	0.688	1.826	0.57	8.3
11	0.699	1.755	0.53	7.2
12	0.708	1.698	0.49	6.3
13	0.717	1.651	0.47	5.6
14	0.725	1.611	0.44	5.1
15	0.732	1.577	0.42	4.6
16	0.739	1.548	0.40	4.3
17	0.745	1.522	0.39	3.9
18	0.750	1.499	0.37	3.7

Note: LCL, lower confidence limit; UCL, upper confidence limit.

Kokinos: In the simulation, s was the standard deviation of the random sample drawn.

Socrates: We don't need to do that since s is the same in both limits. Then, when we calculate the percent change, s or square root of s cancels. To make the math easier, just use $s = 1$.

Agrafos: Easy. Here it is (Table 13.2).

Kokinos: It's slightly different. Now, it says you only need an n of 17.

Agrafos: I checked what it is at $n = 30$. The change in delta from $n = 29$ is 1.9%.

Kokinos: Depending on what we mean by "stabilize" or "good enough," we can get different answers. I think Peismon would still complain if his paycheck were 2% smaller. I would!

Socrates: We've looked at different analyses. We compared the t and the normal distributions. We estimated a standard deviation assuming a normal distribution. What about hypothesis testing?

SAMPLE SIZE AND STATISTICAL SIGNIFICANCE

Peismon: I still think we're missing something. How can so many people say using $n = 30$ is good enough? I've been asked several times not how much data do they need but how much data to be statistically significant. That's different. That specifies a requirement.

Socrates: What do they mean by statistically significant?

Peismon: They usually say at a 95% confidence level, so alpha = 5%.

Socrates: If that is all they care about, then why not say $n = 1$ or 2?

Peismon: That doesn't give 95% confidence!

Socrates: What happens to the margin of error if we decrease the sample size?

Peismon: It gets larger, of course.

Socrates: If they only care about the confidence level and not the margin of error, then $n = 1$ or 2 will be enough to calculate a 95% confidence interval.

Agrafos: In fact, can't we just say plus or minus infinity and not collect any data? I'm 100% confident with that margin of error!

Socrates: Exactly.

Peismon: But that is a useless confidence interval.

Kokinos: I think that's Socrates' point. A person asking simply for the sample size at 95% confidence level doesn't know enough about statistics to know that the width of the interval that is useful or informative must be specified also. As a mentor, you should guide them to specify the interval width or margin of error before you can tell them how much data they need.

Mathitis: We saw that, for hypothesis testing, it's even more complicated.

Kokinos: We seem to be giving inconsistent advice.

Agrafos: And receiving inconsistent advice!

Peismon: Still, $n = 30$ is a rough estimate of the amount of data you typically need. Start there and don't waste time with the formulas.

Mathitis: I think that we've shown that there is nothing typical about $n = 30$.

Socrates: The difference, Peismon, is that we are discussing why that particular advice and trying to understand when it is appropriate and useful guidance. If we are to mentor others, we should have a deeper understanding rather than just blindly following rules that are false or guidelines without understanding when they do not apply or work or there are better ways.

Kokinos: What if each sample costs thousands of dollars? I think it's worthwhile to take all day if you only needed an n of 5 instead of 30?

Mathitis: I would think so. In our course, the instructor also mentioned that you could collect a small sample to estimate sigma. Then, do the calculations to estimate n. If you needed more data, then collect it.

Socrates: What do you think? Should we recommend $n = 30$ when we mentor or teach?

Mathitis: No.

Kokinos: I'll stop using that rule of thumb. I won't give that advice.

Socrates: What should we offer instead since, as Mathitis said, "What sample size do I need?" is probably the most commonly asked statistical question?

Kokinos: I know that the Food and Drug Administration not too long ago started asking for the statistical rationale for sample sizes for studies submitted to them.

Mathitis: If we teach people formulas for sample size or how to use software to determine the sample size, then I would expect them to determine an appropriate sample size. If we don't, then I would teach them to go ask someone who does know for help.

Socrates: What would you teach the helpers and software users?

Mathitis: A better approach, I think, is to teach them a procedure. I would first provide them with the framework for sample sizes. Separate estimation from hypothesis testing, as the first has only one type of error (alpha) and the second has two types of errors (alpha and beta). For hypothesis testing, the framework can be illustrated with software, as it asks for the information needed.

Kokinos: There are numerous online videos and free downloadable software that show how these factors interact. I think I would start there to show them how changing the various factors changes the sample size. I agree with Mathitis that the first step is to ask what analysis they want to do and then go to the simulations.

Mathitis: One possible procedure for confidence intervals might be as follows:

1. Determine the size of the interval desired—the margin of error.
2. Select the confidence level.
3. Determine the sample size.

For hypothesis testing, I might teach them this procedure:

1. Determine the difference they want to detect, delta.
2. Estimate the variation of the population.
   a. A way to approximate the variation if they don't have data is to ask them if they remember the largest and smallest values they have seen and divide that range by six to get a standard deviation estimate.
   b. Another approach is to collect a sample of four or five and use the standard deviation from the sample. They could also be conservative by inflating that estimate, maybe by 10% or 20%.
3. Select the confidence level.
4. Select the power.
5. Determine the sample size.

Socrates: Is that too much information?

Mathitis: As I was thinking of this, I realized that the advice of $n = 30$ is just too cryptic. My thought was to just say "It depends on whether you are estimating or testing or wanting to analyze in some other way" and follow the answer with "That depends on…" and list the factors for those who don't understand the formulas and theory or who have not been taught. For those who do understand the formulas, remind them and assist them with the procedures.

Kokinos: Perhaps we can create a form that standardizes Mathitis' procedure.

14

Myth 14: Can Only Fail to Reject H_o, Can Never Accept H_o

Scene: *Project meeting room (project review with mentor Daskalos)*

BACKGROUND

Daskalos: I see that you have tested three possible root causes using hypothesis tests.

Socrates: That's correct. We were trying to optimize the functionality of a product. Based on some historical data and subject matter experts, we selected these three factors.

Daskalos: What were your conclusions?

Socrates: We identified two factors A and B as significant for optimization and we also found their optimal levels.

Daskalos: And the third factor?

Socrates: We concluded that factor C was not helpful individually or in combination with the other two in improving product performance.

Daskalos: Your report says you accepted the null hypothesis. Remember in class I taught you that, when $p >$ alpha, you fail to reject the null hypothesis.

Peismon: That's what I told Socrates. You can never accept the null hypothesis (Bower and Colton 2003).

Socrates: I'm still not clear why.

Agrafos: Nor I.

Daskalos: The null hypothesis can be viewed as no change or no difference. For example, we might have a null hypothesis that the mean μ is a particular value, for example, $H_0: \mu = 100$. Then, our alternative hypothesis is $H_0: \mu \neq 100$. Let's say we collect data and get a sample mean of 103. Now we assume that H_0 is true and calculate the probability of getting the result 103 or worse.

Kokinos: Why "or worse" and what counts as worse?

Daskalos: We want to consider the worst-case scenario. Values greater than 103 are evidence against H_0, as are values equal to or less than 97. Since the alternative is not equal and the sample mean differed from the hypothesized mean by 3, then we consider all means at least 3 units different from the hypothesized value of 100 as evidence against H_0.

Agrafos: Okay. Then what?

Daskalos: If the probability, p, is less than the risk alpha we were willing to take, then we reject H_0. It means that if H_0 is true, we got results that are so unlikely, we are better off rejecting H_0 and accepting H_a. However, if the probability of the results were not that unlikely, then we fail to reject H_0. We fail to reject because we just may not have enough evidence to support H_0.

Kokinos: The example used in our class was of our legal system. A person is assumed innocent until proven guilty. The null hypothesis is that the person is innocent—they start innocent until they commit a crime. The alternative hypothesis is that the person is guilty. Note that our credo is "innocent until proven guilty." It says we can prove a person guilty not prove a person innocent.

Mathitis: We saw that example in our class also. A verdict of not guilty only means that there may not be enough evidence to convict. There is "reasonable doubt" about the defendant's guilt. Alpha, the probability of a Type I error, indicates the level of doubt that must be exceeded.

Peismon: Another way of looking at it is how scientific theories are tested. We assume a scientific theory and make a prediction on the basis of that theory. We then check whether the prediction was correct or not. If the prediction is right, then we have evidence for it but it doesn't prove the theory. It is only one piece of evidence. We continue making predictions and each correct prediction further supports the theory but does not prove it.

Daskalos: However, if the prediction is incorrect, then it disproves the theory. But it might mean making minor adjustments to the theory or restricting its applicability. For example, a theory might not be true under all conditions. Rather than rejecting it, it may be adjusted to apply except for those conditions. Thus, the principle of the theory is correct but the scope was wrong.

PROVING THEORIES: SUFFICIENT VERSUS NECESSARY

Daskalos: That's why we say "fail to reject." You can't prove the null hypothesis. You can't prove a negative, which is what the null hypothesis states.

Socrates: I can prove that 5–7 is negative.

Daskalos: You know what I mean.

Socrates: Actually, I don't.

Kokinos: The null hypothesis is a statement about what is and the alternative is about what isn't.

Socrates: It sounds like the alternative is the negative—what isn't.

Daskalos: The null hypothesis is about an absence of something. Like in your project, it was the absence of process improvement. You hypothesized that using factor C would *not* improve the process. That's why it is a negative.

Socrates: If it is the use of the word *not* that makes it negative, then why is $\mu = 100$ a negative rather than $\mu \neq 100$ the negative?

Agrafos: That makes sense to me.

Socrates: I get ready to come to work. My null hypothesis is that I don't have my car keys in my pocket and the alternative is I do have them there. I check my pocket and they are not there. Haven't I proven they are not there? Haven't I proven a negative?

Mathitis: They could be there....

Socrates: Consider another case then. I have two jars. Jar A has 100 red marbles and jar B has 100 green marbles. Without you looking, I randomly select a jar and take a marble out. I ask you to guess—hypothesize—which jar it came from. Regardless of the hypothesis, can we not prove which one it came from given the color of the marble we selected?

Mathitis: Yes, in that case, because it is 100% of one thing.

Socrates: Then, the statement that the null hypothesis cannot be accepted or proven doesn't apply to all null hypotheses.

Mathitis: Okay, I accept that.

Socrates: There are other ways to prove hypotheses.

Kokinos: Which ways?

Socrates: Using necessary conditions besides sufficient ones.

Kokinos: What do you mean?

Socrates: The statement "If x then y" says that x is a sufficient condition for y but not necessary. Using the marble example, if all red marbles are large but only half the green marbles are large, then being red is a sufficient condition for a marble being large: If red, then large.

Kokinos: How is that related to hypothesis testing?

Daskalos: It's related in this way. We create an argument by stating an assumption, make a prediction from that assumption, and if that prediction is false, we reject the assumption.

Kokinos: That's what we do to test theories.

Socrates: The method of argument is called *reduction ad absurdum*. If an assumption leads to a contradiction, an "absurd" conclusion, we reject the assumption.

Daskalos: The argument in hypothesis testing can be stated this way:
1. (Assumption): H_0 is true.
2. (Prediction): If H_0 is true, then p will (likely) be greater than alpha.
3. (Inference from 1 and 2): (It's likely that) $p >$ alpha.
4. (Fact based on sample and assumption): $p <$ alpha.
5. (Conclusion: inference 3 contradicts fact 4): Therefore, H_0 is (likely) false.

Peismon: Doesn't that show we can't accept H_0 but only fail to reject? If H_0 is true, p is likely to be greater than alpha but p could be greater than alpha even when H_0 is false. Also, the prediction at step 2 is not 100% true but probably true.

Mathitis: In this example. But in the example of the marbles in the jar, we can prove H_0 is true.

Peismon: But if I hypothesize that the marble is red, and test the size without knowing the color, I won't have conclusive evidence that it is red. Since large marbles are either red or green. Knowing that there are more red large marbles than green large marbles only provides evidence that it probably is red, if randomly selected.

Mathitis: But the reverse is conclusive. If you hypothesize the marble is large without knowing the color, and then check the color, a red marble is conclusive evidence that the size is large.

Kokinos: What is a necessary condition? Why isn't necessary enough?

Socrates: A necessary condition to win the lottery is buying a ticket. But buying a ticket is not enough to win. When we have both necessary and sufficient conditions, we use the phrase "if and only if." With the jars, it is true that "a marble is red if and only if it came from jar A." For the lottery winner, we can say "If you won then you bought a ticket." This is the necessary condition statement. This is the converse of what we had before. Notice that buying a ticket is not sufficient to win. This statement is false: "If you bought a ticket, then you won." However, the reverse is true: "If you won, then you bought a ticket."

Mathitis: Hypotheses that are necessary and sufficient can be proven, then.

Socrates: Consider this. I see an animal that is new to me. I hypothesize it is a mammal. My null hypothesis is that there is no difference between this animal's class and Mammalia. If I see that the animal nurses its young, does that prove that it is a mammal, the null hypothesis?

Mathitis: I guess it would if nursing is a necessary and sufficient condition for being a mammal. Or, just sufficient.

Daskalos: If we choose number of legs, then having four or two legs is necessary but not sufficient for an animal to be a mammal. We know is it not sufficient because reptiles also have four legs. It is supporting but inconclusive evidence.

Socrates: Do we always find sufficient cases to prove scientific theories?

Kokinos: No. Predictions from the theory of relativity are true but none are sufficient.

Peismon: Do we have necessary and sufficient conditions for statistical tests?

Mathitis: We can prove that a population is not normal by checking specific characteristics.

Peismon: The question is can we prove that it *is* normal. I say we can't do that statistically.

Daskalos: We can test for symmetry and unimodality but those are not sufficient as other distributions are also symmetrical with only one mode.

Mathitis: Can we test some unique characteristics, for example, approximately 68% of the distribution is within one standard deviation of the mean, 95% within two, and 99.7% within three?

Peismon: No. That's why we can't prove the null hypothesis. We can't determine exactly those percentages because we use probabilities, the *p* value. It's not like the mammal or the marble examples.

PROVE VERSUS ACCEPT VERSUS FAIL TO REJECT: ACTIONS

Socrates: That's right, Peismon. We do calculate probabilities and make probabilistic decisions when we test statistical hypotheses.

Peismon: You agree then that we cannot prove the null hypothesis?

Socrates: Statistically? No. Do we ever prove the alternative? In other words, you and Daskalos insist I say "fail to reject the null" because the null hypothesis cannot be proven. Then, why not insist I say "fail to reject the alternative" since we cannot prove it either?

Peismon: But when we reject the null, we are accepting the alternative hypothesis.

Socrates: Do we say we accept the alternative? Does accept mean prove?

Mathitis: No. Just because we accept something doesn't mean we have proven it.

Socrates: How do we know we have accepted the alternative hypothesis?

Daskalos: You take an action consistent with or based on H_a.

Peismon: For example, in your project, when you rejected the null hypothesis that factor A was not a cause, you accepted the alternative that it was a cause and your action was to use that factor to improve the process. You determined the setting for that factor and modified the process to use that better setting of that factor.

Mathitis: Look at it this way. If you had told us that you rejected the null hypothesis so you accepted the hypothesis that the factor was critical and its level needed to be changed, we would expect you to make that change. If we see that you make a process change but do not change the setting for that factor, we would ask "I thought you said that was a critical factor?"

Socrates: When I thought that the third factor was not a cause, what action did I take?

Peismon: None.

Socrates: Didn't I act consistent with and based on H$_o$?

Kokinos: What do you mean? How could you if you didn't take any action?

Socrates: As Mathitis said, I will change the setting for the factors that I accepted as critical. Since I didn't find C critical, I won't change the setting of that factor. In practice, I accepted H$_o$.

Kokinos: Not taking an action is different from taking an action.

Mathitis: I don't think that's the case. No action is not the same as an action.

Socrates: In our company, we had the option of contributing to our pension plan. We had to do that by a certain date. I had two choices: contribute or not contribute. By ignoring the communication and doing nothing, didn't I in effect choose not to contribute?

Agrafos: I agree with Socrates. Don't we say not to choose is to choose?

Peismon: That's choosing.

Socrates: By choosing not to change a factor, I am choosing to accept it as not being critical. I have accepted the null hypothesis *in practice*. If not acting is what I should do if a hypothesis is accepted, then I accepted that hypothesis. For example, if the significant factor A—the one where I rejected H$_o$—was set at the optimal level, what action would I have taken?

Kokinos: None. No action is needed, which, in this case, is consistent with accepting H$_a$.

Socrates: If I accept that factor B is not critical, and I change its setting, what would you say?

Kokinos: You're right. I would ask why you are changing it if it isn't critical.

Mathitis: I see what Socrates is saying. We should be doing hypothesis testing to determine what action we take. We take action A$_o$ if we accept H$_o$ and take action A$_a$ if we accept H$_a$ (Tukey 1960). But we need to recognize that an action is defined in a general or broad way to include doing nothing. If doing nothing is consistent with the hypothesis, we claim to be accepting.

Socrates: The reason for including "do nothing" as "taking action" is when it supports the reason for the test. When we want to improve a process, we want to find the conditions under which the process will perform better. The actions are to keep the same the conditions that are irrelevant with respect to performance and change the critical conditions to their optimal settings.

Mathitis: We could change an irrelevant factor for other reasons, for example, cost or time or availability.

Agrafos: Or, we could ignore a factor that isn't critical and let it be whatever it can be—we simply don't control it.

Socrates: Of course, both are right. In the latter case, we call that factor noise.

Mathitis: Then, if H_0 is "this factor is irrelevant to improving the process," then no change might be the action taken. In practice, yes, even when we "fail to reject," we are accepting H_0 by our actions or inactions because we act *as if* H_0 is true.

Socrates: Then even when we act consistent with acceptance of H_a, it doesn't mean that we have proven H_a is true, has it?

Mathitis: I agree.

INNOCENT VERSUS GUILTY: PROBLEMS WITH EXAMPLE

Socrates: Now we can see that the legal example does not accurately illustrate the issues about hypothesis testing (Bower and Colton 2003).

Kokinos: Why not?

Socrates: Let's start with acceptance: innocent versus fail to find innocent. We said that we can determine whether we accept based on our actions. Don't people accept the results of a jury or judge?

Kokinos: Yes, we do. People are found innocent and people accept that.

Socrates: How do you know?

Kokinos: Those who believe the decision was right might celebrate or applaud the decision.

Mathitis: But others may not accept that decision. The victim's family, prosecutor, or detective on the case may continue looking for evidence that shows the person was guilty.

Peismon: All that shows is what I have been saying about failing to reject.

Socrates: Isn't it also true for the decision of guilty? Aren't there people who accept that decision while others reject it? How do we know? We can tell by their actions.

Mathitis: Sure—including their words. Now it's the other side, the defendant's family, defense lawyer, and innocence project that seek to find evidence exonerating the person. They reject the decision. They say the defendant is innocent.

Socrates: We see that we can accept either hypothesis even when the evidence or decision is contrary to that hypothesis.

Agrafos: But isn't the legal system example used to show that we cannot prove the null hypothesis?

Peismon: That's right. We assume a person is innocent until proven otherwise. Not the other way around.

Socrates: That would be easier to believe if it were true that innocence or guilt cannot be proven.

Peismon: What do you mean? It is true. I and many others constantly say that we can never prove someone innocent in our legal system—only guilt.

Socrates: That's flawed reasoning in two ways. Consider the case of the two jars, one with only red marbles and the other with only green marbles. If we prove a marble came from one jar ("guilty"), then we have proven that it did not come from the other jar ("innocent"). If one person is guilty and proven so, isn't everyone else proven innocent?

Kokinos: And the second flaw?

Socrates: Hasn't the innocence project proven people innocent using DNA? Haven't confessions of the actual criminals proven others innocent and themselves guilty? Don't those hundreds of exonerations show that guilt can sometimes not be proven and innocence can be proven?

Kokinos: I think Socrates has a point—or two.

Socrates: Last year, more than a thousand people were murdered worldwide. Did any of you kill any of them? Peismon or Kokinos?

Peismon: Of course I, we, didn't.

Socrates: Can you prove you are innocent?

Mathitis: Alibis are one way of providing evidence that one is innocent.

Socrates: Alibis are ways of saying it was impossible. Every day, there are crimes committed somewhere on this planet. If we learn that a crime was committed in some other country at this moment, do you think no one in this country would be able to prove that he or she is innocent?

Kokinos: They may have hired someone!

Socrates: But they didn't directly commit the crime—they would have abetted.

Daskalos: Socrates is right. We can prove people innocent because they didn't have the opportunity. We learn that from police and detective shows: opportunity, motive, and means. President Washington did not kill President Kennedy—despite all the theories of who killed Kennedy. We can prove that simply because it was impossible for him to have an opportunity.

Socrates: Do all legal systems assume people innocent?

Mathitis: No, in some places today and in the past, people are guilty until proven innocent.

Kokinos: Then why do people always use that example?

TWO-CHOICE TESTING

Socrates: I don't know whether everyone does, but that example is useful for another reason.

Mathitis: What is that?

Socrates: Since there are only two choices, guilty or innocent, it doesn't matter which is the null hypothesis. Thus, in these cases, you can reject either and you can prove either.

Peismon: But we don't. Sometimes, the actual criminal is not found guilty because of a technicality.

Socrates: That is why I said "can be proven" not that they always are. We may fail to prove either hypothesis—or perhaps accept neither.

Daskalos: The legal system example is like the jars. There were only two choices. Rejecting one means the other is it. While we have evidence for one or the other based on the color of the marble chosen, if we reject one jar, we are forced to accept the other jar. If we disprove that the marble came from one jar, then it must have come from the other.

Agrafos: The situation with replacing the lightbulb is then the same. It either works or it doesn't.

Socrates: Exactly. In logic, this is called the law of the excluded middle.

Mathitis: Doesn't that say that either something is true or it is not?

Socrates: Yes. The excluded middle is that it can't be both or neither. So, if you exclude one, it has to be the other.

Peismon: But we may not have enough evidence for either. By default, without the evidence beyond, say, a reasonable doubt, we automatically conclude not guilty. The verdicts are "guilty" or "not guilty." They are not "guilty" or "innocent." That makes it seem more like the probabilistic type of statistical hypothesis testing.

Socrates: I agree with that aspect of it. Think of it this way. Which is the negative: innocent or guilty? Since we showed that either can be proven, then it shows that we can, in fact, prove some negatives. But more importantly, we can have either hypothesis be the null hypothesis. That is why it is not a good example to show you cannot prove the null hypothesis.

SIGNIFICANCE TESTING AND CONFIDENCE INTERVALS

Peismon: But again, for statistical hypotheses, we test using probabilities.

Mathitis: Wouldn't that be one more reason why the innocent versus guilty example may not be the best to illustrate the difference between "accept" and "fail to reject?"

Socrates: Let's consider a simple case of hypothesis testing. We often start with confidence intervals. As Daskalos stated in class, if you have a confidence interval with a lower and upper confidence limit (LCL and UCL, respectively), then we can test a null hypothesis that some parameter like the mean μ is a specific value, for example, H$_0$: $\mu = 100$, by checking whether that value is in the range LCL to UCL; in other words, we ask "Is LCL $\leq 100 \leq$ UCL?" A "no" means we reject the null hypothesis that $\mu = 100$ and a "yes" means that we fail to reject the null hypothesis.

Peismon: But that doesn't prove that the mean has that specific value.

Mathitis: It is evidence that it does.

Socrates: However, notice that the confidence interval also supports the alternative hypothesis that the mean is not that value or, in this example, that $\mu \neq 100$. And apparently, there is more evidence that it isn't than it is.

Agrafos: How does it do that?

Socrates: Is every value in the range LCL to UCL supported by the evidence?

Daskalos: Yes. That's why it is ambiguous.

Socrates: Then, if the confidence interval is (97.0, 104.5), every value in that range is supported by the evidence. But there is only one value in that entire range that equals 100 and an infinite number of values that are not 100. If $\mu = 100$, then only 100 supports that hypothesis while 97.00, 97.01, 97.02, 97.03, and every other value in that range does not support that $\mu = 100$.

Peismon: How does it not support it since by sampling error those are possible values?

Socrates: These other values also support the alternative hypothesis H_a: $\mu \neq 100$. There are more of these other values. Isn't it the case then that using confidence intervals, it doesn't matter what the result is, we always have substantially more evidence that the null hypothesis is false when it hypothesizes a single value?[*]

Mathitis: I hadn't thought of it that way.

Socrates: Now, consider what happens if we reduce the interval. As it gets smaller and smaller, we will eventually exclude $\mu = 100$ except in one case.

Daskalos: When the sample mean equals 100, assuming we have symmetrical confidence limits.

Agrafos: Why is that?

Daskalos: The symmetrical confidence interval is of the form sample value $\pm\Delta$, where Δ depends on the confidence level, population variation, and sample size. If the sample mean $\neq 100$ or μ, then as Δ gets smaller, 100 will eventually be outside the interval.

Socrates: But we don't need to decrease the interval symmetrically. Anyway, the only way a confidence interval can support *only* the null hypothesis $\mu = 100$ is if the sample mean equals μ equals 100 and the interval width is 0.

Kokinos: But in that case, don't we have 0% confidence?

Daskalos: Or, the entire population.

Socrates: And that suggests that any null hypothesis that states a single specific value for the mean or other parameter is probably wrong.

Mathitis: I think Deming said we already know that two populations are not exactly equal, so why test? Or, for single population, we already know that the mean is not exactly equal to μ. It could be $\mu + 0.000000000001$, which makes it different.

---

[*] For example, Carver, R.P., The case against statistical significance testing, revisited, *Journal of Experimental Education*, vol. 61, no. 4, Statistical Significance Testing in Contemporary Practice (Summer, 1993): 287–292.

Daskalos: This is what is called significance testing. Where only the confidence level is provided and there is only one hypothesis: the null. This is Fisher's position and why he stated that the choices were reject or fail to reject H_o.

Peismon: Then Fisher supports my claim.

Daskalos: However, there are other views of hypothesis testing contrary to Fisher's. Neyman and Pearson thought that we should be comparing two hypotheses, not just one (Hager 2013; Lehman 1993).

Socrates: But significance testing at any confidence level has an apparent weakness (Hubbard and Murray Lindsay 2008).

Kokinos: What is that?

Socrates: We never have to ask what sample size to use, because we can always use a sample size of 2 or 3 to calculate a 95% or any % confidence interval with a nonzero width.

Peismon: But that means the interval will be very wide—and useless.

Mathitis: That's why we need to define the interval width before we collect data.

Socrates: Isn't that equivalent to changing our null hypothesis from a single value to a range of values? We are willing to accept any value within the range of say $\pm\Delta$.

Kokinos: That would mean that all the values in that range are equivalent.

Socrates: Yes, practically, legally, statistically, financially, or for whatever reason you decide to not differentiate among them. Without that, then the hypothesis testing table of decisions (Table 10.2) only has one row for Truth. We do not include the power of the test, only the confidence level. When we do significance testing, I agree that we can—not must—conclude "fail to reject."

HYPOTHESIS TESTING AND POWER

Daskalos: This is why we teach hypothesis testing and not significance testing. By adding the second row of Truth to include the alternative hypothesis, we include the power to detect the difference Δ. There was and still is considerable controversy over these two options. Neyman—Pearson proposed the hypothesis testing approach, which Fisher completely disagreed with (Gill 1999). His view is what we are discussing here: you can't prove the null hypothesis.

Socrates: Let's see how the hypotheses are stated when using confidence level, power, and delta. For example, if we are testing a mean, the null hypothesis might be H_0: mean = 100. If we want to detect a difference of delta, what would the alternative hypothesis be?

Daskalos: Actually, we can change the hypothesis to reflect better Kokinos' idea of a negative. We can write the null hypothesis as H_0: |mean − 100| = 0.

Kokinos: Now I see how it is negative or no difference.

Mathitis: Then, the alternative would be H_a: |mean − 100| ≠ 0.

Daskalos: Except that if we calculate the sample size n so we have enough power to detect a difference Δ we consider significant, the alternative hypothesis is H_a: |mean − 100| > Δ.

Socrates: Good. Let's use that alternative hypothesis. Let α be the probability of a Type I error, which is an error when H_0 is true, and β be the probability of a Type II error when H_a is true. Then, $1 − \alpha$ is the confidence level in accepting H_0 when it is true and $1 − \beta$ is the power to detect Δ the difference or accept H_a when it is true. Correct?

Daskalos: Yes, that's correct.

Socrates: When we use power, do we do it for each value of H_a?

Kokinos: Why not?

Daskalos: No, because the power increases as Δ increases. So, we consider the worst-case alternative when considering the power of the test. That's when the difference is exactly Δ.

Kokinos: Then, aren't we doing what you said before, Socrates, testing only two hypotheses?

Socrates: Possibly. Let's see graphically. Consider the case when H_0 is true. Represent that situation—we can assume a normal distribution to keep it simple—showing alpha.

Daskalos: It looks like in Figure 14.1. The dashed line represents alpha. We can consider the one-sided case so the alternative hypothesis is that it is greater than 100 by delta or more.

Socrates: If we had a sample mean that is beyond the dashed line to the right, what *would that mean?*

Kokinos: That p < alpha.

Socrates: We would reject H_0. And accept H_a that the mean is not equal to 100.

Daskalos: That's correct.

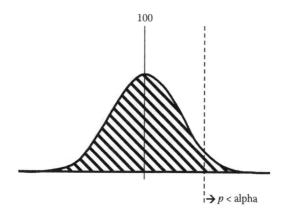

FIGURE 14.1
Testing hypothesis H$_o$ by comparing p to alpha.

Socrates: Just using this graph, when is p > alpha?

Daskalos: When the sample mean is to the left of the dashed line.

Socrates: We shouldn't accept H$_o$, correct?

Peismon: Great! That's what we've been saying.

Socrates: And the reason we cannot accept H$_o$ is…?

Peismon: Because the mean could be any value greater than 100 but less than the value where the dashed line is. Any value to the left of the dashed line results in p > alpha.

Socrates: Even though we said that we would accept values less than delta from 100 as equivalent to 100?

Peismon: But that is not the same as being equal to 100.

Mathitis: I see what you are saying. We are no longer saying the null hypothesis is a single value but a range of values. So, we accept H$_o$ for any sample mean to the left of the dashed line?

Peismon: I don't agree.

Socrates: I agree with you, Peismon.

Peismon: Then I don't understand. What are we discussing if we agree?

Socrates: This is only the confidence level. Let's look at power 1 − β graphically. There are two worst-case scenarios at mean = 100 ± Δ. Let's look only at 100 + Δ to understand the concept.

Daskalos: The curves centered at 100 + Δ and 100 − Δ are worst cases, because of all the values assuming H$_a$, these are the closest to the value assuming H$_o$. With the mean at 100 + Δ, (Figure 14.2), the dashed line on the graph represents β.

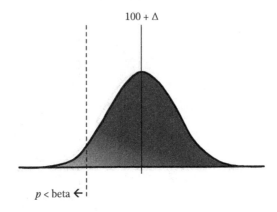

$100 + \Delta$

$p < \text{beta}$ ←

FIGURE 14.2
Testing hypothesis H_a by comparing p to beta.

Socrates: Let's treat this as we did the null hypothesis. Assuming H_a is true, if we get a sample mean to the left of the dashed line, do we reject H_a?

Daskalos: Yes.

Socrates: And if we get a sample mean to the right of the dashed line, do we accept H_a?

Peismon: No, you would fail to reject H_a.

Socrates: Let's see. Put the two graphs together to show confidence and power simultaneously (Figure 14.3) including both curves for H_a. Using the formula for determining sample size on the basis of the four factors alpha, beta, delta, and sigma, the dashed lines D1 and D2 representing beta and alpha align. Where are we when $p < \text{alpha}$?

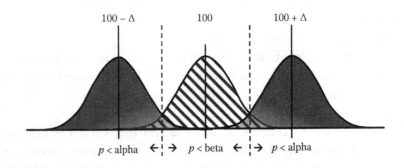

$100 - \Delta$ 100 $100 + \Delta$

$p < \text{alpha}$ ←|→ $p < \text{beta}$ ←|→ $p < \text{alpha}$

FIGURE 14.3
Simultaneously testing hypotheses H_o by comparing p to alpha and testing H_a by comparing p to beta with decision lines D1 (left dashed line) and D2 (right dashed line).

Kokinos: Either to the left of D1 or to the right of D2. We reject H_o.

Daskalos: We do this because we assumed H_o is true and this result is too unlikely if it was true.

Mathitis: That's right.

Socrates: When $p >$ alpha, where are we and what is our decision?

Peismon: What I have said from the beginning: we fail to reject H_o.

Socrates: However, from the perspective of H_a, we have the equivalent of "$p <$ beta." We are between D1 and D2. When we assume H_a, and calculate p, it will be less than beta.

Mathitis: I see it now. If assuming H_o we rejected H_o when $p <$ alpha and therefore accepted H_a, then it makes sense that if assuming H_a we get $p <$ beta, we should reject H_a and accept H_o.

Daskalos: There are actually two p values. There is one calculated assuming H_o, which we can label p_α, and another calculated assuming H_a, which we can label p_β. If $p_\alpha <$ alpha, we reject H_o and accept H_a. If $p_\beta <$ beta, we reject H_a and accept H_o.

Kokinos: The graph tells it all. We don't need to do either calculation. Depending on which side of the dashed line the sample mean occurs, we make our decision.

Socrates: If we have software that will do that for us. But we can always calculate just the one p value, what Daskalos called p_α, when n is based on all four factors: alpha, beta, delta, and sigma. When the sample size is determined for all four factors, $p_\alpha = p_\beta$ and the dashed lines coincide.

Kokinos: Why?

Socrates: Because we are either to the right of the dashed line or to the left. One side represents $p <$ alpha and the other represents $p <$ beta. If to the right, then we reject H_o, and if to the left, we reject H_a.

Agrafos: Alpha, beta, delta, and sigma. I now see that we have all those factors.

Kokinos: What if p equals alpha?

Socrates: We can arbitrarily add equality to either one.

Mathitis: Traditionally, we reject H_o if $p <$ alpha. So, $p =$ alpha goes with failing to reject H_o.

Daskalos: But in any case, under these conditions, we can accept or reject either hypothesis.

Kokinos: But I was taught that, in statistical testing, we can't have either hypothesis as the null.

Daskalos: The calculation of p depends on a specific hypothesis. That means that the hypothesis used to calculate p must have the equal sign. The two p values I suggested are calculated for each hypothesis using the equal sign for that specific value.

Mathitis: So, what Socrates says is true. If we believe we can reject the null when $p <$ alpha, then we should believe we can reject the alternative when $p <$ beta.

Socrates: Not exactly. When we determine the sample size to have a certain power to detect a difference, we always have a gray area.

Agrafos: What gray area?

Socrates: See the area between D1 and H_a, and between H_a and D2? A true population mean in these areas is neither H_o nor H_a. To recognize a difference Δ means we can't recognize differences less than it.

Peismon: I was right then. That's why we fail to reject H_o.

NULL HYPOTHESIS OF ≥ OR ≤

Socrates: Yes, because H_o hypothesizes only one value. We looked at two other null hypotheses in class.

Daskalos: Yes. We considered H_o: $\mu = 100$ versus H_a: $\mu \neq 100$, but we could have H_o: $\mu \leq 100$ versus H_a: $\mu > 100$ or H_o: $\mu \geq 100$ versus H_a: $\mu < 100$.

Peismon: I don't think that makes a difference in failing to reject the null hypothesis.

Socrates: It might, because now we have a range of possible values for the null hypothesis.

Peismon: How would that make a difference?

Socrates: Consider this example. Suppose I suspect that a coin is biased toward heads. What is my null hypothesis?

Mathitis: It can't be that it is biased because we don't know how much of a bias it has.

Agrafos: Couldn't we use the approach Kokinos described? Make the null there is a bias of at least a certain amount versus the alternative hypothesis that the bias is less than that amount?

Socrates: That's true, Agrafos. Let's start with the simple case first.

Peismon: Then, H_o must be that the coin is fair.

Socrates: That's the practical problem. Statistically, it is that the probability π of heads = 0.5 is H_o versus H_a: $\pi > 0.5$. If we get 55% heads and $p = 0.2$, what would you conclude?

Peismon: That we cannot reject H_o. We don't have enough information that the coin is fair.

Socrates: What if we got 10% heads and $p = 0.000$, to three decimal places?

Kokinos: We have to reject the null hypothesis because $p < 0.05$ or other alpha values.

Peismon: I agree.

Mathitis: Except that rejecting H_o that the coin is fair means we would have to conclude that it is biased toward heads. But if that were the case, why only 10% heads?

Socrates: It's because our null hypothesis was wrong.

Mathitis: What should it be?

Socrates: The complement to biased heads is fair or biased tails.

Mathitis: Then, the null hypothesis should have been H_o: $\pi \leq 0.5$ versus H_a: $\pi > 0.5$.

Socrates: Exactly. Do we now have enough evidence to reject H_a, and therefore, accept H_o?

Daskalos: We do. Interestingly, in this case, the power of the test might not be important. We don't necessarily need to calculate a p value assuming H_a is true as we did before. If the results are extreme enough from the assumed equal value—0.5 in this case—and in the direction of the inequality of the null hypothesis—in this case less than—then we reject H_a.

Mathitis: That would be true where the null hypothesis includes $\geq$ versus $<$ in the alternative.

Socrates: Returning to Agrafos' point, we don't have to include a delta that allows for bias.

Daskalos: If we do, then our hypotheses are H_o: $\pi \leq 0.5$ versus H_a: $\pi > 0.5 + $ delta. We still have the same situation as before with a gray area.

Mathitis: So, sometimes we can only fail to reject H_o but other times we can accept H_o.

PRACTICAL CASES

Socrates: Let's consider a practical application of statistical hypothesis testing. We develop a new product and want to show that it is better than the old. What is the null hypothesis?

Mathitis: The null hypothesis is H_0: new = old and the alternative is H_a: new is better than old.

Peismon: We specifically set them that way because we cannot accept H_0 only fail to reject. The objective is to reject H_0 so we can prove that the new is better. That's why we make the alternative hypothesis what we want to show.

Socrates: Do we prove that the new is better than the old or do we only reject that the new is the same as the old?

Mathitis: I think we reject the null, H_0, but we accept H_a by acting *as if* H_a is true.

Daskalos: That's right, we don't prove H_a.

Socrates: Suppose rather than two products, it is a new product against a placebo, for example, a vaccine against nothing? Can we not prove that a vaccine works?

Mathitis: I would say yes, since the facts prove that for many vaccines versus doing nothing.

Daskalos: I agree and hadn't thought of that. The null hypothesis is H_0: vaccine is as effective as doing nothing and the alternative is H_a: the vaccine is effective (to whatever degree we wish).

Socrates: We also have the reverse. We need to show generics are equivalent or to show work transfer from one site to another has the same performance or we switch one ingredient for another and we want to show there is no difference. What is the null in each case?

Mathitis: Given Peismon's argument about what we want to prove, we make H_a there is no difference or they are equivalent. But that doesn't make sense because then we are saying that the "negative" or "no difference" hypothesis is the alternative.

Socrates: Don't we still make H_0: the two are equivalent and the alternative H_a: they differ?

Daskalos: I think we do in practice. Certainly for some Food and Drug Administration (FDA)-regulated products we do, or they accept or insist in that structure of the hypotheses.

Socrates: Then, do we not accept H$_o$ if the evidence is strong enough to support equivalence?

Mathitis: Or, perhaps if the evidence against H$_o$ is too weak.

Kokinos: But, in the pharmaceutical industry, they do the opposite, as Mathitis said. The null hypothesis is that the difference is greater than a certain amount and the alternative is that the difference is not more than that amount. It's called a two one-sided t test (TOST) (Richter and Richter 2002; Schuirmann 1987).

Agrafos: I'm not sure I understand what the hypotheses are.

Kokinos: Rather than H$_o$: |mean − delta| = 0 and H$_a$: |mean − delta| > 0, there are two sets of hypotheses. One set is H$_{o1}$: mean − delta < 0 versus H$_{a1}$: mean − delta ≥ 0 and the other is H$_{o2}$: mean − delta > 0 versus H$_{a2}$: mean − delta ≤ 0. Then, two one-sided t tests are done.

Agrafos: Originally, we said that the null hypothesis was that the difference was less than delta, and using the TOST, it's the alternative hypothesis that states the difference is less than delta.

Peismon: They do this because of what I said: you can't accept the null hypothesis.

Socrates: I think they do it because they believe as you do Peismon, not because what they believe is true.

Peismon: But it is true.

Kokinos: The FDA, for example, provides guidance on this and states that testing the first way using equality in H$_o$ is not the way to show equivalence. Since equivalence doesn't mean equal, there is an acceptable amount of difference, delta. The alternative hypothesis—what we want to prove, accept, show—is that there is equivalence. That's why H$_a$ is two treatments, drugs, processes, laboratories, and so on are equivalent if the difference between them is no greater than delta.

Peismon: In these cases, the null hypothesis must be H$_o$: not equivalent.

Socrates: However, we could keep the original structure, determine the confidence level and power we want, and then switch them. That is, if we typically have 95% confidence and 80% power, we switch to have 80% confidence and 95% power.

Peismon: Why would that work?

Socrates: We have included power. Also, we don't have equality in the hypotheses.

Mathitis: How does that affect the results?

WHICH HYPOTHESIS HAS THE EQUAL SIGN?

Socrates: Recall when we talked about distribution-free analyses (Myth 10) and we said that, to calculate any probability, we need to assume some probability distribution?

Kokinos: Yes, I recall. We calculated the probability of 3 heads in a row, assuming a fair coin.

Socrates: If I said the probability of heads is less than one-half, you could only calculate a range of probabilities for three heads in a row based on the probability of a single head.

Kokinos: That makes sense.

Socrates: Understanding that, does the null or the alternative hypothesis have the equal sign?

Mathitis: When we have, for example, H_o: $\mu = 100$ versus H_a: $\mu \neq 100$, then we can't calculate a p for the alternative because there is no specific value to assume. So, we say that the null hypothesis must contain the equal sign.

Socrates: Then, Peismon is right in saying we should conclude "fail to reject" H_o when $p >$ alpha.

Peismon: Finally, we agree.

Socrates: But we saw that if we have a difference Δ to detect with power $1 - \beta$, both the null and the alternative hypotheses have an equal sign. The null has it for a specific value and the alternative for that value plus Δ. Now, we can reject either hypothesis. If rejecting one hypothesis means accepting the other, in principle or in practice by our actions, then we can accept the null.

Mathitis: I see that now.

Daskalos: As do I. In fact, we cannot calculate a sample size for a given alpha and beta, without having the equal sign for both the null and alternative hypotheses.

Mathitis: Unless we do a range of possible sample sizes, which is a power curve. But you are right, Daskalos, each sample size in that power curve requires us to assume a specific hypothesis.

Socrates: And we see one more instance of when the null hypothesis can be accepted.

Daskalos: When the null hypothesis is a range of values, it is of the form $\geq$ or $\leq$. In these cases, we can have values that are so unlikely if the alternative is assumed, that we can reject the alternative.

Mathitis: That was something I hadn't realized. And why the legal system with only two choices also falls into this type of hypothesis testing where the equal sign occurs with both hypotheses and we can therefore reject either, not just the null.

Daskalos: Also, that either one can be the null hypothesis when there are only two cases.

Socrates: From these discoveries, we should also see that the equal sign can often be added to the alternative. For example, we can test H_o: $\mu > 100$ versus H_a: $\mu \leq 100$ or H_o: $\mu < 100$ versus H_a: $\mu \geq 100$.

Peismon: Why can we do that?

Socrates: Because "<100" could mean simply 99.99999999999.... And when we calculate a p value, we will always do it when $\mu = 100$ regardless of which hypothesis has the equal sign. We can calculate a p for H_a as we showed before.

Daskalos: Then, compare that p to beta rather than alpha. But that's the same as the traditional case but switching the hypothesis, as Kokinos showed us with TOST.

Socrates: That's why I was wondering why not use traditional testing for equivalence—H_o is equivalent and H_a is not equivalent and switch the levels of alpha and beta. If traditionally, alpha = 0.05 and beta = 0.2, then use alpha = 0.2 and beta = 0.05.

Mathitis: We were right about only being able to "fail to reject" but for specific conditions. But you were right in there are more conditions where we can, in fact, accept the null and also reject the alternative. That's especially the case when you determine the sample size needed to detect a difference delta at a specified confidence level and power. Then, you can reject either hypothesis, and if rejecting one means accepting the other, we can accept the null hypothesis.

Daskalos: I agree with you now, Socrates. And with your project.

Kokinos: To me, the lesson is that if we spend so much time in class learning about the relationship of confidence, power, delta, and sigma as variation, then we need to be more rigorous in using this learning in determining sample sizes, designing the studies appropriately, and interpreting the conclusions correctly, logically.

Mathitis: Not just us, we need to ensure those whom we mentor are just as rigorous.

BAYESIAN STATISTICS: PROBABILITY OF HYPOTHESIS

Socrates: We have also forgotten that there is another approach. Recall our discussion about what p meant and what question we really wanted to answer (Myth 10).

Mathitis: Yes. We really want to know the probability of H_o given the data rather than the probability of the data given H_o. We talked about how, assuming that the parameter of a distribution is a random variable, we can change the question and get a probability.

Kokinos: However, that's how we test scientific theories. We assume a theory is true, make a prediction on the basis of that assumption, and then check whether the prediction occurred.

Socrates: However, statistical hypotheses about the value of parameters can be treated differently. As we said earlier, we don't think a mean is exactly a specific value, like 10.

Kokinos: Then, we don't have to do hypothesis tests on the mean, do we?

Peismon: I still think you have to.

Socrates: Consider TOST.

Agrafos: Can you explain that again?

Kokinos: That's when we want to show equivalence between two methods, for example, between two assays or production lines. Rather than test whether the two means are exactly equal, we test whether they differ by no more than a certain amount delta, Δ. Because of the belief that you cannot accept the null hypothesis, we make the alternative hypothesis that the difference is not more than Δ. Then, we do two tests to reject the null that it is greater than Δ. It is equivalent to checking whether a confidence interval of the difference is in the interval $(-\Delta, +\Delta)$.

Socrates: Rather than test whether that difference is no more than Δ, can we calculate the probability that it is no more than Δ?

Daskalos: We could, using Bayesian statistics. We only need to assume that each mean is a random variable. That would make the difference a random variable. We would then empirically develop a probability distribution for the difference and calculate a probability of the difference being in the interval $(-\Delta, +\Delta)$.

Agrafos: How is that different from the confidence interval of the difference being in the interval $(-\Delta, +\Delta)$?

Daskalos: The confidence level of the interval is determined before the analysis and, more importantly, it is not a probability once the sample is taken because we assume the difference is constant.

Socrates: Then, this is similar to determining the distribution for the tossing of coin. Right?

Daskalos: Yes, I see that it is with the two differences I just noted.

Mathitis: When we flip a coin, we assume that the probability of heads is fixed. Now, we assume that it is not fixed. That makes more sense to me because while the coin is fixed, the process of flipping the coin or getting the results varies.

Socrates: If the mean is a random variable, we answer a different question. Rather than asking "Is the mean = mu?" we ask "What is the probability that mean is in the range $(-\Delta, +\Delta)$?"

Mathitis: I prefer the second question.

Peismon: I still like the first.

Socrates: Notice two issues about asking if the mean is a specific value. First, we thought that we could only fail to reject the null hypothesis. But we learned conditions when we also accept the null hypothesis. Consider the former. What do you know about the mean, then?

Mathitis: You can't know anything since you failed to reject. It could be equal to that value or not. We don't know.

Peismon: But when we reject the null hypothesis, we do know.

Socrates: What do we know when we reject H_o?

Mathitis: We only know it is not that one value—we probably know that anyway.

Peismon: Using Bayesian statistics, we don't know either.

Socrates: True. However, we have an estimate of the probability that the mean is within a certain range. If being within that range is not an error, we have the probability of no error.

Mathitis: I like that better.

Kokinos: Does that mean we avoid the issue of whether we accept or fail to reject the null hypothesis?

Mathitis: I think so.

Socrates: There seems to be another advantage. We avoid the gray area.

Mathitis: Because we merely calculate the probability of being within a prescribed range.

Agrafos: I like that it answers the question I really want to answer. It's more informative. Really, it models how we think more closely.

Kokinos: I see the advantage for testing equivalence, which is a common question in science and product submissions to regulatory agencies. It seems a better approach than TOST since we don't have to worry about which hypothesis to make the null and which to make the alternative.

15

Myth 15: Control Limits Are ±3 Standard Deviations from the Center Line

---

Scene: *Break room (after introduction to control charts lesson)*

BACKGROUND

Agrafos: Could someone explain the construction of a control chart? I didn't understand why the limits are set at three standard deviations from the center line?

Peismon: Let's start with what a control chart is and does. All processes have variation that can be detected if the measures are granular enough.

Agrafos: We talked about measurement error before. Now, we are discussing the process, right?

Peismon: Yes, although control charts are also used in a measurement system analysis. Control charts represent the voice of the process through process data. The control limits represent the inherent variation of the process. The inherent variation is calculated as three times the standard deviation. So, the control limits are set at three standard deviations from the center line of the chart, which is the average of the points. The line below the center line is the lower control limit (LCL) and the line above it is the upper control limit (UCL) as in this chart (Figure 15.1).

Mathitis: The vertical axis is in the measurement units, like centimeters or counts, while the horizontal line represents time. It could be a label, like a serial number, but it is in time order.

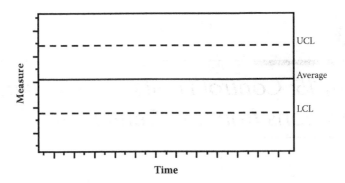

FIGURE 15.1
General structure of a control chart.

Socrates: And this is how limits for all control charts are calculated—the center line or average plus or minus three standard deviations?

Peismon: Yes.

Kokinos: There are warning limits you can add to a chart, but the decision limits are the average plus or minus three standard deviations.

Socrates: And it applies to all data appropriate for the type of control chart?

Peismon: Yes.

Socrates: That makes it easy to calculate then: calculate the standard deviation of all the data, multiply by three, and add and subtract from the center line.

Peismon: You can't use the standard deviation of all the data.

Socrates: I'm confused. You said plus or minus three standard deviations.

Peismon: There are different kinds of standard deviations.

Socrates: What other standard deviations are there?

Peismon: Because you collect data in subgroups, there are actually three standard deviations. There is the standard deviation of all the data, the standard deviation within subgroups, and the standard deviation between subgroups.

Socrates: Saying three times *the* standard deviation implies there is only one standard deviation. Perhaps it would be better saying three times a standard deviation.

Mathitis: Or, better, would be to say which standard deviation.

Socrates: Which is used for setting the control chart limits?

Peismon: The within-subgroup standard deviation.

Socrates: So, the control limits are three times the within-subgroup standard deviation, which should be easy to calculate—although not as easy as the standard deviation of all the data.

Agrafos: But each subgroup has a standard deviation. Which is used?

Peismon: It isn't straightforward—it is often miscalculated. Each subgroup's standard deviation is an estimate of the within-subgroup variation. But there are different ways of combining them to produce a final within-subgroup variation. There are formulas that use factors from tables. But today, with so much software, you don't have to do it manually.

Socrates: Can you give me an example?

Peismon: Sure. Let's use the $\bar{X} - R$ charts. The mean $\bar{x}$ and range r of each subgroup is calculated. The $\bar{X}$ chart displays all the subgroup means. The R chart shows all the subgroup ranges. For the $\bar{X}$ chart, $\text{UCL} = \bar{\bar{x}} + A_2 \times \bar{R}$ and $\text{LCL} = \bar{\bar{x}} - A_2 \times \bar{R}$, where $\bar{\bar{x}}$ is the mean of all the subgroup means, A_2 is found in the tables I mentioned, and $\bar{R}$ is the average of the ranges.

Kokinos: The $\bar{X}$ chart assesses process centering and the R chart assesses process variation.

Socrates: Why do you need tables if it is just three times a standard deviation?

Mathitis: Before software, using the range was much easier for manual calculations. For small subgroup sizes, the range chart is used for the variation chart. So, there is a factor that is used to convert the range to a standard deviation.

Socrates: Is A_2 that factor?

Mathitis: Yes it is.

Socrates: So, you think A_2 includes 3 and the conversion factor?

Mathitis: It must.

Peismon: And for large subgroup sizes, we use the standard deviation or S chart.

Mathitis: When the sample size is 8 or more. Although, with software, you might as well use the S chart all the time as it is more efficient. You want to estimate the standard deviation. Using ranges means converting them to standard deviations, which will increase the error.

STANDARD ERROR VERSUS STANDARD DEVIATION

Socrates: Ok. What is the formula for the $\bar{X}$ chart UCL when the standard deviation is used?

Peismon: That formula is similar but uses a different factor: $\bar{x} + A_3 \times \bar{s}$, where A_3 is a factor similar to A_2 and $\bar{s}$ is the average subgroup standard deviation.

Socrates: Since it's just three times the standard deviation, then why call it A_3 if it's just 3?

Peismon: No, it isn't just 3. A_3 depends on the subgroup size n.

Mathitis: As does A_2.

Socrates: Why do they depend on n if it is just three times the standard deviation?

Peismon: There are tables that show the values of A_2 and A_3 for each subgroup size.

Mathitis: For example, $A_3 = 3 / \left[c_4 \times \sqrt{n} \right]$. So, we have the c_4 factor and the $\sqrt{n}$ factor. The c_4 factor removes the bias from s.

Socrates: That's strange. Before, we were discussing the equation for s, and you thought it was a correct formula without bias. But now, we are clearly stating that s is biased and that the equation for s must be modified, suggesting that it isn't correct.

Mathitis: You're right. In one context, we say one thing, and in another context, we say the opposite and don't realize we are doing that.

Peismon: Do you also see the factor $\sqrt{n}$? See how A_3 depends on the sample size?

Socrates: Then, it isn't three times the standard deviation but $3/\sqrt{n}$ times the within-group standard deviation adjusted for the bias. If I said the circumference of a circle is twice the radius, what would you calculate the circumference to be if the radius is 5?

Peismon: Ten, of course. But the circumference is 2π times the radius. It's not twice the radius—you forgot the factor π.

Socrates: Do you see my confusion? Agrafos asked about control limits. You and Peismon said they were three times the standard deviation away from the center line. I expected to be able to multiple a standard deviation by three to get how far away they are from the center line.

Agrafos: I did too.

Socrates: Now, you are saying that the standard deviation is divided by the product of the square root of n and factor c_4 before it is multiplied by three. How is that three times?

Agrafos: I certainly didn't expect to be using tables if all we had to do was multiply by three.

Mathitis: You're right. I see what you mean. But the c_4 factor removes the bias so it is explainable. Besides, we are estimating, and you taught me there is no correct way to estimate.

Kokinos: Then why does everyone say that the control limits are three times the standard deviation away from the center line?

Mathitis: It could be a shortcut label. Possibly just to indicate they are not twice or some other multiple including these other factors.

Peismon: I can explain. Do you know the difference between standard deviation and standard error?

Kokinos: Yes, the software I use produces some basic statistics that include the mean, standard deviation, minimum, maximum, quartiles, and the standard error. I was taught that the standard error was the sampling error. Here is one explanation:

Std Dev
is the sample standard deviation computed for each level of a categorical variable. It is the square root of the variance of the level values.
Std Err
is the standard error of the mean of each level of a categorical variable. It is the standard deviation, *Std Dev*, divided by the square root of N for each level.[*]

Peismon: Yes, that's one application. For a single sample, the sampling error of the sample mean, $\bar{x}$, is the standard deviation divided by the square root of the sample size n. Standard deviation is of the original data, while standard error is the standard deviation of a statistic.

Agrafos: What does the standard error represent?

Peismon: If we take all possible samples of size n and calculate the mean of each of those samples, then we have a distribution for the sample means. The standard deviation of those means, of the distribution of the sample means, is called the standard error.

---

[*] Minitab version 16, Help Glossary.

186 · *Understanding Statistics and Statistical Myths*

Mathitis: Except for individual value chart or *I* chart, all control charts are graphs of statistics, for example, mean, range, standard deviation, proportion, and count. So, the control limits represent the variation of those statistics. They are three times the standard deviation of the statistic plotted. For example, for the $\bar{X}$ chart, the control limits are three standard deviations of the sample mean from the center line; for the *S* chart, they are three standard deviations of the sample standard deviation from the center line, and so on.

Agrafos: Except they aren't three times the standard deviation because of the other quantities.

Socrates: Since the standard deviation of a statistic is the standard error and we distinguish between standard error and standard deviation, then it is more accurate to say that the control limits are three standard errors of the plotted statistic from the average of that statistic.

Mathitis: I would agree with that. In fact, by stating it that way, it removes the question of what standard deviation to use.

Kokinos: But the *S* chart and *R* chart do not have symmetrical lines, so they can't be three standard deviations or standard errors from the center line.

Mathitis: You're right, but only when the LCL is negative. Then it is replaced by zero. We usually talk about the $\bar{X}$ chart as if everything true of it applies to all control charts when it doesn't. The charts with asymmetrical limits are based on having approximately the same probability of being within the control limits as the $\bar{X}$ chart when the process is in control, replacing for negative limits when the statistic cannot be negative. For example, ranges and standard deviations can't be negative, so zero replaces the calculated value.

Peismon: But the limits are not probability limits.

WITHIN- VERSUS BETWEEN-SUBGROUP VARIATION: HOW CONTROL CHARTS WORK

Socrates: Let's postpone that discussion. Is the standard error of the statistic used for the control chart limits for all charts?

Mathitis: That's right.

Socrates: And that is the average within-subgroup variation divided by the square root of *n*?

Mathitis: Yes, for the $\bar{X}$ chart.

Socrates: But the standard error is the between-subgroup variation of the statistic plotted. For example, for the $\bar{X}$ chart, it is the variation of the subgroup means.

Agrafos: Why?

Socrates: According to Peismon's explanation, it is the standard deviation of the sample means. Each subgroup represents a sample. We calculated the mean of each subgroup. So, we can now calculate the standard deviation of those sample or subgroup means and multiply by three.

Peismon: That's right. That's what I explained earlier.

Socrates: But the variation of the subgroup means is not within-subgroup variation. It is between-subgroup variation.

Peismon: No, it can't be because we use the within-subgroup variation.

Mathitis: There seems to be a contradiction. I agree with what you each say. Variation of the means is the between-subgroup variation but the calculation of the limits uses within-subgroup variation. If the standard error represents variation of the subgroup means, why are we using within-subgroup variation? Each subgroup has a mean and then we calculate the standard deviation of those means. So, $\bar{s}\,/\sqrt{n}$ is the standard deviation of the means but it is based on within-subgroup variation.

Socrates: Which is it? Are the limits based on the between- or within-subgroup variation?

Mathitis: Within since it includes the square root of *n*.

Socrates: This should apply to all $\bar{X}$ charts and all data, right?

Mathitis: Yes.

Socrates: Let's look at an example. Assume you have collected 25 subgroups of size 4 and the four numbers are exactly the same for each subgroup: 1, 2, 3, and 4. What are the subgroup means and standard deviations?

Kokinos: Each subgroup has an average $\bar{x} = 2.5$ and a standard deviation $s = 1.29$.

Socrates: Is there within-subgroup variation?

Kokinos: Yes, because the within-subgroup variation $s > 0$.

Socrates: Is there between-subgroup variation?

Kokinos: No, all the subgroup means are the same: 2.5.

Socrates: If there is no between-subgroup variation, then where should the limits be if they are 3 times the between-subgroup variation?

Kokinos: Without any variation, the center line = LCL = UCL.

Mathitis: That doesn't make sense. According to the formula, they will be at $2.5 \pm A_3 \times \bar{s}$. We know that both A_3 and $\bar{s}$ are greater than zero; so, the center line will not equal the control limits.

Peismon: We must have done something wrong.

Socrates: Let's try it another way. If the UCL is three standard deviations away from the center line, then take the difference between the two and divide by three. What should that equal?

Kokinos: That should equal *some* standard deviation. We just aren't sure which it is.

Socrates: Using your formula with $\bar{s}$, what is UCL − center line?

Mathitis: That's just $A_3 \times \bar{s}$. Each subgroup standard deviation is 1.29, which is also their average $\bar{s}$. For $n = 4$, $A_3 = 1.628$, so the difference is $1.628 \times 1.29 = 2.100$.

Socrates: If we divide by three?

Mathitis: 0.7. So, some standard deviation should equal 0.7. We'll have to adjust for the bias.

Socrates: Now, let's calculate all three standard deviations, adjust them for bias, and see which standard deviation to use.

Mathitis: We already know that, for within-subgroup variation, $n = 4$, so $c_4 = 0.9213$. For total variation, $s = 1.124$ and $n = 100$, so $c_4 = 0.99748$, and for between-subgroup variation, $n = 25$ subgroups, so $c_4 = 0.9892$.

Socrates: Does any standard deviation equal 0.700 (Table 15.1)?

Mathitis: No, none.

Socrates: Try this. You said that since we are plotting the means, we should be looking at the standard error of the means. What is the standard error of the numbers 1, 2, 3, and 4?

TABLE 15.1

Comparison of Three Standard Deviations to ±3 Standard Deviation Control Limits

Standard Deviation	Biased	Unbiased
Total (all the data)	1.124	1.1265
Within ($\bar{s}$)	1.291	1.401
Between (standard error)	0	0

Mathitis: That's just the standard deviation divided by the square root of the sample size. So, that's $1.29/\sqrt{4} = 0.645$. That's close!

Socrates: And if we remove the bias?

Mathitis: For $n = 4$, $c_4 = 0.9213$, so $0.645/0.9213 = 0.700$!

Socrates: Which standard deviation did we use?

Mathitis: The within-subgroup variation.

Kokinos: But why is the between-subgroup variation zero? If there is no variation of the means, why aren't the control limits zero distance from the center line?

Socrates: Good question. If we take all possible samples from the same population of size n and calculate the mean of each sample, what is the standard deviation of the sample means?

Kokinos: It will be the population standard deviation σ divided by the square root of n.

Socrates: We can check it two ways.

Kokinos: We calculate all the sample means, then the standard deviation of all those sample means and see if it is $\sigma/\sqrt{n}$.

Mathitis: The other way is what we do to calculate control limits. We estimate σ by taking the standard deviation of each sample, average them, and divide by $\sqrt{n}$. That should be equal to the way Kokinos calculated it.

Socrates: In the $\bar{X}$ chart, what represents Mathitis' way of calculating the standard error?

Peismon: The control limits, as I said.

Socrates: In the $\bar{X}$ chart, what represents Kokinos' way of calculating the standard error using within-subgroup variation?

Peismon: We don't use that approach.

Socrates: Don't the points, the sample means, represent that?

Mathitis: I guess so. The $\bar{X}$ chart has the subgroup averages. They represent the actual process variation between subgroups. Their variation represents the actual standard error of the means.

Socrates: Independently, we estimate what that would be assuming all the subgroups came from the same population. That is shown, with a margin of ±3 standard errors, by the control limits.

Mathitis: I get it. Since there is within-subgroup variation, then there should be between-subgroup variation equal to the within divided by the square root of n. The control limits represent this expected variation under the assumption of sampling from the same population.

Socrates: The plotted points represent what actually occurred. When do we consider them equal?

Mathitis: If the assumption is true. Then the points—all the subgroup means—should be within the control limits.

Kokinos: Then, if points fall outside the control limits, that is evidence against the assumption.

Mathitis: That's right.

Kokinos: But if there is variation within, why is there no variation between the means?

Socrates: I made up this example to understand the difference between within and between variation, and how a control chart works. Control charts use two estimates of the standard error and compares them: the points versus the limits. If the example consisted of each subgroup having zero variation, for example, all n values are the same, but subgroup means differed, for example, subgroup 1 had 4 threes and subgroup 2 had 4 nines, what would the control limits be?

Kokinos: Then, we would have $s = 0$, no within-subgroup variation. The control limits would equal the center line: $UCL = LCL = \bar{\bar{x}}$. But no points would be on the center line.

Mathitis: That would also show that control limits estimate the standard error one way and the points represent that same standard error in a different way.

Kokinos: Why is the total less than within?

Socrates: Because this was an artificial example. If the samples really did come from a population with nonzero variation, then random samples would vary.

I CHART OF INDIVIDUALS

Kokinos: But what about the individual values chart, or *I* chart? There is no within-subgroup variation to use.

Mathitis: It depends on what we mean by standard deviation. If the data are assumed to come from a random variable X, so the data are $x_1, x_2, \ldots, x_n$, then an *I* chart of these individual values will have control limits that are a multiple of the standard deviations of the xs. Traditionally, these will be $\bar{x} \pm 3\bar{s}$.

Peismon: That's not right. It will not be the standard deviation of all the data. It will be based on an estimate of the within-subgroup variation. If the subgroup size is one, then subgroup ranges are artificially created by taking the range of consecutive values, typically two consecutive values. If we have *n* data, then we have *n* – 1 "moving" ranges. As we discussed before, these are converted into a standard deviation.

Socrates: But it is the standard error of the statistic. In this case, the statistic of the subgroup is the single value. In other words, the standard deviation divided by the square root of the sample size equals the standard deviation because $n = 1$: $\sigma/\sqrt{n} = \sigma$.

Peismon: That makes sense.

Mathitis: For all other control charts, $n > 1$, so it is not a multiple of *s* but of *s* divided by the square root of *n* or the standard error. Since all other control charts are charts of statistics, then for the control limits to represent the variation of these statistics, they must be the standard deviation of the statistic. These are called standard errors. Since we make a distinction between standard deviation and standard error, then we should correctly say ±3 standard errors.

Kokinos: More importantly, even if we say ±3 standard deviations, the statement is incomplete at best. It doesn't specify which standard deviation.

Mathitis: Kokinos is right. The standard error is a function of the standard deviation. But there are several standard deviations: standard deviation within subgroups, standard deviation between subgroups, and standard deviation total.

Socrates: And the critical one: standard deviation of the statistic, or the standard error. What is the practical risk of continuing to say ±3 standard deviations without specifying which standard deviation and making a distinction between standard error and standard deviation?

Mathitis: I know that sometimes—although less and less I imagine—people calculate control limits manually. In other words, they enter formulas in software like spreadsheets. If they do not know the difference, they will calculate the wrong limits. They may use the wrong standard deviation to begin with. Since standard deviation is greater than standard error, they may think their process is in control when it isn't. Examples of the difference would be helpful in getting the message across.

Socrates: What's the benefit of saying it more precisely?

Kokinos: It provides a learning opportunity in distinguishing between standard deviation and standard error. This helps explain the difference between an initial distribution from which the data come and a sampling distribution.

Mathitis: It explains the difference between parameters and statistics. In addition, the connection to confidence limits might make the broader concept of sampling error easier to explain.

Kokinos: If software programs distinguish between standard deviation and standard error when calculating statistics of a sample by listing them both, it doesn't make sense to call it one when it is the other.

Socrates: Doesn't it also show us how a control chart works?

Peismon: We know how—by comparing the points to the control chart limits.

Socrates: Yes, but what do each represent?

Mathitis: I see. We also learned that we estimate the standard error in two ways. The points represent the actual between group variation while the control limits represent the estimated between-group variation assuming all the data came from a single distribution.

16

Myth 16: Control Chart Limits Are Empirical Limits

Scene: *Break room (continuation of control chart discussion)*

BACKGROUND

Peismon: In one sense, it doesn't matter how the limits are calculated because the control chart limits are empirical limits.

Agrafos: That's true. Everything I've read and heard says the same thing, that Shewhart derived the limits empirically.

Mathitis: I would agree based on online discussions I've been involved in with others.

Peismon: Read Deming and Wheeler; they say that Shewhart said they were empirical limits.

Socrates: Are you saying that Deming, Wheeler, and others believe it is true because they think Shewhart says it is true? Or, are you saying that they believe it is true, independently of what Shewhart thought or said?

Agrafos: What difference does that make?

Socrates: If we want to have a deeper understanding, we need to get at the reason. If it is Deming or Wheeler saying something is true, then we can get at their reason from what they write or say. On the other hand, if Deming, for example, is saying that Shewhart claims something is true, then we check for the reason from Shewhart's writings—not from Deming's.

Kokinos: Even if they quote Shewhart?

Socrates: Possibly, if the quote contains reasons and do not just repeat the claim. We can illustrate that with the discussion so far. You say control chart limits are empirical limits. We don't know why because you didn't provide an explanation. Are you saying that because you believe it or because someone else believes it—Deming or Wheeler?

Kokinos: I accept it because the experts say it is true.

Socrates: Then we will not get at a reason for it being true from you, but from Deming or Wheeler. If Deming accepts it because Shewhart says it, then we need to get it from Shewhart.

Mathitis: Since Kokinos says that Wheeler quotes Shewhart, then we must look at what Shewhart said, and search for his reasons.

Socrates: Exactly. This reminds me of a story in a chapter on design of experiments in a graduate school textbook I used. The story was to show the value of data—just as we are claiming that the limits are empirical—that empirical justification is needed, not just theory.

The setting is outside an inn in Italy during the Renaissance. Several wealthy men are waiting for the stable boy to bring their horses. When the stable boy arrived with the horses, he heard the men discussing the number of teeth a horse has—one saying 38, another saying 42—and the others choosing sides. After a while, he interrupted, "Excuse me, master, but here are four horses. I would be pleased to count their teeth for you."

The eldest laughed. "See how uneducated these peasants are. Dear boy, we don't need to count them. Aristotle was the greatest thinker of all time. We only need to remember what he said was the number of teeth."

Kokinos: Great story.

Agrafos: Is that where the saying "straight from the horse's mouth" comes from?

Socrates: (laughing) I don't know, Agrafos, maybe. If we want to know what Shewhart meant, go straight to Shewhart. Others can only give their interpretation of Shewhart. Although it's a worthy discussion: which interpretation is most useful, whether Shewhart meant that or not.

Peismon: Maybe. But control chart limits were derived empirically. It's obvious from Shewhart the limits must be empirical (Shewhart 1980, 276). "Hence the method for establishing allowable limits of variation in a statistic $\bar{\theta}$ depends upon theory to furnish the expected value $\bar{\theta}$ and the standard deviation σ_θ of the statistic θ and upon empirical data to justify the choice of limits $\bar{\theta} \pm t \times \sigma_\theta$" (Shewhart 1980, 277).

Mathitis: That convinces me that Shewhart intended the limits to be empirical.

Kokinos: I agree.

DEFINITION OF EMPIRICAL

Socrates: What does empirical mean?

Peismon: They are based on data as a result of experience.

Kokinos: The conclusions are based on observation or experimentation or study. That's what I do. It's the opposite of theoretical determinations.

Agrafos: Can you give me an example?

Peismon: I walk across the room and count how many strides it takes. I do this for several room sizes. Then, knowing the actual length of the rooms, I estimate that my stride is approximately 2.75 feet long. I use that factor to convert strides into feet and determine the length of rooms or other distances. That's an empirical derivation. Similar to what Shewhart did in determining where to set the control limits.

Socrates: The formula, 1 Peismon stride = 2.75 feet, is an empirical formula. So is the result from using that formula. Is the formula 3 feet equals 1 yard empirical?

Peismon: No, that is true by definition. Derived formulas from theory or assumptions or definitions are not empirical.

Socrates: In statistical analyses, the calculation of a p value based on the assumption of the null hypothesis is not empirical and neither is the formula used to calculate it. Is that correct?

Peismon: I would say yes, that's true.

Mathitis: An example of theoretical, not empirical, conclusion is the t distribution. Gosset (1908) derived the Student t distribution for the sample mean assuming a normal distribution with mean and standard deviation unknown. It was a mathematical derivation.

Socrates: What about alpha, the probability of an error if the null hypothesis is true?

Peismon: That's given, like 3 feet equals 1 yard.

Mathitis: We choose that, but we didn't define it. I don't think it is empirical, although it could be similar to what Peismon is saying about the 3 standard error limits. Somebody selected it, and it seemed to work and continued using it and now it's commonly used and accepted.

Agrafos: Is it empirical or not? I'm still confused.

Socrates: Let's put it this way. The original selection could have been arbitrary, supported empirically, and then passed along and used for the original and other reasons.

Agrafos: I use it because I was taught to use it.

Socrates: Exactly. And, some people learn that they can choose other values for alpha than 5% and do so.

Mathitis: Then, we have at least four cases: (1) completely empirical like Peismon's conversion of 1 stride to 2.75 feet, (2) theoretical or academic or both so not at all empirical like the derivation of the *t* distribution, (3) neither, like definitions, and (4) initially empirical and then supported by theory.

Kokinos: Then, the 3 standard error limits seem to be an example of the fourth case.

Peismon: The decision to use 3 standard error limits was determined empirically, not theoretically.

Socrates: Today, no one empirically derives the limits they use. They just use the formulas. So, they are not *your* empirical limits in the sense that you used data to determine your limits.

Mathitis: You mean it would be similar to me using Peismon's stride formula for my stride even though we have clearly different heights and have different stride lengths.

Socrates: You wouldn't get the same results and that may have unwanted consequences.

Kokinos: In other words, Shewhart developed them for processes he was working on and we accepted them blindly for our processes without checking whether they were the best for our processes. In that sense, our limits are not empirically derived—they are just passed on.

Peismon: People don't have to develop their own empirically because Shewhart did initially. And Wheeler has done so since then.

Mathitis: But with my accepting your empirical stride rule, unless the empirical derivation is for all cases and conditions, we don't know if they work as well or at all. The inference from a few cases is not empirical.

EMPIRICAL LIMITS VERSUS LIMITS JUSTIFIED EMPIRICALLY

Socrates: Note that Shewhart says the basis for and justification of the choice of limits is empirical. Is that the same as empirical limits?

Agrafos: If the limits are justified empirically, are they not empirical limits?

Socrates: If we empirically justify speed limits of 55 mph, are the limits empirical or are they speed limits?

Agrafos: Both, I guess. They are empirical speed limits.

Peismon: The control chart limits are justified empirically. That's what I think Shewhart and the others mean.

Socrates: But to say something is justified means it is justified for a purpose, criterion, or objective. How is the speed limit of 55 mph justified empirically?

Mathitis: I think I understand. For example, if we want to reduce fatal traffic accidents, we might study various factors related to such accidents, making observations, doing experiments, and so on. The U.S. federal government did that. We conclude that if cars travel at 55 mph or less, there is a significant decrease in fatal traffic accidents. We established the limit of 55 mph empirically, but we did it for the purpose of reducing fatal traffic accidents.

Kokinos: The same thing was done with seat belt use and making drunk driving illegal.

Socrates: What is the purpose of the limits—regardless of whether they are determined theoretically or empirically or guessing?

Peismon: Shewhart says it is to achieve an economic balance of chasing false alarms and failing to respond to true signals by not addressing true problems (Shewhart 1980, 276).

Socrates: Now, we can better understand the difference between empirical limits and empirically justified limits. The limits may be economically based or justified. We understand this to mean that they are determined to achieve some economic benefit. But when we say they are empirically based, we don't mean that they are determined to achieve some empirical benefit. After all, the economic benefit is also determined empirically.

Kokinos: Okay, I see the difference.

SHEWHART'S EVIDENCE OF LIMITS BEING EMPIRICAL

Socrates: What is Shewhart's empirical evidence of that economic balance?

Peismon: He says experience shows that 3 standard error limits are best economically. Thus, they are empirical limits.

Socrates: It reminds me of another story—this one true. Remember the Pythagorean theorem, that $c^2 = a^2 + b^2$. In the 17th century, the mathematician Fermat conjectured that there were no positive integers a, b, and c for which $c^n = a^n + b^n$ is true. He made this conjecture in his copy of the book *Arithmetica*.* He also wrote he had a wonderful proof but too long to fit in the margin.

Mathitis: I know the story. It took nearly 400 years and the development of significant new mathematical tools for anyone to finally prove his conjecture. Andrew Wiles took years to develop the proof and hundreds of pages to write it (Kolata 1993; Wiles 1995).

Peismon: What's the connection?

Socrates: We're merely asking for actual data from real processes provided in Shewhart's book.

Kokinos: I've read Shewhart and recall some examples he used as evidence to support this claim. That would be empirical. Let me get the book.

Socrates: Let's review them. But first, let's discuss the structure of a study to empirically determine what control chart limits best achieve the purpose of economic balance of false alarms and true signals. What tests should be done and under what conditions?

Mathitis: To show that something is best requires at least a comparison of different options.

Socrates: In this case, it would be comparing different limits, right?

---

* Pierre de Fermat, in 1637 wrote on a translated copy of *Arithmetica* that he had a proof too long to fit in the margin.

Mathitis: Yes. And also we need to calculate or estimate the costs using each set of limits we compared. To check the false alarm rate, false nonalarm rate, and their costs both when the process is in control and when it is out of control.

Agrafos: Do we need to look at all the charts and different subgroup sizes?

Mathitis: Yes, good point.

Socrates: Then, we have these four tests corresponding to the four combinations of stable or not versus alarms or not.

> Test 1. When the process is stable (no assignable causes), are there no alarms?
> Test 2. When there are no alarms, is the process stable (no assignable causes)?
> Test 3. When the process is unstable (assignable causes), are there alarms?
> Test 4. When there are alarms, is the process unstable (assignable causes)?

> To do the tests on assignable causes, such a study would have to include changes to be detected.
> We have the following conditions:

> Condition 1. Shewhart's five control charts
> Condition 2. Various limits
> Condition 3. Various subgroup sizes

Mathitis: That looks good. If we think of others along the way, we can add them.

Socrates: One more thing. The purpose of the empirical studies was to determine the best economic limits. So, there is an analysis that needs to be done for each study:

> Analysis: Compare costs to determine which limits achieve the lowest cost (economic balance).

Kokinos: Here are the examples on Shewhart (1980), pages 18–22. The first example was the history over 3 years showing that as assignable causes were found and eliminated, the percent of points out of control decreased.

Peismon: He used this as "Evidence that Criteria Exist for Detecting Assignable Causes."

Mathitis: That's a study of Test 4: If alarms occur, are there assignable causes?

Kokinos: He also noted that as assignable causes were eliminated, the percent defective decreased.

Socrates: One of the benefits of having controlled product. This example showed that points beyond the control limits were good indicators of assignable causes existing.

Peismon: That's empirical evidence for setting the limits.

Socrates: Not completely. Does he say whether he chose the limits ahead of time or determined them based on what assignable causes they found and could associate them specifically with the points beyond the limits?

Kokinos: No, he doesn't explain in his book.

Mathitis: I see what you are saying. It could be that not all the points beyond the control limits had assignable causes and that some points within the control limits had assignable causes. This example only does Test 4.

Kokinos: He also didn't say what the limits were for that example. And it is only for an $\bar{X}$ chart, so it's a limited study with respect to Condition 1.

Agrafos: But there's another issue. He doesn't say that he checked various limits to find which limits achieved the economic balance he sought. This example is an incomplete study of Test 4 since we don't know the limits, only one limit was used, and there is no mention of the cost. Conditions 2 and 3 are not addressed even for this control chart and test and there is no analysis.

Kokinos: But he does recognize the limitations: "Such evidence is, of course, one-sided. It shows that when points fall outside the limits, experience indicates that we can find assignable causes, but it does not indicate that when points fall within such limits, we cannot find causes of variability" (Shewhart 1980, 19).

Mathitis: Does he address the other case?

Kokinos: Yes. He provides data on 204 megohm readings and constructs a control chart showing points outside the control limits.

Mathitis: Does he say what those limits are?

Kokinos: No, only that the subgroup size is four. There were 51 averages plotted on an $\bar{X}$ chart.

Peismon: That's a familiar study that others have also analyzed. The limits he used were 3 standard error limits. You can confirm that since he has a table with all the data.

Socrates: What does he conclude?

Kokinos: He says they investigated and found several causes of variability and eliminated them. Another 16 sets of 4 values each were taken and the control chart showed no points—no subgroup averages— beyond the limits. I assume he used 3 standard error limits but he doesn't say so. He then addresses the other side of the argument. But he says that despite much work, they did not find any assignable causes and that proved that the lack of signals meant there was only common cause or random variation (Shewhart 1980, 21).

Socrates: Does he say it proves it?

Kokinos: That's my interpretation—not his word.

Socrates: Isn't that because failure to find no assignable causes is different from there not being assignable causes? My failure to find my keys in the house doesn't mean they are not in the house. How many times have we searched for something and not found it but someone else has?

Peismon: But that is the evidence and shows why 3 standard error limits work.

Socrates: When we discussed hypothesis testing, you insisted that you cannot accept the null hypothesis because failure to reject does not mean proof of or acceptance of a hypothesis.

Mathitis: Those are contradictory positions. Why treat hypothesis testing decisions and control chart decisions differently?

Peismon: That was hypothesis testing and control charts are not tests of hypotheses.

Socrates: Let's address that issue later. Now the discussion is whether 3 standard error limits are empirical. How well does this study meet the requirements?

Kokinos: Again, only one control chart, only one subgroup size, one set of limits—but he doesn't even state what they are, we had to determine them.

Peismon: He does all four tests.

Socrates: I'm not sure of that. He clearly does Test 4. There were alarms and he checked for and found assignable causes. He also did Test 2: there were no alarms, and he checked for and could not find assignable causes.

Mathitis: Since we don't know for sure that there weren't any, it is only circumstantial support of Test 2.

Socrates: What other evidence was there, Kokinos? You did say pages 18–22?

Kokinos: That's right. He has one more example. This graph shows that all the points are within the control limits. He concludes that, as before, without special cause signals, the control chart indicates there is only common cause and all assignable causes have been eliminated (Shewhart 1980, 21–22).

Socrates: Does he say they looked for assignable causes? Or, only that they concluded it wasn't necessary?

Kokinos: No, they didn't look.

Socrates: In other words, he doesn't actually test that assumption. Then, this is not even a study of any of the tests under any conditions we listed.

Mathitis: Is that all the evidence that the 3 standard error limits are empirically derived?

Peismon: Yes, but he says there were other studies.

Kokinos: But after page 22, for the next 250+ pages, he does not provide evidence for selecting the limits he used. He describes the statistical theory use to determine which statistics to use.

Peismon: That leads to the two quotes saying the limits must be justified empirically and that 3 standard error limits seem acceptable.

Socrates: What happens after that, in the remaining part of the book?

Peismon: He does provide additional empirical support for the limits. In Chapters XIX and XX entitled "Detection of Lack of Control," he looks at various actual data sets to illustrate and develop the five criteria or control charts to use.

Socrates: Does each study of a control chart do all four tests for Conditions 2–4?

Peismon: No. He didn't need to since he already concluded that 3 standard error limits were acceptable. He confirms empirically that they work.

Socrates: Did he at least show the economic benefit of using 3 standard error limits?

Kokinos: He didn't. I also read the very next section. The title provides some insight: "Role Played by Statistical Theory." Here he says while mathematical statistics may appear to have a minor role, it doesn't (Shewhart 1980, 22). He explains that all physical phenomena are statistical distributions and therefore the sampling theory is invaluable.

Socrates: Let me see. Note that the next 200+ pages deal with that statistical theory: statistical definition of quality, presentation of distributions, statistical laws, and sampling fluctuations of different statistics, ending with a quote we looked at before where he says experience and many observations would not be feasible (Shewhart 1980, 243).

Mathitis: Let me summarize what we have discovered:

- Shewhart states in one sentence that 3 standard error limits "seem acceptable."
- He shows three control chart examples, but makes no mention of what the limits are or how they were selected or determined for any of them.
- He has only one example that was studied under both states of having alarms and not having alarms (Tests 2 and 4).
- No example includes any subgroup size but $n = 4$, so he doesn't study Condition 3.
- None of the three actual examples covers all control chart types (no study of Condition 1), only the $\bar{X}$ chart is used.
- Between the three examples and the statement that 3 standard error limits seem acceptable, there are 250+ pages all on statistical theory. Those pages also don't state any specific limits.
- The next 100 pages use actual data to develop his five control charts and criteria for detecting lack of control but the empirical studies only use 3 standard error limits and do not compare false alarm rates, true signal rates, or economic benefits using other sigma limits—analysis of Conditions 2 and 3 are missing.
- There are no examples of economic analyses: the analysis critical to showing empirically an appropriate economic balance is achieved.

Kokinos: That's what I see also.

Peismon: That's a rather harsh assessment of this great work.

Socrates: It does seem harsh. But remember that this is *only* an evaluation of the empirical data Shewhart provided in his book. It is not an evaluation of the value of control charts or of whether any other justification is provided for the control limits, let alone of the contribution of his work.

Peismon: Then why have this discussion?

Socrates: One reason is to see what theory we can add to the empirical work. You're a follower of Deming. You are far more familiar with his work, especially his work on profound knowledge. What does he say about knowledge without theory?

Peismon: There isn't any or very little.

Socrates: That's all we are seeking: a profound understanding. I believe much of Shewhart's book is about theory not empirical evidence. I believe that because we can't find many studies. He may have had other studies that were more complete. They are just not in his book.

Agrafos: Why do you think he didn't include them?

Socrates: My opinion? He anticipated Deming and focused on the theory.

WHEELER'S EMPIRICAL RULE

Peismon: Okay, but Wheeler also has studies showing the limits to be empirical.

Mathitis: You mean his Empirical Rule?

Peismon: Exactly. Why call it empirical if it isn't?

Agrafos: What is the Empirical Rule?

Peismon: The Empirical Rule states that there is a certain percentage of data that will fall within one, two, and three standard deviations of the average regardless of the distribution.

Agrafos: What are those percentages?

Peismon: About 60%–75% within one standard deviation of the mean, 90%–98% within two standard deviations of the mean, and 99%–100% within three standard deviations of the mean.

Socrates: How did he show this?

Peismon: His used six distributions but I understand he has evaluated many more distributions (Wheeler and Chambers 1992).

Socrates: These are distributions of real data? From actual processes?

Peismon: Well, no.

Mathitis: Isn't empirical using data from real processes? What did he use?

Peismon: He looked at uniform, triangle, normal, chi-square, exponential, and Burr distributions.

Socrates: These are theoretical probability distributions.

Peismon: I suppose so.

Mathitis: Did he sample from them like Shewhart did from a bowl?

Peismon: Not exactly. He did simulations, which we can do now. I'm sure Shewhart would have done the same if he could have.

Socrates: That's fine. In a certain sense, it is better. It is more rigorous.

Mathitis: It doesn't seem empirical to me as they are theoretical probability distributions.

Socrates: He could, as you explained Gosset did for the *t* distribution for the sample mean, have derived the sampling statistics and then determined exact calculations.

Peismon: But not all sampling statistics are derivable or even calculable if derived.

Socrates: I agree. Some are intractable and that is why people have developed approximations—even to the normal distribution.

Agrafos: But if we said that mathematical derivations and their resulting numbers are not empirical, then these results are not empirical either, right?

Mathitis: Good point.

Socrates: I agree, Agrafos. Still, let's understand what Wheeler showed because it mimics what Shewhart did with the bowl of chips to simulate the normal distribution. Peismon?

Peismon: First, he showed how much of these distributions complied with each part of the Empirical Rule. This is just the individual values, as if you would use the *I* control chart.

Kokinos: And, the results were?

Peismon: Here, see? Four of six matched the percentages for one standard deviation, five of six for two standard deviations, and all six for three standard deviations.

Agrafos: This suggests that using three standard deviations is better for more distributions.

Kokinos: But your conclusions are not right. For three standard deviations, the percentage should be between 99% and 100%. Two of the six were below 99%: 98.6% and 98.2% (Wheeler and Chambers 1992, Figure 4.6, 64).

Peismon: To be fair, in the Empirical Rule, Wheeler uses the words "roughly," "usually," and "approximately" to describe the percentages expected.

Agrafos: How far off can it be and still be considered meeting the Rule?

Socrates: Before we address that, let's see what else was done.

Peismon: He then took 1000 samples of subgroups of size $n = 2, 4,$ and 10 from each distribution and graphed the histogram of the subgroup averages and ranges.

Kokinos: How did those compare to Empirical Rule percentages?

Peismon: He compared to the three standard deviation part, requiring 99% or more. For $n = 2$, only the exponential sampling distribution for averages fell below 99%, at 98.8%. For ranges, the Burr (98.7%), chi-square with two degrees of freedom (97.4%), and the exponential (97.4%) fell slightly below 99%. For $n = 4$, the chi-square (97.4%) and exponential (97.4%) range distributions fell below 99%. For $n = 10$, the Burr (98.9%), chi-square (97.9%), and exponential (96%) fell below 99%.

Mathitis: He didn't have to sample to find the distribution of averages. He could estimate them using the central limit theorem.

Peismon: But Wheeler calls that a myth. You don't need to use the central limit theorem.

Socrates: It isn't used for other charts, but it certainly applies to the $\bar{X}$ chart. It's disingenuous to say it doesn't apply when his simulation is a manual application of it.

Kokinos: You're right, Socrates. Here's what Wheeler says, "The central limit theorem does apply to Subgroup Averages: as the subgroup size increases, the histogram of the Subgroup Averages will, in the limit, become more 'normal' regardless of how the individual measurements are distributed" (Wheeler and Chambers 1992, 78).

Peismon: Yes, but he also says "Even though the central limit theorem applies to the Subgroup Averages, it is not the reason why control charts work" (Wheeler and Chambers 1992, 79).

Kokinos: But it's what Socrates said, he continues by saying "the central limit does *not* apply to Subgroup Ranges, therefore, if the central limit theorem was the basis for control charts, then the Range Chart would not work" (Wheeler and Chambers 1992, 79).

Socrates: I agree with Peismon. Wheeler is making the valid point that the central limit theorem applies to $\bar{X}$ charts but that does not mean the central limit theorem explains why all control charts work.

Peismon: Perhaps not even any.

Mathitis: There are different types of control charts, some for continuous measurements and some for discrete. In addition, the ones we use now are not the ones Shewhart derived in his first book. He had five charts for different statistics: average, proportion, standard deviation, and correlation coefficient. We don't use two of these five, the ones on correlation coefficient. Regardless of the criterion, each had the same signal: a point beyond the control limits.

Socrates: Yet, each could have a different explanation for why it works.

Peismon: But he says that, by "brute forcing," at least 99% of the points will fall within the limits assuming the process is in control; no theoretical argument is needed.

Socrates: I don't know what he means by brute force. The fact is that there are proven theorems—and Shewhart used these theorems, repeatedly referring to them—that show the percent, or probability, of points within symmetrical limits.

Mathitis: Tchebycheff's theorem says the probability of being within $t\sigma$ of the mean is $1 - 1/t^2$.

Peismon: But that only assures us of 89% using three standard deviations.

Mathitis: But there is a difference. These theorems provide the exact minimum probability while the Empirical Rule provides "roughly" or "usually." This suggests that, for specific distributions, there are limits that are more optimal.

Socrates: He also mentions the Camp–Meidell theorem. That brings us to $1 - 1/(2.25t^2)$, or 95% for $t = 3$, the 3 standard error limits.

Peismon: That's still not 99%.

Mathitis: No, but neither did all cases he show achieve 99%. The range sampling distribution from an exponential distribution only achieved 96%. That's not far from the 95% of the Camp–Meidell theorem. In addition, Wheeler's calculation is a point estimate from a sample. How low would a lower 95% confidence interval go?

Socrates: Also, other conditions such as unimodality or finiteness (e.g., the uniform or triangle) increase the percentages. That is why some of the distributions Wheeler simulated did better than others. On the other hand, for the chi-square with two degrees of freedom and the exponential distributions, he found more than 2% of the ranges beyond the 3 standard error limits.

More importantly, your claim was that Shewhart established the limits empirically. Whatever Wheeler did, whether empirical or theoretical, provides no evidence that the limits were derived empirically by Shewhart. This evidence, Wheeler's evidence, comes after the fact.

Peismon: But they support the rationale for using them. Wheeler says so.

Socrates: I want to understand why control charts work. Empirical derivation alone does not provide that understanding. Theory is necessary for understanding.

I think Wheeler's work is wonderful. He is providing theoretical support by manually trying numerous distributions. But the original theory came from Shewhart. Mathitis and I lamented the brevity and sparseness of examples and studies to show why 3 standard error limits. However, the hundreds of pages that show theory provide a deeper understanding. Should we oppose such efforts to understand? Theory helps improve and develop better methods. Isn't that why we are applying control charts to begin with? To improve?

Agrafos: I agree, Socrates. Isn't that what Deming said, without theory there is no knowledge?

Peismon: True.

Socrates: We can have knowledge without theory. Knowledge of facts, for example, a 2-year-old may be able to recite the value of pi or her address. What is lacking is understanding.

Peismon: Deming's profound knowledge.

Socrates: Wheeler's work is on that theory—shown in a straightforward and clear manner.

Peismon: His theory shows that as you widen the limits, the percentage within them varies less and less. That seems to be true for the initial universe or population distribution and for sampling distributions of averages and ranges. However, the least amount of variation of the percent within the limits occurs with 3 standard error limits and that's why those are used.

Agrafos: I thought the reason for choosing the 3 standard error limits was to achieve an economic balance not to reduce the variation of the percentage within those limits.

Kokinos: Isn't that obvious? As you widen the limits, more and more of the distribution is included regardless of the distribution. Why not use even wider limits? If we went to 4 standard errors, would there be no distributions that fall outside the predicted Empirical range?

Mathitis: That's my point. The Empirical Rule says at least 99% and his own study contradicts that.

Peismon: The Empirical Rule doesn't mean all cases fit it, just the vast majority—roughly.

Mathitis: Doesn't that show that sometimes 3 standard error limits are not empirically derived?

Peismon: But that's not his point. Wheeler's point is that if you make the limits wide enough, then a point outside the limits is very likely to be a true signal. So, you don't need theory to tell you that—it's logic.

Kokinos: I thought you said it was empirical.

Peismon: He confirmed it with his studies.

Kokinos: Of theoretical distributions using theoretical sampling—simulation. Besides, as I asked before, on the basis of that logic, why not make the limits even wider, say 4 standard errors or 5 standard errors? To cover every possible distribution, use 10 standard error limits since, by Tchebycheff's inequality, that would guarantee at least 99%, since $1 - 1/10^2 = 0.99$.

Socrates: As with Shewhart, Wheeler also doesn't consider other limits beyond three standard deviations. As Kokinos asks, would wider limits define a high percentage that all distributions meet? But critically, not one theoretical simulation tells us which limits are economically best by achieving the economic balance that was the intent of choosing 3 standard error limits.

Mathitis: I agree. As with Shewhart's examples, Wheeler's simulations do not do all four tests under all three conditions and make a conclusion based on the economic analysis.

Peismon: He does look at different sample sizes, different limits, and different control charts; those are Conditions 1–3.

Agrafos: But only of Test 1 and no economic analysis.

Kokinos: But also why does he use only integer multiples and no limit beyond 3? If wider limits are better, then he should be comparing 3 standard error limits to limits with greater multiples of standard errors, for example, 3.1, 3.2, and so on.

Mathitis: Wheeler's evidence is more than what Shewhart showed in his book but is also limited. And Wheeler's evidence is theoretical, despite the naming of the rule as empirical.

EMPIRICAL JUSTIFICATION FOR A PURPOSE

Socrates: That puzzles me in two ways. One, why is there so much focus on a couple of statements that the limits are empirically justified or based? Two, why ignore the significant theoretical work comprising most of Shewhart's book? Is being theoretically based a disadvantage or weakness or flaw?

Agrafos: Listening to the discussion, I can see why you say that. It seems that, because Shewhart said the limits are empirical but offered little evidence and Deming repeats it, everyone else views it as settled matter—no matter whether it is useful or beneficial to also see them in other ways or to support their empirical derivation with theory.

Kokinos: I tend to agree. For example, I started playing golf as a hobby and developed my own way of swinging empirically. I was able to reach a certain level of success that I enjoyed. But when I wanted to get better and couldn't, I began studying swings of professional golfers, reading books, and eventually took lessons. Changing my golf swing from empirically based to theory based made a difference.

Socrates: That's a good point, Kokinos. It's tied to the discussion we had initially about the difference between empirical limits versus limits empirically justified. We concluded that to justify, empirically or theoretically, there had to be a purpose. In your case, it was to improve your performance. Shewhart said that the purpose of process control was to achieve an economic balance of chasing false alarms and failing to address true alarms. Now that we have looked at Wheeler's evidence, what has he shown?

Kokinos: As the limits get wider, the percentage of false alarms decreases. That seems to achieve one purpose: reducing the false alarm rate.

Mathitis: But that is checking only one of the four tests.

Peismon: It's something.

Socrates: Aren't those percentages probabilities?

Kokinos: Yes, of course.

Socrates: Then, are the limits in Wheeler's Empirical Rule probability limits?

Mathitis: I hadn't thought of that but I would have to agree.

Peismon: But Wheeler says the limits are empirical and not probability limits.

Agrafos: Why can't they be both?

Peismon: They can't be both if they can't be probability limits.

Socrates: Let's discuss that.

17

Myth 17: Control Chart Limits Are Not Probability Limits

Scene: *Break room (continuation of discussion on control charts)*

BACKGROUND

Socrates: Peismon, why aren't control chart limits probability limits?

Peismon: Besides that everyone says they are not?

Socrates: Everyone?

Peismon: Shewhart, Deming, Wheeler, and other experts have said control chart limits are not probability limits.

Mathitis: But others have written articles and books using them as probability limits. Juran, Duncan, Grant, and others consider them to be probability limits.

Socrates: It sounds like you two are arguing that because someone—especially if that person is an expert—says it's true, then it must be true. This is an argument from authority: Authority A says X is true. Whatever A says is true. Therefore, X is true.

Peismon: They must have a reason for saying it is true.

Socrates: And that is what we want to know.

Peismon: Wheeler quotes Shewhart.

Kokinos: Shewhart said control chart limits aren't probability limits but empirical limits.

Mathitis: We just discussed that topic (Myth 16). Shewhart is saying that empirical evidence is needed to justify that whatever limits are chosen the probability of false alarms is not economically burdensome.

Peismon: I've never used them as probability limits and they work for me. Using them as probability limits doesn't serve any purpose. It just confuses things, makes it more complicated. Deming said treating control chart limits as probability limits would be an obstacle to their use.

Socrates: That's another issue that doesn't preclude them being probability limits. Maybe others find them useful.

Mathitis: I can see that. I view them as probability limits in a sense. We teach control charts by showing a picture of a normal distribution on the vertical axis suggesting that the probability of being beyond the limits is 0.27%. It hasn't stopped people from using them.

Peismon: But that's not true for all control charts.

Mathitis: But it does counter Deming's view.

ASSOCIATION OF PROBABILITIES AND CONTROL CHART LIMITS

Socrates: We said that Wheeler's Empirical Rule showed the percentages of being within various multiples of the standard error. Are those percentages probabilities?

Kokinos: Of course. I said they are the probability of being within the limits.

Mathitis: That is exactly what Wheeler's simulations showed. His Empirical Rule states for various limits the probability range of being inside those limits. He is justifying theoretically or empirically those probabilities.

Socrates: Then, we have a probability associated with each set of limits. How does Shewhart use the limits? He goes through at least four steps: he provides a definition of control, followed by theory, he then suggests limits, and finally describes criteria including examples. Let's see if, in each step, he mentions the limits alone or with a probability.

Mathitis: That sounds like a good approach.

Socrates: Kokinos, will you find the passages?

Kokinos: Let me know what to search for.

Socrates: Find his definition of control. It should be at the beginning. Use the Glossary.

Kokinos: It's on page 6. He writes, "a phenomenon will be said to be controlled when, through the use of past experience, we can predict, at least within limits, how the phenomenon may be expected to vary in the future. Here it is understood that prediction with limits means that we can state, at least approximately, the probability that the observed phenomenon will fall within the given limits" (Shewhart 1980, 6). Later, related to the basic laws of control, he repeats that saying a process in control is being able to predict within limits with a given probability (Shewhart 1980, 121).

Mathitis: He clearly associates a probability with the limits in defining process control.

Socrates: How about the theory, Kokinos?

Kokinos: For theory, in Chapter II "Scientific Basis for Control," he writes that removing assignable causes leaving only common cause variation puts the process in a "state where the probability that the deviations in quality remain within any two fixed limits (Fig. 5) is constant" (Shewhart 1980, 17).

In Section 7 of Chapter XIII "Sampling Fluctuations" entitled "The Problem of Determining the Allowable Variability in Quality from a Statistical Viewpoint," he writes that if a measure of product quality is statistically controlled, there is a probability associated with an output unit being within a certain range determined by functions of the measure (Shewhart 1980, 73).

For each statistic he considers in Chapter XIV, he says, "we shall need to know the probability P that a statistic of a sample of size n produced by this constant system of causes will fall within the range θ_1 to θ_2..." (Shewhart 1980, 175).

Then, in Chapter XVII "Design Limits on Variability," he writes that the determination of limits depends on knowledge about the distributional functions. This allows the "determination of the probability associated with any interval X_1 to X_2" (Shewhart 1980, 268).

Mathitis: Again, probability and limits are connected.

Socrates: He also refers to Tchebycheff's inequality, which relates a probability to an interval. Now, Kokinos, tell us how he determines the limits.

Kokinos: That's in Chapter XIX "Detection of Lack of Control in Respect to Standard Quality" in Section 2 "The Basis for Establishing Control Limits." Again, he says the distributional function of a statistic based on a sample size *n* "gives the probability that the statistic θ will have a value lying within the limits θ₁ to θ₂" (Shewhart 1980, 275).

Peismon: But he has also clearly stated the limits must be empirically based.

Mathitis: Yes, but for what purpose?

Kokinos: And the limits are always associated with a probability. That seems to be the overwhelming purpose, to reduce the probability of unnecessary costs.

Socrates: Let's see how he applies these limits. He has five criteria, each one for a different statistic and a different control chart. What does he say about the limits?

Kokinos: In Chapter XX "Detection of Lack of Control" in Section 8 "Use of Criterion I—Some Comments," he states that we must first answer several questions. One question is the number of subgroups of size 4, since that is the subgroup size he seems to prefer.

Socrates: Why is that question important?

Kokinos: He states that it is important because the "expected probability of a statistic falling within the ranges established by Criterion I approaches the economic limiting value only as the total number *n* of observations approaches infinity" (Shewhart 1980, 315). He also states that even for only two subgroups of size 4, the difference in the expected probability is negligible. Oh, there is a footnote. He cites an unpublished manuscript by F.W. Winters researching Criterion I. Winters calculated the probability of false alarms to be 0.0085 under one condition and 0.00014 under another (Shewhart 1980, 315).

Mathitis: He now calculates a specific probability.

Kokinos: He also uses a specific probability regarding Criterion II, using the statistic $|d|/\sigma_d$ where $|d|$ is calculated using this equation:

$$\frac{n}{n-1}\overline{\sigma^2} - \frac{m}{m-1}\sigma_{\bar{x}}^2 \qquad (17.1)$$

where *m* is the number of subgroups, *n* is the size of each subgroup, and the other factors involve the subgroup variances and squares of the subgroup averages (Shewhart 1980, 319).

The probability that $|d|/\sigma_d > 3$ is approximately 99% (Shewhart 1980, 320), based on a normal distribution.

For Criterion III on the correlation coefficient *r*, he uses the 3 standard error limits of $0 \pm 3/\sqrt{(n-1)}$. So, if *r* is outside the limits, it signals special cause or assignable cause variation (Shewhart 1980, 323). While he doesn't state a probability, he provides evidence that the distribution of the correlation between two controlled variables is approximately normal.

Mathitis: That makes these limits approximately 99.73% limits.

Socrates: What about the fourth and fifth criteria?

Kokinos: Criterion IV also uses three times the standard error of a difference statistic based on two sets of subgroups, one using *X* only and the other using an uncorrelated *Y* (Shewhart 1980, 323). The idea is to check whether the independent variable *Y* is a possible assignable cause for variation in *X*.

Socrates: Okay, what about Criterion V?

Kokinos: He again states a specific probability criterion of 0.001. Less than that signals special cause variation or instability (Shewhart 1980, 329).

Socrates: His five criteria for detecting lack of control are each based on probabilities. It appears that Shewhart intended the control limits to be probability limits in the sense that there is an associated probability of the plotted statistic being within the limits when in control.

Mathitis: I would certainly agree.

Peismon: But that doesn't exclude them from being empirically derived, which is what is meant by being empirical limits.

Mathitis: Even if you are right, it doesn't mean that being empirical excludes being probabilistic. They could be empirically derived probability limits, as Agrafos suggested.

Agrafos: Why do Deming and others claim they are not probability limits? It doesn't make sense given that in every situation Shewhart associates a general or specific probability with the limits.

Peismon: What if we can't calculate the probability? Then, they can't be real probability limits.

218 • *Understanding Statistics and Statistical Myths*

CAN CONTROL LIMITS BE PROBABILITY LIMITS?

Socrates: What is needed to be able to calculate probabilities?

Mathitis: The probability distributions. If we know the probability distribution, we can calculate probabilities. If we know one for the statistic—the sampling distribution—then it's great; if we know it for the universe or population from which we sample, then we can derive the sampling distribution of the statistic, or use simulation as Wheeler did.

Peismon: That's my point—you can't know the distributions. Certainly not the universe probability distribution, and therefore, we can't calculate the probabilities. That's why Wheeler checked numerous distributions with different shapes.

Socrates: Why do you think we can't know them?

Peismon: Neave and Wheeler wrote an article discussing this very issue to answer both questions of whether we can know the probability distributions and whether we can calculate probabilities for the limits. They claim that there is too much uncertainty. They paraphrase Shewhart: "even if the process were exactly stable, and if the normal distribution were appropriate (neither of which we would ever know), we would still never know the value of its mean. And even if we did, we would never know the value of its standard deviation. We could only estimate these from the data. Since the probability calculations depend upon all of these things, it will always be impossible in practice to compute the required probabilities" (Neave and Wheeler 1996).

Mathitis: We have Peismon quoting Neave and Wheeler quoting Shewhart.

Kokinos: But Shewhart does say in those pages we talked about before that we can't know the functional distribution well enough to establish the limits (Shewhart 1980, 275).

Socrates: That's good. It clearly identifies the prerequisite to calculating probabilities: knowing the probability distributions that can depend on knowing the parameters of the distribution, for example, the mean and standard deviation of a normal distribution.

Mathitis: I am now thinking that Peismon might be right. It does seem that we can't know the probability distributions because of what Neave and Wheeler said.

Peismon: Shewhart did not base control limits on precise calculations of probabilities. That's the point I am making.

Kokinos: Although the sections I read on the five criteria he calculated specific probabilities.

Peismon: But those are theoretical distributions.

Socrates: Let's say you are correct, Peismon. Then there are contradictions with two other assertions. The first is with hypothesis testing and estimation.

Peismon: What contradictions?

Socrates: Do Neave and Wheeler and you think that probabilities can be calculated when doing statistical inferences of testing hypotheses and estimating probabilities? Did Shewhart believe the probability p of hypothesis testing can be calculated?

Peismon: Yes, we all do.

Socrates: When testing whether two population means are equal, do we know the population distributions, the population means, and the population standard deviations?

Kokinos: If we knew them, then why test?

Mathitis: We don't know the mean and standard deviation when using the t test. It was developed specifically for that case. And we calculate the probability p.

Socrates: And how is that p calculated?

Kokinos: We assume that the null hypothesis is true, and that assumption is about a probability distribution.

Mathitis: Even when we use nonparametric or distribution-free tests, there is an assumed probability distribution.

Peismon: And that's the point they are making. That assumed probability distribution could be anything so we don't know exactly what the probability is.

Socrates: Could it not be anything when we do hypothesis testing? Isn't that why often people do a test on the distribution to validate the distributional assumption before doing the original test, say, on the difference between two population means?

Mathitis: And then we choose to do other tests that do not require that particular distribution when the distributional test fails.

Peismon: But we often assume that the sample we take comes from a much larger population. Say, it is a continuous production process. We assume we sample from the population of possible units that could have been manufactured. Then we don't know what it is.

Socrates: But we do calculate probabilities in those cases. Like for flipping a coin. We don't have a finite number of results from which we sample. We assume the finite results come from an infinite number of possible results. We don't have those and we don't know what they are. Yet, we calculate probabilities.

Peismon: So what do we use other than a control chart if our assumed distribution is not correct?

FALSE ALARM RATES FOR ALL SPECIAL CAUSE PATTERNS

Agrafos: But wasn't a point exceeding the limits a false alarm in Wheeler's simulations?

Socrates: Yes. Peismon, we want to keep the false alarm rate low, correct?

Peismon: Yes.

Socrates: How would we know if you had kept the rate low, or low enough if you can't calculate it? You might want it less than 1%, but because of the lack of precision, it could be 5% or more.

Peismon: That's where the empirical justification applies.

Mathitis: So that justification *is* for the probability! Meaning the limits are probability limits.

Socrates: We also have another reason for believing that probabilities can be calculated. That relates to the false alarm rates.

Agrafos: You mean that they are probabilities?

Socrates: Yes, but more than that. Some years after Shewhart's book, other special cause patterns were created called the Westinghouse Rules. What was the basis for these rules?

Mathitis: They were designed to detect other kinds of special causes that a point beyond the control chart limit rule would not detect.

Socrates: What is one of those rules?

Mathitis: There is the rule of 2-out-of-3 consecutive points beyond 2 standard errors on the same side of the center line.

Socrates: Why two out of three and not two out of five or seven out of nine?

Peismon: They are meant to have the same false alarm rate as a point beyond the 3 standard error limits.

Socrates: Now, Peismon, you can't have it both ways. How can you know that the other rules have the same false alarm rates as a point beyond the 3 standard error rule if they are not probability limits or the probabilities can't be calculated?

Kokinos: I think Socrates has a point, Peismon.

Peismon: But those probabilities can't be calculated exactly.

Mathitis: They can, assuming a probability distribution. Assuming a normal distribution, we know that the probability of a point beyond the 2 standard error limits is 0.05. Hence, the two-out-of-three rule has a probability based on two out of three and three out of three: $3 \times 0.025 \times 0.025 \times 0.975 + 0.025 \times 0.025 \times 0.025 = 0.001828 + 0.000015625 = 0.00184375$. That's approximately what we get for one point beyond the 3 standard error limits assuming a normal distribution or the 0.001 probability that Shewhart used in his criteria.

Mathitis: The same issues seem to apply to both control chart limits and hypothesis testing but Neave and Wheeler don't seem concerned with them when testing hypotheses. That seems contradictory.

Agrafos: But didn't we discuss previously that Wheeler looked at a variety of distributions to support his Empirical Rule?

Socrates: That's true. And we said those were percentages and, thus, probabilities.

Agrafos: That's what I thought. So isn't Wheeler saying that the 3 standard error limits have a limited range of probabilities of points occurring beyond them if the process is in control?

Kokinos: Now I think they are probability limits.

Mathitis: I think there is a contradiction or inconsistency in saying they are not probability limits and then showing that they have certain probabilities. By calculating the percentage of averages and ranges within each set of limits, he calculated the probability of being within each set of limits. He is showing that the 1, 2, and 3 standard error limits are probability limits.

Peismon: I think he is showing that you reduce the false alarm rate as the limits widen.

Mathitis: I thought the reason for choosing the 3 standard error limits was to achieve a balance not to reduce the variation of the probability within those limits.

WHEELER USES PROBABILITY LIMITS

Agrafos: So, where are we in answering the question?

Socrates: The claim was that we cannot calculate probabilities because we cannot know the probability distribution. The counterpoint was that all that is needed is an assumption of a probability distribution for either the sampling distribution of the statistic or the universe from which the samples came.

Peismon: Still, even if true, the assumed probability distribution may be so far off from the true distribution to be useless.

Socrates: Unfortunately, if you are relying on Wheeler, you should be aware that he actually does calculate probabilities in the manner we have suggested.

Kokinos: Really, he does?

Mathitis: That I would like to see!

Socrates: He says this: "Another aspect of the c-chart concerns the fact that the Poisson Distribution is severely skewed when the average is small. This skewness changes the chances of getting a false alarm. For average counts that are less than 1.0 per sample, there is an approximate chance of 3 to 4 percent that a point will exceed the upper 3 sigma limit. For Average Counts between 1.0 and 3.0, the chance of a false alarm is about 2 percent, and for averages between 3.0 and 7.0, the chance of a false alarm is about 1 percent" (Wheeler and Chambers 1992, 274).

Mathitis: It sure looks like he is calculating probabilities.

Kokinos: And assuming a probability distribution.

Socrates: Note also that he is saying that, with asymmetrical distributions, the Empirical Rule for 3 standard error limits may be violated significantly. The difference of 96% is close to what the Camp–Meidell theorem calculates and also close to what Wheeler estimated for the chi-square distribution. We have one answer to Agrafos' question of what does "approximately" mean.

Peismon: But this is a theoretical determination.

Socrates: Thus, it provides support for our discussion of the value of profound learning. If we don't seek to understand why things work, then it is more difficult to learn when they will not work. That limits our ability to develop methods that do work or work better.

Kokinos: Wheeler is also showing with this example that the 3 standard error limits are not the best ones to use. And he is identifying a condition when that occurs: substantial skewness or asymmetry in the distribution.

Socrates: And why did he determine that better limits were needed?

Kokinos: He looked at the false alarm rate.

Socrates: That's all?

Mathitis: I see where you are headed. We discussed the four tests to determine the best limits and Wheeler is using only Test 2 (Myth 16): false alarms when no assignable causes exist. So, the decision to change limits is not based on achieving the "right balance" since he didn't look at the effect on identifying true signals.

Agrafos: He did calculate the probability of a false alarm.

Kokinos: And to do that, he had to assume a probability distribution. In this case, it was the Poisson distribution.

Socrates: He also says that the formulas "for the 3 sigma limits do not produce a lower control limit until the Average Count per sample exceeds 9.0" (Wheeler and Chambers 1992, 274). Note that his 3 sigma limits refer to 3 standard error limits. Because of the two problems of no lower control limit (LCL) and skewness varying false alarm rates per the average count $\bar{c}$, he recommends different limits: "However, since the assumptions justifying the use of a c-chart also justify the use of the Poisson Distribution, there is an easy way to remedy both of the shortcomings of 3 sigma limits for Poisson Counts. This may be done by using the 0.005 LCL_c and the 0.995 UCL_c probability limits shown in Table 10.7" (Wheeler and Chambers 1992, 275).

Mathitis: I didn't know that. The issue of no LCL occurs for all variation charts. The R chart for ranges has LCL = 0 for $n \le 6$ and the S chart for standard deviations has LCL = 0 for $n \le 5$. Zero is used because, otherwise, the calculation would yield a negative value. The same thing occurs for other averages of other distributions that can only have positive values.

Kokinos: Why does that occur?

Mathitis: We talked about this before. If the limits are average statistic ± 3 standard errors, then when the average is near zero, LCL = average statistic − 3 standard errors may be less than zero. So, the constants used in the formulas are zero to avoid negative numbers.

Socrates: He also says probability limits could be used for the u chart.

Kokinos: What do you think, Peismon?

Peismon: I don't know… I didn't know that.

Socrates: Let's take this one step further. He calculated probability limits by assuming a probability distribution. We can work backward to determine the probability of the 3 standard error limits assuming the same probability distribution.

Peismon: How?

Socrates: Let's use the standard normal distribution z as an example, with mean = 0 and standard deviation = 1. What z value corresponds to a cumulative probability of 0.975?

Peismon: We look up that probability and find the z value. That's an easy one—it's 1.96.

Socrates: What is the probability of getting a value $z \leq 1.5$?

Peismon: We look it up in the table to find the area to the left of 1.5. It's 0.9332.

Socrates: You can do both: start with a probability and look up the z value corresponding to it, or given the z value, look up the probability corresponding to it.

Peismon: Yes, of course.

Socrates: Wheeler calculated the c chart control limit values for 0.005 and 0.995, and they were slightly different from the 3 standard error values. Now, we just need to use the 3 standard error values to reverse the lookup to find the corresponding probabilities.

Kokinos: Show me.

Socrates: Let's use the formula for the c chart control limits that Mathitis said. If the average count, $\bar{c}$, is 9, to make the math easier, then the 3 standard error limits are UCL = $9 + 3\sqrt{9} = 18$ and LCL = $9 - 3\sqrt{9} = 0$. Right?

Kokinos: Okay.

Socrates: Wheeler's probability limits for the c chart when $\bar{c} = 9$ are 0.995 UCL = 18.5 and 0.005 LCL = 1.5. These limits are for what intended probability?

Peismon: Obviously, 99%: 0.995 − 0.005 = 0.99.

Kokinos: The probability of a false alarm is 1%. This range is 17, which is narrower than the 3 standard error limits of 0 to 18.

Socrates: Yes, but don't forget that this is a discrete variable. The possible results are the nonnegative integers: 0, 1, 2, and so on. So, the probability limits will not cover exactly 0.99 since you can't get a count of 1.5 or 18.5. The only difference is that the 3 standard error limits allow a count of 0 and 1 (both less than 1.5). The probability of those counts occurring is 0.001 if the average count is 9.

Peismon: There is very little difference.

Socrates: And Wheeler says as much, saying that the difference between the probability limits and the 3 standard error limits results in "the probability of a point falling below (above) the 0.005 LCL_c (0.995 UCL_c) is never greater than 0.005 when the process is in control" (Wheeler and Chambers 1992, 275).

Mathitis: Making control chart 3 standard error limits probabilities limits!

Kokinos: If he recommends probability limits, doesn't that make them useful?

Socrates: He states what useful means: "Thus, the [probability] limits provide a balance between sensitivity to process improvements and process deterioration, while minimizing the chance of a false alarm. They are reasonable alternatives to the 3 standard error limits for c-charts" (Wheeler and Chambers 1992, 275).

Mathitis: However, he doesn't address why this balance of false alarm rate and correctly identifying true signals (which he does not calculate) is better than before.

Peismon: That's only for these two charts, which Shewhart did not include in his book.

Socrates: But other authors do the same thing with the $\bar{X}$ chart. Assuming a normal distribution, what does a 3 standard error limit equate to probabilistically?

Mathitis: Clearly, 99.73%.

Socrates: If we assume we don't know the mean and standard deviation, then rather than using the normal distribution, what would we use?

Mathitis: The *t* distribution.

Socrates: What is the equation for the standard error using the *t* distribution?

Mathitis: It's $t_{1-\alpha, n-1} \cdot s/\sqrt{n}$.

Socrates: The task is to find $t_{1-\alpha, n-1}$ knowing that $t_{1-\alpha, n-1} \cdot s/\sqrt{n}$ is equal to the difference between the upper control limit and the grand average. In other words, $t_{1-\alpha, n-1} \cdot s/\sqrt{n} = UCL - \bar{\bar{x}}$. We can solve for $t_{1-\alpha, n-1}$ knowing s, n, UCL, and $\bar{\bar{x}}$. Once we know $t_{1-\alpha, n-1}$, then we can find alpha.

Kokinos: You are saying that regardless of whether the limits are economically based, nothing stops us from assuming an appropriate or reasonable probability distribution and determining the alpha that corresponds to those control limits.

Socrates: That's correct. You can assume any distribution—appropriate or inappropriate. After all, that's what we do when we calculate p for hypothesis testing. The null hypothesis could be false in many ways and quite significantly.

Mathitis: So, for $n = 4$, 3 standard error limits correspond to 94.2% limits. I see. We can do this for every control chart to determine what probability the 3 standard error limits equate to.

OTHER USES OF PROBABILITY LIMITS

Socrates: There are other uses for probability limits. The first is related to other tests for control.

Kokinos: You mean the Western Electric rules.

Socrates: Yes. Shewhart only used the criterion of a point beyond the control limits as an indication of assignable causes. The other indicators came into use in the mid-1950s.

Mathitis: Those rules were designed to have approximately the same false alarm rate as the 3 sigma rule as we showed.

Socrates: Peismon, do you agree?

Peismon: Yes.

Socrates: The only thing that was required was an assumption of a probability distribution. Are these other rules useful in the same way that the 3 standard error limits are?

Kokinos: Yes. I use them.

Mathitis: If they provide the same balance as Wheeler said about the probability limits for c charts, then you would have to agree they are useful.

Kokinos: You said there were other uses. What else?

Socrates: The calculation of ARLs, the average run lengths, before a false alarm assuming a controlled process.

Kokinos: They also require a probability. The ARL is the reciprocal of the probability of a false alarm, or $1/p$.

Mathitis: ARLs can be useful—and they depend on the 3 standard error limits or whatever limits are used as being probability limits. How can you calculate that without knowing *p*?

Socrates: Can the 3 standard error limits and control chart limits in general be probability limits?

Kokinos: Yes, they can, just by assuming a probability distribution.

Mathitis: Either the sampling distribution of the statistic or the distribution of the universe.

Socrates: Are they probability limits?

Peismon: Not necessarily.

Mathitis: They are if you assume a probability distribution; otherwise, not.

Peismon: I'm not convinced.

Socrates: Consider a different application. Where does the *p* value come from when using the *t* test for testing hypotheses?

Mathitis: There is a formula for computing the *t* value. For example, the formula for a one-sample test of the mean is $t = \bar{x}/(s/\sqrt{n})$. That *t* value is then converted to a probability.

Socrates: Does the unconverted *t* value correspond to a probability?

Mathitis: Yes. In my class, we explained that the formulas produce a *t* value for the data, which is compared to a critical *t* value that corresponded to alpha. We could make that comparison instead of comparing *p* to alpha.

Kokinos: It is still a probability.

Mathitis: Yes, but only if you assume the *t* distribution. Otherwise, it's just a number. The 3 standard error limits have similar formulas. They are just numbers until we assume a probability distribution and then we can equate them to probabilities.

Socrates: Is it necessary to assume a probability distribution to compute control chart limits?

Peismon: No. You can just calculate the 3 standard error limits using tables for the constants and the formulas or letting software do the calculations for you.

Mathitis: If you want to have specific probability limits, you do need to assume a probability distribution.

Socrates: Is it useful to know what the probability is—for 3 standard error limits?

Peismon: I would say no.

Kokinos: I'm inclined to say maybe, sometimes it is. After all, Wheeler switches to probability limits because he knew that the 3 standard error limits for the c chart had significantly varying false alarm chances or probabilities depending on the average count. That sure looks like knowing the probabilities was a critical factor for making the change.

Socrates: And being able to calculate them.

Mathitis: As I am learning more about statistics, the answer to your question, Socrates, is "it depends."

18

Myth 18: ±3 Sigma Limits Are the Most Economical Control Chart Limits

Scene: *Break room (continuation of control charts)*

BACKGROUND

Socrates: We agreed that Shewhart talked about establishing the 3 standard error limits for economic reasons, as the title of his book states. But are they the best limits for this purpose?

Peismon: Everyone knows they work—they've been used successfully for more than 80 years.

Socrates: But they "work" doesn't mean they are the most economical limits. Trial-and-error works, flat earth theory works…. The geocentric theory that the sun orbits the earth worked with respect to predicting solar and lunar eclipses.

Peismon: Right after Shewhart said that the purpose was to balance the cost of investigating false alarms against the loss of ignoring problems (Shewhart 1980, 276), he said to use 3 sigma limits for this purpose.

Socrates: However, he also mentions another economic advantage, which is more aligned with the title of his book. He says that being in a state of statistical control enables one to reap the benefits of reduced inspection and scrap costs, higher and uniform quality, and reduced tolerance limits (Shewhart 1980, 34).

Kokinos: What does that mean?

Socrates: I take it to mean that the economic value is not only in minimizing false alarms and missed opportunities but also in maintaining a state of control. Thus, resources to manage a process are reduced while the quality of the product is increased. We don't know which is greater and it's not clear which he means from all the pages of his book we have discussed.

Peismon: I know it works for me.

EVIDENCE FOR 3 STANDARD ERROR LIMITS BEING ECONOMICALLY BEST

Mathitis: Are you saying that 3 standard error limits are the most economical ones for all situations, for example, products, services, environmental conditions, different sites, different currencies, different costs of wages, and all costs of searching for false alarms and missing opportunities?

Peismon: Why not?

Socrates: The question isn't why not but why, especially given the absence of economic studies and analyses in Shewhart's and Wheeler's writings.

Agrafos: Can we describe such a study as we did before for evaluating studies that tested whether control chart limits were empirically supported or not?

Socrates: Yes. This time, focus on a new Condition 4: Various costs of false alarms and failing to detect true alarms (Myth 16).

Kokinos: And add another analysis to those studies: the selection of the most economical limits and not just to compare the false alarm and true signal rates.

Mathitis: In that case, there is no evidence in Shewhart's book. His examples not only did not have complete studies of all four tests under all four conditions, there were no cost analyses done. That's simply because there were no comparisons.

Kokinos: Then it's also true for Wheeler's simulations. While his simulations covered different control charts (Condition 1), different limits (Condition 2), and various sample sizes (Condition 3), he did not include any costs (Condition 4) or do any economic analysis (Analysis).

Mathitis: Even in his example of the *c* chart, where he changed from 3 standard error limits to probability limits, the reason was not because of a better economic value but solely to reduce false alarm rates.

Peismon: He explains that "While they have a basis in probability theory, three sigma limits were essentially chosen because they provide reasonable action limits. They strike an economical balance between the two types of errors that one can make in interpreting data from a continuing process. They neither result in too many false alarms, nor do they miss too many signals" (Wheeler and Chambers 1992, 82–83).

Mathitis: How does he know if he did not assess the probability of detecting a true signal for either the 3 standard error limits or the probability limits? And by replacing them with probability limits, he is admitting the 3 standard error limits are not the best for the *c* and the *u* charts.

Socrates: One thing is to show in a study that any other limits but 3 sigma will never have fewer false alarms or fewer missed signals. This shows that it is best when best is defined as the minimal total probability of false alarms and missed true alarms. With respect to best economical limits, 3 standard error limits may have more false alarms or missed opportunities, but the combined cost of each should be the least. Does Shewhart or Wheeler or anyone refer to actual costs of false alarm searches and missed true signals?

Kokinos: Neither do.

Socrates: There are other factors that Shewhart did not mention. While Shewhart was the father of control charts, there is much more that is understood about them since he created them.

Mathitis: What are those factors?

Socrates: Not just subgroup size, but how frequently one should sample, what kinds of assignable causes. Recall that he did not use other rules for detecting assignable causes. All these conditions will affect the economic value. It seems incredibly fortunate that the 3 standard error limits are the most economical regardless of the chart used, subgroup size, number of subgroups, time between subgroups, what process changes are critical, costs of scrap, and so on.

Mathitis: Stating it that way, I would agree. And, neither Shewhart's nor Wheeler's examples cover all these conditions.

Socrates: Consider this. A company has two production lines, one in a high-cost area and one in a low-cost area. Or, one produces a cheap product with high profit margin and another produces an expensive product with low profit margin. Does it really make sense that 3 standard error limits would be economically optimal for both lines—regardless of the control chart used?

Kokinos: Then, why does everyone say 3 standard error limits are the most economical?

EVIDENCE AGAINST 3 STANDARD ERROR LIMITS BEING THE BEST ECONOMICALLY

Mathitis: I have said it because that was what I read and learned. I repeated what others said.

Socrates: Let's do a search. Here, it seems that others have looked at the economic design of control charts. The earliest I find and the basis for other such studies was Duncan's article (Duncan 1956).

Mathitis: What did he show?

Socrates: He looked at these conditions: subgroup size, sampling interval, shift in mean to be detected, costs, and standard error limits. His conclusion is that as the first four conditions varied, the most economical limits also varied.

Kokinos: I guess it makes sense.

Mathitis: I found another. In 1963, Bather published an article on minimizing costs (Bather 1963).

Kokinos: Searching the bibliography of each one we find others. There is also the fine-tuning of Duncan's model (Goel et al. 1968). Ten years later, Chiu published an article showing that different limits had different economic benefits (Chiu 1973). In 1998, Tagaras and Lee published another article also showing the same under other conditions (Tagaras and Lee 1988).

Mathitis: They all found that depending on conditions, 3 standard error limits are not always the most economical limits?

Socrates: Montgomery has a review of these articles (Montgomery 1980). Here are three more references:

1. Juran: "While ±3σ limits are the most widely used, some situations call for different limits."

...other multiples may be chosen for different statistical risks. For example, ±3σ limits involve a very small risk...of looking for trouble that does not exist. However, these 3σ limits may have a larger risk of failing to detect trouble when it does exist (Footnote: These are called Type I and Type I errors). Limits set at ±2σ limits would have a larger risk of the first type of error but a smaller risk of the second type. These risks depend on the sample size and other factors. A study by Duncan (Ref. 2, p. 398) found that charts using 2σ or even 1.5σ limits are more economical than charts using the conventional 3σ limits. (Juran 1951, 23–25)

2. Acheson J. Duncan: "It will be pointed out in Chapter 21, however, that in certain circumstances 2 sigma and even 1.5 sigma limits may be more economical than 3 sigma limits" (Duncan 1974, 382).

   Also: "Under certain circumstances charts using 2 sigma or even 1.5 sigma limits are more economical than charts using the conventional 3 sigma limits. This is true if it is possible to decide very quickly and inexpensively that nothing is wrong with the process when a point (just by chance) happens to fall outside the control limits, i.e., when the cost of looking for trouble when none exists is low. Contrariwise, it will be more economical to use charts with 3.5 sigma to 4 sigma limits if the cost of looking for trouble is very high." (Duncan 1974, 449)

3. Grant and Leavenworth: "As in the case of the Xbar and R charts, the use of 3 sigma limits rather than narrower or wider limits is a matter of experience as to the economic balance between the cost of hunting for assignable causes when they are absent and the cost of not hunting for them when they are present...special cases arise in which the use of narrower limits, such as 2 sigma, are desirable" (Grant and Leavenworth 1980, 225).

Agrafos: Aren't Grant and Leavenworth recommending doing exactly what we said Shewhart did: use experience to determine which limits are best for a particular situation?

Peismon: And that would make the empirical limits. Socrates, you asked for actual studies. These seem to be just theoretical analyses.

Kokinos: Juran mentions an actual study by Duncan.

Peismon: Where are the details?

Mathitis: Now you want details on the studies! I suggest you look at these references.

COUNTEREXAMPLES: SIMPLE COST MODEL

Socrates: From what we have read, we don't see the details. These, however, are more relevant and possibly more complete than Shewhart's and Wheeler's examples in that they do assess the economic value of using different limits, different subgroup sizes, and frequency of subgroups, and conclude that other limits are more economical in certain cases.

Mathitis: That covers at least three of the four conditions, which is more than Shewhart and Wheeler did. They only say 3 sigma limits are more economical without addressing costs or comparing to other limits or considering any of the other issues we generated.

Peismon: It's easy to address costs. Look, if you consider all possible things that can change in the process, you would be spending a long and costly time looking for their causes. Most would not be worth the effort of eliminating. Yet, you will still have that wasted cost. Using narrower limits is equivalent to looking for the items on that long list. Wider limits means looking for items on a much smaller list. The cost is much less. Basically, we are using the Pareto principle. Forget the trivial many and search for the critical few.

Kokinos: From that reasoning, then Wheeler's c chart example with probability limits shows why 3 standard error limits were not the most economical—the limits were narrower. Also, as I said before, why not even use wider limits then if you want restrict yourself to just the critical few?

Socrates: I agree in principle. The point is that there is no evidence that 3 standard error limits are the cutoff of the trivial many from the critical few for all situations. Yes, you can say it has "worked" for decades but not in the sense of 3 standard error limits are always more economical than any other limits—that is the discussion.

On the contrary, we have theory and examples that show other limits are more economical under certain conditions.

Mathitis: How do they assess costs to make this comparison?

Socrates: Let's see if we can create a framework for the analysis.

Kokinos: How?

Socrates: To determine the economic value of selecting symmetrical limits $\pm t\ \sigma$, we can compare the expected "cost of looking for trouble when it does not exist" (false alarm, or error I) plus the cost of "overlooking troubles that do exist" (false nonalarm, or error II). Let

$$E[C \mid t = x] = \text{expected total cost given } \pm x \times \sigma \text{ limits} = c_I \times p_{tI} + c_{II} \times p_{tII}$$
(18.1)

where p_{tI} = Probability(false alarm|multiple t), p_{tII} = Probability (false nonalarm|multiple t), c_I = total cost of a false alarm, and c_{II} = total cost of a false nonalarm.

For $t = 3$ to be the best economical multiple, then

$$E[C \mid t = 3] < E[C \mid t = k \neq 3], \quad \text{or} \quad p_{3I} \times c_I + p_{3II} \times c_{II} < p_{kI} \times c_I + p_{kII} \times c_{II}$$
(18.2)

Mathitis: Then, it is not true if we reverse the inequality.

Socrates: That's right. Let's do that and solve for the cost ratio:

$$c_{II}/c_I > (p_{3II} - p_{kII})/(p_{kI} - p_{3I})$$
(18.3)

Mathitis: Do we use an average chart?

Socrates: Let's make the math easy and use an I chart, assume the data come from a normal distribution, and consider $t = 3$ versus $t = 2$ for a mean shift of +1 standard deviation during the time one individual value was taken. Then,

$$E[C \mid t = 3] = p_{3I} \times c_I + p_{3II} \times c_{II} = 0.0027 c_I + 0.9772 c_I$$
(18.4)

$$E[C \mid t = 2] = p_{2I} \times c_I + p_{2II} \times c_{II} = 0.0455 c_I + 0.8400 c_{II}$$
(18.5)

Socrates: Shewhart's claim is false if $E[C|t = 3] > E[C|t = 2]$ or $c_{II}/c_I >$ (0.9772 – 0.8400)/(0.0455 – 0.0027) = 0.3119.

Mathitis: Then, $t = 2$ is more economical whenever the cost related to false nonalarm or error II is greater than 31% of the cost related to a false alarm or error I.

Peismon: Again, this is just theory.

Socrates: So far, but the theory lets us develop cases that match—just as Shewhart did in developing the concept of statistical control.

Peismon: Do you have actual cases?

Socrates: Peismon, you are getting to be just like me! Before, I was asking for actual examples that support Shewhart's and Wheeler's claims. Now you are asking the same of me.

Peismon: (smirking) And?

Socrates: I will give you two actual cases. If the cost of failing to find an assignable cause is very large relative to the cost of searching for false alarms, then we have a counterexample. Contact lenses are one such example.

Peismon: Explain.

Socrates: Contact lenses are mostly water so the cost is very low—pennies. Manufacturers throw them away. However, the cost of not finding a defect on a lens is far greater and can be overwhelmingly high. Eye infection, temporary and permanent eye damage, lawsuits, hosting and responding to Food and Drug Administration (FDA) site visits, loss of reputation and market share all add up to a significant amount even when spread across millions of lenses.

Peismon: How does that relate to control chart limits?

Socrates: Manufacturers are willing to pay for the cost of 100% inspection—manual and automated—because the ratio of c_{II}/c_I is greater than 1 let alone 0.31.

Peismon: Then they don't use control charts so it's not a relevant example.

Socrates: If they used a control chart and wanted—needed—to detect every possible signal, where would they set the limits? What multiple of the standard error would they use?

Peismon: I don't know.

Mathitis: I get it. They would use 0 standard error limits to detect every possible signal.

Kokinos: Then why have a control chart?

Mathitis: That's Socrates' point. They don't. Zero standard error limits are more economical than 3 standard error limits.

Peismon: I'm not sure I accept that example since they don't use control charts.

Mathitis: That's the point, Peismon. It's not economical to use them if the limits they would use are 0 standard error limits. Why bother collecting the data and plotting them if every point is a signal? It also matches what Duncan and Acheson and others have said.

Socrates: Have you heard of the Higgs boson experiment?

Peismon: Yes. A European group of scientists used a particle accelerator to determine whether this theoretical particle exists.

Socrates: The problem is that if it exists, it would have an exceedingly short life span. But, if it exists, it would cause an increase in gamma rays. While the experimenters didn't appear to use control charts, the approach was based on the same concept. Graph gamma ray quantities wait for an extra boost—any amount greater than normal or background noise.

Kokinos: Like a point beyond a control limit.

Socrates: Exactly. To quote an article with an apropos subtitle "The Game of Bumps": "So the signature of a Higgs boson or any other paradigm-shattering new particle would be an unexpected excess of gamma rays or some other particles—an anomalous bump on a graph. Dr. Tonelli said this happened about once a month now that the collider was running, but random flukes would also produce bumps" (Overbye 2013).

Mathitis: You're going to say that they didn't use 3 sigma limits.

Socrates: Before we get to that, there is one more point. The theory is that if it is not Higgs boson causing the excess gamma rays or the "bump on a graph," then it would be a single "bump." If it is the Higgs boson, it would persist and even grow.

Kokinos: Like more than one point beyond the control limits.

Socrates: Yes. To answer Mathitis, let me quote from the article again: "To physicists, the gold standard for a discovery is '5 sigma,' a term meaning that the odds it occurred by chance are less than 1 in 3.5 million" (Overbye 2013). The article is using a different sigma than with control charts and is based on a normal distribution. But the point is that they could analyze the data on control charts and set 5 sigma limits and see if the excess gamma rays persist beyond them.

Peismon: That doesn't show it's more economical.

Socrates: Depends on how you define economical—what costs or how you measure costs. Clearly, for this, they thought the cost of chasing false alarms with a narrow limit was greater than the cost of missing true signals: billions of dollars over many years by continuing the study.

Kokinos: I can understand that.

Mathitis: Then, we have two actual cases, one with the narrowest limits possible (0 standard errors so no control chart is needed) and another with much wider limits (5 standard errors).

OTHER OUT-OF-CONTROL RULES—ASSIGNABLE CAUSES SHEWHART DIDN'T FIND BUT EXIST

Socrates: Let's look at another issue related to Shewhart's megohms example.

Kokinos: This is the one where the first control chart had several points beyond the 3 standard error limits, they searched for assignable causes, found them, and eliminated them. Then, the second control chart didn't show any points beyond the limits, they searched for assignable causes, but didn't find any.

Socrates: That's the one.

Mathitis: You said that failure to find assignable causes didn't mean there weren't any.

Peismon: And I thought the example was strong enough to show the value of 3 sigma limits.

Socrates: Shewhart used the 3 standard error rule for the average and standard deviation control charts. It wasn't until a quarter century later that other rules were published.

Mathitis: The Western Electric rules (Western Electric Company 1956, v).

Socrates: That's right. We all know today and have seen these criteria used successfully to indicate possible assignable causes. And we have found assignable causes.

Peismon: They serve their purpose. I use them.

Mathitis: But those rules don't use the 3 standard error limits.

Peismon: That doesn't mean the limits shouldn't be 3 standard errors.

Socrates: But these other rules signal true process changes with true assignable causes that the 3 standard error limits may not find.

Mathitis: Yes.

Peismon: And?

Socrates: Did Shewhart use these other rules in 1931?

Peismon: Of course not. They weren't invented then.

Socrates: The additional rules indicate smaller or different process changes as the applicable points are all within the 3 standard error limits.

Mathitis: True.

Socrates: Now, Shewhart claims his experience was that no assignable causes existed when there were no points beyond the 3 standard error control limits. That would mean that either (a) except for the megohms, he didn't really check other cases to confirm both alarms and nonalarms; or (b) that not one of his studies ever had these other patterns defined by the Western Electric rules.

Peismon: We have no evidence for either case.

Socrates: Actually, we can show that some of the other special cause patterns occurred with the original megohm data (Figure 18.1). We see that by applying seven other rules than a point beyond the control limits, the control chart exhibits four other alarm types:

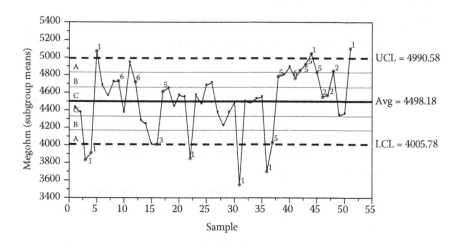

FIGURE 18.1

$\bar{X}$ chart of Shewhart's megohm data with $n = 4$ and 3 standard error limits with alarms for eight types of assignable cause patterns (numbered points) using the zones A, B, and C.

- 9 consecutive points on the same side of the center line
- 6 consecutive points steadily decreasing (or increasing)
- 2 out of 3 consecutive points beyond 2 sigma limits
- 4 out of 5 consecutive points beyond 1 sigma limits

Mathitis: We also find those patterns in our processes today.

Socrates: Clearly, the other rules indicate that the 3 standard error limits alone are not always the most economical since they fail to signal when true assignable causes occur that meet these other criteria or rules while not exceeding the 3 standard error limits.

Mathitis: That makes sense. Why use them, otherwise?

Socrates: They may signal other kinds of assignable causes besides the ones you would expect with 3 standard error limits. What does a point beyond the 3 standard error limits likely indicate?

Mathitis: A shift in the mean or an increase in variability.

Kokinos: Or both.

Socrates: But the Western Electric rules also signal trends, mixed distributions.

SMALL CHANGES ARE NOT CRITICAL TO DETECT VERSUS TAGUCHI'S LOSS FUNCTION

Peismon: The purpose of having 3 standard error limits is to ensure we don't have many false alarms. With many false alarms, we would be unnecessarily searching for unassignable causes. That's a big waste.

Socrates: What about changes that do have assignable causes? Changes signaled by the other rules but not identifiable by the rule beyond the 3 standard error limits?

Peismon: We don't want to look for every change, only the big ones with significant economic consequences.

Mathitis: How do you determine which ones are significant?

Socrates: What about small, economically significant changes that have assignable causes?

Kokinos: Like the contact lens example. It seems that the size of the change is dependent on the industry and product. So, we can't just say that all small changes for every situation are insignificant.

Mathitis: I agree with Kokinos. A deviation of a centimeter or less when doing surgery could mean death. That cost is high. But such a deviation between two pieces of wood that form part of an ornament has minimal or no cost.

Kokinos: What about Taguchi's loss function? The idea is that you should be aiming for as little deviation as possible around the appropriate target. Any deviation from that target results in a "loss to society."

Peismon: It's a great way to understand the flaws of specifications. It's a point Deming has made. I use that concept to motivate continuous improvement.

Agrafos: How does it work?

Mathitis: The loss function uses a quadratic function, meaning that a deviation of 4 units would have a 16-multiple loss while a 3-unit deviation would have a 9-multiple loss. If the length of a table should be 6 feet, then every deviation from that has a "loss to society," that increases as the square of the deviation from 6 feet.

Socrates: Is that a net loss?

Mathitis: Taguchi doesn't explicitly say so, but I think we all assume it is a net loss. Otherwise, we'd have to calculate the benefits and combine to net it.

Socrates: Peismon, if you accept Taguchi's idea of the loss function and that it is a total net loss, then every deviation has an economical consequence. If the idea is to minimize those losses, then clearly you want to identify all changes, even the small ones.

Peismon: But that isn't practical.

Socrates: That's why people use specifications—because the loss owing to a small deviation in some cases is less than the gross "loss to society" owing to the deviation—like a deviation in an ornament. However, that is not the case for all products and services.

Mathitis: In other words, Taguchi's loss function is inconsistent with 3 standard error or wider limits, just as it is with specifications. The former says that all changes from target incur net losses while the latter says only the "big" ones—without defining what they are.

Kokinos: So, those are contradictory beliefs we learned in isolation: all deviations from a target are significant versus disregard all deviations except the big ones when monitoring processes.

Socrates: Peismon, would you want to detect a change that increased the defect rate 20-fold?

Peismon: Of course.

Socrates: Assume $C_{pL} = 1.33$ and $C_{pU} = 2.33$, then $C_{pk} = 1.33$. What is the defect rate for a process with a $C_{pk} = 1.33$?

Agrafos: I can look it up. It's 63 per million.

Socrates: Suppose the process mean shifts one standard deviation toward the lower specification limit. Then $C_{pL} = 1$, $C_{pU} = 2.67$, and $C_{pk} = 1$. Now what is the defect rate?

Agrafos: It's 1350 per million. That's more than a 20-fold increase.

Socrates: How likely are 3 standard error limits to detect that? And how likely versus 2.5 or 2 standard error limits?

Agrafos: Much less likely.

Kokinos: I remember now that we were told that 3 standard error limits were designed to detect large changes, as Peismon said, of approximately 1.5 standard deviations or more.

Mathitis: Then, it isn't designed to detect a 20-fold increase in defects in certain conditions.

IMPORTANCE OF SUBGROUP SIZE AND FREQUENCY ON ECONOMIC VALUE OF CONTROL CHART LIMITS

Mathitis: Another example is medical emergency responders who need to have fast turnaround times. Minutes, sometimes even seconds, delayed and the consequence could be permanent damage or death. If they were to use a control chart to monitor their response times, I hope they would look at small increases in response time—not like 5 standard error bumps of the Higgs boson—and search for assignable causes.

Kokinos: In other cases, seconds or minutes are too small to even measure, let alone worry about. For example, most of us don't worry about the loss of a few seconds every hundred years, creating the need to adjust the calendar every 400 years for leap days.

Socrates: The issue is that the 3 standard error limits are set for few false alarms, a very low probability of false alarms. For the same subgroup size, the lower the false alarm rate, the higher the chances of missing true alarms.

TABLE 18.1

Probability of Detecting a 1.5 Standard Deviation Shift of the Mean Using Different Subgroup Sizes

N (Subgroup Size)	4	5	10
Mean	100	100	100
Standard deviation	5	5	5
LCL	92.5	91.3	89.4
UCL	107.5	108.7	110.6
1.5 standard deviation shift in mean	7.5	7.5	7.5
Prob(detecting shift)	0.50	0.57	0.72

Mathitis: Also, we know that control chart limits widen as subgroup size decreases.

Peismon: This means that there are other ways of detecting these small changes and still use 3 standard error limits.

Agrafos: How is that?

Peismon: Increase the subgroup size. That makes the control chart more powerful. For example, look at detecting a 1.5 standard deviation shift in the mean using an $\overline{X}$ chart (Table 18.1). See what happens when we use the same mean and standard deviation but only change the subgroup size.

Kokinos: The probability increases.

Peismon: So, we can still use 3 standard error limits. Even for detecting the change from $C_{pk} = 1.33$ to $C_{pk} = 1.0$ because of a shift in the mean.

Socrates: Or, use the same sample size but narrower limits. Using 2 standard error limits with $n = 4$, the probability of detecting the shift is 0.70, almost the same as with $n = 10$ but less data.

Mathitis: Or, we can decrease the sample size and also the limits. For $n = 2$, 2.1 standard error limits yield a probability of 0.51 to detect a 1.5 standard deviation shift in the mean. For $n = 3$, 2.6 standard error limits produce a 0.5 probability.

Peismon: But using 2.5 standard error limits, and worse 1.83 standard error limits, increases your chances of going after false alarms. That increases your overall costs. That's why we use 3 standard error limits.

Mathitis: But if we don't know the costs of chasing false alarms and the costs of scrapping defects, then how can we tell which is better economically? Recall that others, such as Duncan, Chiu, and

Tagaras, show that there are other factors and their combinations that determine more economical limits than 3 standard errors.

Kokinos: Your argument is fallacious. Pick any multiple and we can then say that any other multiple will increase or decrease the false alarm or true alarm rates relative to that.

Socrates: Consider two dice, A and B. A has the numbers 1–6 and B has the numbers 2–7. What is the population standard deviation and mean for A?

Kokinos: The standard deviation of the numbers 1–6 is 1.7 and the mean is 3.5.

Socrates: What are the 3 standard deviation limits then?

Kokinos: LCL = 3.5 – 3 × 1.7 = –1.6 and UCL = 8.6.

Mathitis: I see it already. No matter how many times you toss die B, it will never exceed those limits. You will never detect that you changed the die using this out-of-control rule.

Peismon: But, as soon as I get a seven, I know the process changed.

Socrates: If both dice are possible inputs to the process, it may not have changed.

Peismon: But you can increase the subgroup size. Use an $\bar{X}$ chart with $n = 2$?

Kokinos: Now, the sampling standard deviation or standard error is $1.7/\sqrt{2} = 1.2$. So, LCL = 3.5 – 3 × 1.2 = –0.1 and UCL = 3.5 + 3 × 1.2 = 7.1. It still won't. We'll need at least $n = 3$.

Socrates: Now, we can better understand that making the limits wider does not necessarily lead to enough power to detect changes. In this example, an *I* chart or $\bar{X}$ chart with $n = 2$ has zero power with respect to detecting a change in average. But using narrow limits does.

Kokinos: We could use other rules or alarms than just points beyond the limits.

Socrates: That's true. But recall that Shewhart only used a point beyond the limits. Yet, it shows us the benefits of understanding how a control chart functions to increase its utility.

Peismon: What about using a different chart?

Socrates: What is the standard deviation of die B?

Kokinos: It's the same as die A. Then that won't work either.

Peismon: These are just theoretical examples.

Socrates: Let's look at a real example. We'll use Shewhart's since you accept those results (Shewhart 1980, 19).

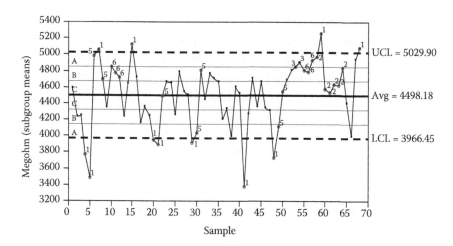

FIGURE 18.2
$\bar{X}$ chart of Shewhart's megohm data with $n = 3$ and 3 standard error limits with alarms for eight types of assignable cause patterns (numbered points) using the zones A, B, and C.

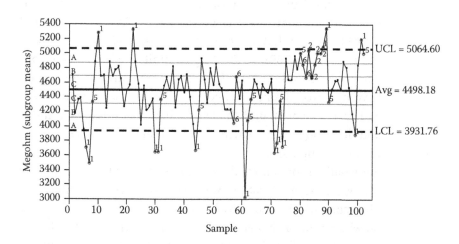

FIGURE 18.3
$\bar{X}$ chart of Shewhart's megohm data with $n = 2$ and 3 standard error limits with alarms for eight types of assignable cause patterns (numbered points) using the zones A, B, and C.

Kokinos: Which example?

Socrates: The megohm one. The subgroup size was $n = 4$ with 3 sigma limits. In the first set of data, there were eight points beyond the control limits, I believe is what you said. While Shewhart didn't state which points had assignable causes, his point was that some were true alarms. Let's see what happens if we had used subgroups of size $n = 2$ and $n = 3$.

Agrafos: Here are the control charts for $n = 3$ (Figure 18.2) and $n = 2$ (Figure 18.3).

Kokinos: This does make a difference. For $n = 4$, we only had eight points beyond the control limits. With $n = 3$, we have 11, and with $n = 2$, we have 15. The extra alarms were changes we didn't see with $n = 4$. The 121st through 124th data produced no alarms when averaged but produce an alarm when separated into two groups: {3075, 2965} and {4080, 4080}.

Peismon: But that is just an analysis of Test 4 and it uses 3 standard error limits. Shewhart also showed that after removing the assignable causes, the control chart showed no more assignable causes and none were found.

Socrates: We can do the same here. Using 2 standard error limits would have resulted in even more alarms for both $n = 2$ and $n = 3$ subgroup sizes. Now let's use the estimate of the standard deviation from the chart that was in control. The $UCL_{n=4} \approx 4700$ and $\bar{x} = 4425$. For $n = 4$, $A_3 = 1.628$, so $\bar{s} \approx (4700 - 4425)/1.628 = 168.9$. For $n = 2$, $A_3 = 2.659$, so the 2 sigma limits are $UCL_{n=2} = 4425 + 2.659 \times 168.9 = 4874 > UCL_{n=4}$. Hence, the same results would have occurred. Using 2 standard error limits, we would have had true alarms, found the assignable causes, removed them, and then had no false alarms. The economic advantage?

Kokinos: Half the sample size.

Socrates: For small out-of-control deviations, we need a larger n, and for changes of short duration, we need more frequent subgroups. The economic value of which limits to use will depend at least on which type of change you want to detect and how much data you want to collect. You can change the limits to accommodate changes in subgroup size and frequency.

Mathitis: Then, if we decide to collect 200 data, we can divide into 50 groups of 4 or 100 groups of 2. But it's not clear that both are economically the same using 3 standard error limits for both.

Socrates: The point is that he didn't make this comparison. Therefore, as others including Wheeler determined, there are other limits that are more economical. You could benefit from studying your specific situation, just as we did when we discussed whether always using $n = 30$ as the sample size was a good rule of thumb.

Agrafos: That seems to make it much too complicated.

Peismon: Stick with 3 standard error limits.

Socrates: Only if you don't care about the economics—which contradicts the original claim. More importantly, if the original claim is true, 3 standard error limits are the most economical limits regardless of the subgroup size, frequency of subgroups or time interval between subgroups, amount and type and duration of changes, cost of false alarms and missing true changes, type of control chart, and all the other factors we discussed; there is a simple solution.

Peismon: What's that?

Socrates: Use $n = 1$ for an *I* chart with 3 standard error limits with the longest interval possible between each subgroup—say, years. That simplifies everything.

Peismon: You know that won't work.

Agrafos: Why not?

Peismon: You will assuredly miss something.

Mathitis: Because 3 standard error limits are not the most economical ones for all cases regardless of the frequency of subgroups.

PURPOSE TO DETECT LACK OF CONTROL— 3 STANDARD ERROR LIMITS MISPLACED

Kokinos: Besides, it doesn't make sense.

Peismon: And defeats the purpose of a control chart. Wheeler says "[t]he express purpose of the control chart is the detection of a lack of control. If a control chart cannot be used to detect a lack of control, then what is the reason for ever using a chart?" (Wheeler and Chambers 1992, 82).

Socrates: If you wanted to detect an intruder in your home while sleeping, would you assume that no intruder would ever come in?

Peismon: Of course not. I would assume the opposite.

Socrates: Would you have an alarm system?

Peismon: Of course.

Kokinos: We have more than that, to also include fire alarms, motion detectors, glass breakage, beams to detect someone entering a room.

Peismon: Just as art galleries, museums, banks, and jewelry stores do to guard their valuables.

Socrates: Would you err on the side of more false alarms rather than less true signals?

Mathitis: More false alarms. Look at security at the airport. Do you want a detection system that waits until you are almost certain someone is a terrorist before signaling? I don't.

Peismon: You aren't frustrated going through security?

Mathitis: That's a different issue. No one wants the process to be cumbersome, time-consuming, and intrusive. That's the false alarm cost. But, if those same people discovered that a terrorist did get through and crashed a plane killing everyone, you can be sure their relatives would be complaining about the opposite.

Agrafos: And they did.

Kokinos: The point also applies in other areas. Do you want a toothache and swollen jaw to be the signal of an infection or do you want a signal before the swelling and infection? Or, which do you prefer: finding out in the emergency room that you really had a heart attack or identifying a symptom of a possible heart attack that may sometimes be wrong?

Peismon: Of course, I want the earliest signal possible.

Kokinos: Even if you have false signals of heart attacks.

Peismon: Especially of the heart attack.

Socrates: Consider the two recent health situations: enterovirus D68 and Ebola. A man flew from Liberia to the United States having been in physical contact with a person infected with Ebola. He presented with fever, abdominal pain, and vomiting. All symptoms of Ebola. Which alarm do you want: all three must occur, any two must occur, or any one must occur?

Peismon: I want any one. I don't want to wait until all three occur—that may be too late.

Socrates: Because the man died, some airports will now take passengers' temperature. What temperature do you want for the alarm: a

	higher temperature to avoid false alarms or a lower temperature to avoid false nonalarms?
Peismon:	Obviously, a lower temperature.
Socrates:	Do you see that one symptom versus three and lower temperature versus higher are both equivalent to narrow control chart limits versus wider ones?
Kokinos:	By lack of control, we mean a true alarm, not a false alarm, right?
Mathitis:	Right.
Peismon:	We use 3 standard error limits to significantly reduce the false alarm rate.
Mathitis:	I see that. We can equate one symptom with 1 standard error limits, two symptoms with 2 standard error limits, and three symptoms with 3 standard error limits since more symptoms decreases the false alarm rate just as more standard errors does. But we want to decrease the false nonalarm rate.
Socrates:	Do you see the contradiction? With all the other cases, you were not interested in reducing the *false* alarm rate, you were interested in increasing the *true* alarm rate.
Mathitis:	The wider the limits, the fewer true assignable causes you will detect. Having wider control chart limits is equivalent to requiring all three symptoms to flag someone who might have Ebola. If you want to be sure you detect an infected person, you don't mind false alarms—better than failing to detect one.
Agrafos:	That was what happened to the man from Liberia. The hospital and Centers for Disease Control and Prevention (CDC) appear to have taken the approach of reducing false alarm rates when the purpose was to identify Ebola-infected people. If the purpose of a control chart is to identify special-cause infected processes, we want a low false nonalarm rate.

TABLE 18.2

Healthy versus Ebola Infection Errors

	Signal	
Truth	**Alarm (Fever, Vomit, or Abdominal Pain)**	**Silence (No Fever, Vomiting, and Abdominal Pain)**
Ebola	No error (true alarm)	Error (false silence)
Healthy	Error (false alarm)	No error (true silence)

Socrates: Or, equivalently, a higher true alarm rate. How can we increase the true alarm rate?

Mathitis: One way is by having a higher false alarm rate.

Kokinos: That wasn't my understanding of false alarms.

Peismon: I didn't understand that from class either.

Kokinos: Isn't the risk crying wolf too much? Soon, we ignore the cries and the alarms.

Socrates: Consider this decision table (Table 18.2). What is the purpose of the alarm?

Kokinos: To detect Ebola-infected people.

Socrates: Which error do you want to minimize then?

Kokinos: False silence.

Socrates: To minimize that error, are you willing to accept more alarms rather than silences?

Kokinos: Yes, I am and so are my family, community, and country.

Mathitis: With more alarms, there will most likely be more false alarms.

Socrates: That's the price for the purpose of detecting Ebola-infected people.

Mathitis: Or, more scrutiny of more people—which also has a price.

Socrates: By reducing the false silence rate, what rate are we increasing?

Kokinos: The true alarm rate.

Socrates: What happens if we focus on reducing the false alarm rate? Remember that rows list the truth and we can only be in one row at a time.

Mathitis: For false alarms, then we have no Ebola-infected people. To reduce that rate to zero, we would never have an alarm, so no alarm system.

Socrates: If you did that, what would be your purpose: to detect Ebola-infected people or to detect healthy people?

Mathitis: It would be to detect healthy people since that is the row that has false alarms.

TABLE 18.3

Control Chart Errors

| Truth | Signal (Special Cause Pattern) | |
	Alarm	Silence
Lack of control	No error (true alarm)	Error (false silence)
Control	Error (false alarm)	No error (true silence)

Socrates: A special cause is an intrusion into the process just like an infection is an intrusion into a person. When we don't want that special cause—lack of control or infection, we want to be notified. The symptoms of Ebola are the alarms just as special cause patterns are alarms. This is the decision table (Table 18.3) for control charts. What is the purpose of control charts?

Peismon: To detect changes, lack of control.

Mathitis: If the purpose of control charts is to detect lack of control, we are in the first row. We want to increase the true alarm rate and reduce the false silence or nonalarm rate—not reduce the false alarm rate since it doesn't apply.

Socrates: How can we reduce the false alarm rate to zero?

Kokinos: You would not have a control chart.

Agrafos: Or use limits as wide as possible and use only the rule of a point beyond the control limits is an alarm. In other words, discard the other rules.

Socrates: Exactly.

Mathitis: But that would decrease the true alarm rate.

Socrates: Almost assuredly. Then, if the purpose is to detect lack of control—remember we are in the first row of Truth—which means increase the true alarm rate, how should we do it?

Kokinos: We should be using narrower limits then.

Agrafos: Or, use more rules than a point-beyond-the-limits rule.

Socrates: These other rules are equivalent to the other detection systems for intruders.

Peismon: So, we can keep the 3 standard error limits but also use the other rules. The 3 standard error limits might be like the alarm on the doors to the outside of the house signaling when it is open. The other rules might be like detecting glass breaking, someone entering a room, motion.

Socrates: Exactly.

Peismon: Or rather than just checking for fever, we also check for nausea and abdominal pain but we only require one symptom to occur to signal an alarm.

Mathitis: Or, we could also use narrower limits, for example, 2.5 standard error limits. In fact, another development for control charts were the warning limits.

Kokinos: Doesn't Shewhart say we should strike the right balance?

Socrates: In our discussion of whether the control limits were probability limits, we quoted Shewhart. We identified several places where he associated a probability P with the limits. That probability was the probability of being *within* the limits—not beyond.

Kokinos: I don't understand the connection.

Socrates: He says "Even when no trouble exists, we shall look for trouble $(1 - P)N$ times on the average after inspecting N samples of size n. On the other hand, the smaller the probability P the more often in the long run we may expect to catch trouble if it exists" (Shewhart 1980, 276). For $P = 0$, then we would look for trouble every time: $(1 - 0)N = N$. We do this to ensure we detect all intruders.

Kokinos: Now I get the contact lens example. A defect on a lens is an intrusion—we want zero defects. We want $P = 0$.

Mathitis: He uses P in the opposite sense we did for false alarms. His P is our $1 - p$.

Agrafos: The same is true for Ebola. We want zero failures to detect an infected person.

Socrates: Exactly.

Kokinos: He is saying what we just recognized. To detect more lack of control situations, we need to make P smaller and one way of doing that is with narrower limits.

Peismon: Or using the other rules.

Kokinos: Then, 3 standard error limits are not the most economical for all the conditions we talked about, and if the purpose of control charts is to detect changes, they aren't the best limits to use either as they focus on the wrong error, false alarms rather than the true alarms.

Agrafos: The Ebola example convinces me that there is much more to determining the appropriate control chart limits than just accepting what Shewhart said since he did not provide any supporting evidence from what his experience was. Clearly, if failing to detect a special cause results in a single death or an epidemic, we shouldn't be using wide limits like 3 standard error limits.

19

Myth 19: Statistical Inferences Are Inductive Inferences

Scene: *Project team meeting (day after control chart discussions)*

BACKGROUND

Agrafos: Why is the use of control charts to manage and improve processes called *statistical* process control if the limits are not probability limits?

Peismon: Control charts use statistics, that's why. We discussed this before (Myth 3). There are descriptive statistics that don't use probabilities and inferential statistics that do. Wheeler and Stauffer recently described four statistical questions (Wheeler and Stauffer 2014). First, start with a known universe or population. Their example is a bowl with 4000 white beads and 1000 black beads. Or, it could be the results of a game. They are all known universes. We can characterize the universes using statistics to answer the first question: (Description) "Given a collection of numbers, is there a meaningful way to represent all of the information in that collection using one or two summary values?" Statistics can answer these questions, for example, the average or percent of red marbles.

Mathitis: In that statistical application, there are no probabilities and no inferences.

Kokinos: But with a known universe, we can also calculate probabilities when *selecting* members of the universe. For example, we have a deck of cards with known suits and values. Those are descriptive statistics summarizing what we know. But we don't know the result of randomly selecting a card. Since we know the universe, we can calculate the probability of selecting one card or combinations of cards of the same suit or value, or of dealing a particular hand.

Peismon: That's the second application of statistics. The question is about the probabilities of events drawn from that known universe: (Probability) "Given a known universe, what can be said (probabilistically) about samples drawn from it?"

Socrates: These calculations are mathematical or deductive inferences.

Mathitis: True. But sometimes we don't know every member of the universe. For example, a customer buying a lot wants to know the percent defective. Or, we don't know how many of each color in a jar of marbles.

Peismon: That's the third statistical application. We still have a single universe but we do not know all its characteristics. We use inferential statistics from the sample to estimate the population characteristics. Wheeler and Stauffer state the question we want to answer is (Inference): "Given a sample from an unknown universe, and given that everything about the nature of the sample is known, what can be said about the nature of the unknown universe?"

Socrates: Then, in the second application, we know the distribution, while in the third application, we assume one. Then, both use deductive reasoning.

Peismon: No, that's not right. When we make an inference from the sample to the population, we are using inductive reasoning. We seek to know as much as possible about this universe knowing only what the sample tells us.

Socrates: Could it still be deductive reasoning?

Kokinos: My understanding of inductive logic is that the conclusion is a generalization from specific instances, which is the reverse of deductive logic.

Peismon: Since the sample represents specific instances but we are interested in the population, which is the generalization, we are using inductive reasoning.

Agrafos: For example, we see 50 white swans and conclude that all swans are white. That's inductive reasoning. But if we assume all swans are white, we can conclude that if the bird in the pond is a swan, it is white. That's deductive reasoning.

Socrates: Would either of these be statistical inferences?

Peismon: Those are just examples. Statistical inferences are generalizations but use probabilities to quantify the likelihood of the conclusion.

REASONING: VALIDITY AND SOUNDNESS

Socrates: It seems that there are two issues. One is whether statistical inferences use inductive or deductive reasoning and the other is when inductive reasoning is statistical. So that I can understand these distinctions, tell me what you mean by an argument or reasoning?

Mathitis: In an argument, we provide reasons for a position or view.

Kokinos: We want the view or position to be based on or supported by the reasons. There has to be a connection, however weak or strong.

Socrates: Are the reasons always true? Are they facts?

Kokinos: We would like them to be, but no, they can be true most of the time. The more often they are true, the stronger the support for the view. This is where statistics plays a role. Rather than seeing 50 white swans, we might see that most swans are white and therefore conclude that the next swan we see will probably be white.

Mathitis: Sometimes, we make assumptions, perhaps to either test the reasons or to support a view. In hypothesis testing, we make assumptions about the values of distributions, for example, the mean or variance. Then, we test those assumptions using statistical theory.

Kokinos: Often, we assume something that we want to show to be false. We use a *reduction ad absurdum* argument. We can reject the assumption if it leads to a contradiction of a fact or two contradictory statements.

Socrates: Then, can we define an argument as having two components: premises and conclusions? Premises may be facts, may be statements that are true in varying degrees and conditions, or they may be assumptions, but they are the reasons supporting the conclusion.

Mathitis: Yes, and we want those premises to be true all or most of the time or under the relevant conditions. Probabilities allow us to quantify the strength of that connection or relevancy.

Kokinos: We want arguments to be valid. That means that it is impossible for the conclusion to be false if the premises are true.

Agrafos: Don't we also want the premises to be true?

Socrates: Those are the two key characteristics, true premises and valid conclusions, which together make an argument sound.

Agrafos: Then, every argument or reasoning tries to be sound.

Peismon: But we argue and reason from premises that are not always true. And, as Kokinos said, sometimes we don't have enough information or knowledge to have strong reasons or a strong connection between the premises and the conclusion.

Mathitis: That's why we argue, disagree.

INDUCTION VERSUS DEDUCTION

Agrafos: Does soundness and validity depend on whether it is a deductive or inductive argument?

Mathitis: What do you mean?

Agrafos: Deductive arguments are valid, aren't they?

Socrates: Not all. We can still create invalid deductive arguments, for the reason Kokinos said—the conclusion is not connected appropriately to the premises. For example:

- Premise 1: All swans are white.
- Premise 2: My pet is white.
- Conclusion: Therefore, my pet is a swan.

Agrafos: You're right. This is a deductive argument, but it is not valid since the conclusion can be false when both premises are true.

Mathitis: It's not even sound. Socrates doesn't have any pets.

Socrates: Why is it a deductive argument?

Peismon: Because the argument is from a general statement to a specific instance. From all swans to a specific swan or pet.

Kokinos: The other type of argument is inductive, where we argue from specific instances to a generalization as we said earlier about generalizing to all swans from some swans.

Agrafos: Then, inductive arguments have the advantage of having premises that are facts. So, they are halfway to being sound.

Mathitis: But they are not valid since it is possible for each premise to be
true and the conclusion false. No matter how many swans you
see that are white, how do you know all swans are white?

Agrafos: It's like the discussion we had about accepting or failing to reject
a hypothesis. You can't prove all swans are white by simply seeing only white swans.

Socrates: Perhaps that suggests another way to distinguish between
deductive and inductive arguments that can help resolve this
issue. Consider this. If there are 100 marbles in a jar and I
see that each marble in the jar is green, is not the generalization that all marbles in the jar are green true and a valid
conclusion?

Peismon: But that wouldn't be an inductive argument as you have seen the
entire universe.

Socrates: That's the point. Rather than differentiating between deductive
and inductive reasoning by whether a generalization is a premise or conclusion, we can distinguish the two types of reasoning
by whether the conclusion is about the members of a class referenced in a premise or about members not in the class referenced
in a premise.

Agrafos: But we are still talking about swans whether it's inductive or
deductive. How is one conclusion about swans and the other not
about swans?

Socrates: The distinctive characteristic of inductive arguments is not
whether it is a generalization from particulars but whether the
conclusion is about other things than the particulars. I see 50
swans and all are white. My conclusion is that the next swan I
see will be white. This is not a conclusion about all swans but
about one swan in particular. Is it inductive?

Mathitis: Yes, and I see what you are saying. The difference is whether the
conclusion is about just the 50 swans you have seen or about
other swans you haven't seen. If you concluded all marbles in all
jars were green, that would be an inference beyond the observed
instances.

Peismon: But isn't that also true about deductive arguments? We don't see
all the swans and yet make inferences about the ones we don't
see.

Socrates: The difference is in the premises. In the deductive argument, there is a premise about all swans. Thus, any conclusion about any swan is already included in that premise.

Kokinos: But in the inductive argument, we make a conclusion about other swans not in the premises. Isn't that a generalization?

Mathitis: Perhaps not if the conclusion is about just the next swan.

FOUR CASES OF INDUCTIVE INFERENCES

Socrates: Swans make up a reference class. Both types of arguments refer to this reference class. But we also have another class about whiteness.

Agrafos: You mean things that are white. Not just swans are white but other things are white.

Socrates: That's right. This is the attribute class because we are saying that swans have this attribute of being white. What is the relationship between things that are swans and things that are white? What is the relationship between these two classes for each argument?

Mathitis: In the deductive argument, the generalization statement is that the reference class of swans is a subset of the attribute class of white things. Using Venn diagrams, we have this in Figure 19.1.

Socrates: Exactly. What is the other premise in the deductive argument?

Mathitis: That we have selected one member of the reference class—in this case, a swan.

Socrates: Now, you can see why selecting something from the "Swan" circle must necessarily be in the "White" circle. The "Swan" class is a subset of the "White" class.

Kokinos: That's the conclusion: Selecting something from "Swan" also means selecting something from "White."

Socrates: Does this diagram represent a premise or conclusion in the deductive argument?

Kokinos: A premise. But in the inductive argument, it is the conclusion. Or, we have one circle that contains all the swans we have seen but our conclusion is about a swan outside that circle.

Mathitis: I see. This helps clarify the two types of arguments. That makes the inference an extrapolation. We know that is not a valid conclusion.

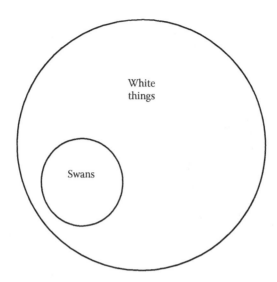

FIGURE 19.1
Venn diagram showing how deductive reasoning works.

Agrafos: Then, in the case when we view every member of the reference class and see that each member is a member of the attribute class, there is no extrapolation. It's just a summary of the facts, of all the members.

Socrates: Exactly.

Agrafos: What does the argument look like using a Venn diagram?

Socrates: We have two choices for the conclusion. We can illustrate it like in Figure 19.2. We observed some members of "Swan" and saw that they were also members of "White." But we don't know about other members of "Swan."

Mathitis: Is the "Swan" circle all within the "White" circle omitting the dotted arc or does the "Swan" circle extend beyond the "White" circle as shown by the dotted arc? That's the question. That's where the extrapolation can occur. We don't know which is true.

Socrates: Consider two jars with marbles. One has 900 black and 100 white marbles. Suppose all the black marbles are on top of all the white marbles in this jar. I draw one marble at a time from this jar starting at the top. The first 800 marbles will be what color?

Agrafos: Black. So you would inductively conclude that either all marbles in the jar are black or that the next marble selected is black. That would be an inductive inference because it's about something

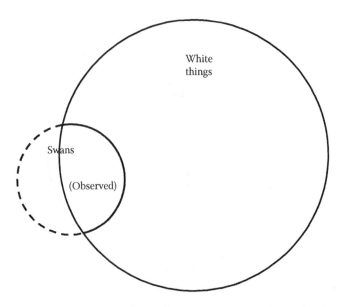

FIGURE 19.2
Venn diagram illustrating inductive reasoning: Are all swans within the "White" circle or are some swans outside the "White" circle?

other than the 800 marbles. That seems reasonable—as reasonable about the swans.

Mathitis: But it would be false about all the marbles. And it would be false about the next marble after you withdrew all 900 black marbles. That's why inductive reasoning is invalid.

Peismon: But when we use statistics, we don't just select the marbles from the top. We take a random sample. Now you are not likely to get the first 800 marbles to all be black.

Socrates: Excellent point, Peismon. Note the two ways of selecting marbles: nonrandom and random. In both cases, we extrapolate to all marbles or to the next marble. But there is more.

Kokinos: Like what?

Socrates: Regardless of whether I select the marbles randomly or not, I extrapolate to the same jar or to the second jar and make an inference about the color of the marbles in that jar.

Kokinos: That's not reasonable extrapolating to the second jar since you don't have any information about that jar or the marbles in that jar, only about the first jar and its marbles.

Mathitis: Or you need some additional information, which would be another premise in the argument. For example, all marbles came from the same process and were randomly stored in jars containing 1000 marbles each.

Socrates: Then, the four inductive cases are (1) nonrandomly sample from a population and extrapolate to the population sampled, (2) randomly sample from a population and extrapolate to the population sampled, (3) nonrandomly sample from a population and extrapolate beyond the population sampled, and (4) randomly sample from a population and extrapolate beyond the population sampled.

Mathitis: I don't think there is any practical difference between the third and fourth cases.

STATISTICAL INFERENCES: PROBABILITY DISTRIBUTIONS

Agrafos: Which of these are statistical inferences?

Socrates: To answer that question, we need to know what a statistical inference is.

Mathitis: I think the second and fourth cases describe statistical inferences. We use random samples to make inferences.

Socrates: Why must the samples be random?

Mathitis: We want to make sure that the sample represents the population. If the sample is not random, then it could be biased. We talked about this regarding estimation and hypothesis testing. Nonrandom sampling gets you the 800 black marbles in a row while the random sampling will get approximately 80% black and 20% white—at least in theory.

Kokinos: We insist that the samples be random so that the conclusion will be valid.

Socrates: Like calculating *p* values. For valid conclusions, we must be reasoning deductively, not inductively. In general, whether estimating or hypothesis testing or inferring other conclusions, are they not statistical inferences if statistical theory is applied?

Peismon: Yes, obviously.

Socrates: What do we mean by statistical theory?

Mathitis: Statistical theory is application of probability theory to inferences. Application of statistical theory is simply applying probabilities to real-world phenomena.

Socrates: How do we do that?

Mathitis: For example, we assume a particular distribution, for example, a normal distribution or a binomial distribution, and make inferences on the basis of that assumption.

Socrates: What is a probability distribution?

Mathitis: That's a theoretical model used to describe real-world phenomena.

Peismon: But it could also be the results of actual processes. For example, data plotted on a histogram describes actual distributions. The description application and the question we want to answer are about actual distributions.

Socrates: Why a histogram?

Peismon: Any graph that has specific two dimensions will work. A histogram has one axis showing the data scale. For example, if we measure the diameter of a marble in centimeters, the horizontal axis would be centimeters. The other axis shows the frequency of each result.

Mathitis: So, you're saying that scales and probabilities define a probability distribution. Those are the two dimensions.

Socrates: Yes, a list of possible results and a function that assigns to each member of the list its probability. For example, the results of tossing a fair, six-sided die produces a probability distribution because the list of possible results is {1, 2, 3, 4, 5, 6} and the probability function assigns to each member in that list the probability 1/6. A fair coin's probability distribution is the possible results {head, tail} with a probability of 0.5 for each. A biased coin also produces a probability distribution: the same list {head, tail} but the probability function might be Prob(head) = 0.9 and Prob(tail) = 0.1. That's straightforward for discrete distributions whether quantitative or qualitative. For continuous probability distributions, we must use calculus to define probabilities for the possible values.

Kokinos: I get it. A deck of cards is the list of choices and the number of suits and cards of different values are the distribution.

Mathitis: Then Shewhart's bowl of chips representing the normal distribution was the list. He shuffled them and drew them one at a time and thereby assigned probabilities to the values.

Peismon: Again, these are all examples of the first and second applications when the universe is known. When we don't know the universe, then we make statistical inferences.

Socrates: But there are different cases of not knowing the universe. I produce a lot of 1000 items. I don't know how many are defective but I can randomly draw a sample and statistically estimate the defect rate or test whether it is less than a certain amount.

Mathitis: Which means you must use theoretical probability distributions.

Socrates: Consider a die with the numbers 1–6. That's the list. If the die is fair, the frequencies are 1/6 for each number. Replace the 6 on the die with 60. What changes?

Mathitis: Only the list, if it is still a fair die. If we keep the original numbers but load it so it favors the number 6, we have the same list but the assigned probabilities change.

Socrates: It's critical to understand that the deck of cards or the bowl of chips or the list {heads, tails} are not probability distributions until probabilities are assigned to them. The key question is how are probabilities assigned? In the case of Shewhart's bowl of chips, the probability of each chip was the same but the number of chips with the same value was in approximate proportion to what the normal distribution would produce. That's one way of assigning probabilities by the distribution. That's the frequentist definition—the probability equals the frequency or proportion. So, Mathitis, the shuffling and selecting was not what assigned the probabilities to the chips. That does assign probabilities to the process results or selections.

Agrafos: How is this related to what we were discussing?

Socrates: A random variable is defined as having a probability distribution. If we think the results appear "random" or vary, then they are candidates for applying statistical theory assuming probability models.

Kokinos: But if we just make calculations using only the data, we are not making any inferences. Those calculations would be simply descriptive statistics, not inferential statistics.

Socrates: Then, to use statistical theory, we would use theoretical probability distributions. Those, like real ones, would still consist of two components: the possible results and the probabilities (or relative frequencies) of each possible result.*

---

* The complete definition of a probability measure has three components: the set of outcomes called the sample space; the set of events, which are subsets of the sample space; and the probabilities of each event. See, for example, Woodroofe, M., *Probability with Applications* (New York: McGraw-Hill, 1975): 40–41. The triplet is also referred to as a set of elementary events, random events (subsets of elementary events), and probabilities, for example, Fisz, M., *Probability Theory and Mathematical Statistics* (New York: John Wiley & Sons, Inc., 1963): 5–13.

Peismon: That's right.

Socrates: Why do we need probability distributions?

Kokinos: Because we make inferences using probabilities. The purpose of inductive arguments is to provide credence or strength to the conclusion since it is an invalid conclusion. One way to show the strength of the argument is to use probabilities. The more probable the conclusion, the stronger the argument and more reasonable to accept the conclusion.

Peismon: Rather than using a point estimate, we use a confidence interval estimate. The inductive inference is that the distribution mean is within the confidence interval estimate. The confidence level provides a measure of the strength of that conclusion. The strength of the conclusion is an indication of how reasonable the induction is.

Mathitis: The same is true for hypothesis testing. The calculation of a probability, p, tells us the strength of our conclusion.

Socrates: Then, to apply statistical theory, we need random samples and the application of a probability distribution model, consisting of the possible results and their probabilities.

Peismon: I agree.

Socrates: In that case, statistical inferences can be deductive, not inductive.

INFERENCES ABOUT POPULATION PARAMETERS

Peismon: *That* is not a valid conclusion. It's often the case that neither the individual values nor the sample have the characteristic the population does. Unlike the case of seeing n white swans and inductively inferring that all swans are white, we are making an inference statistically about the population mean, for example, from a sample where none of the individual members nor the sample mean equals the population mean. That cannot be deductive reasoning, especially according to your distinction of extrapolation, it is inductive.

Agrafos: Like a fair die. The possible results are the numbers 1 through 6 but the average is 3.5, which no number equals.

Peismon: That's what I mean. We don't toss the die 25 times, note that each result has a mean of 3.5, and then infer that the population mean is 3.5. Most will not have a mean of 3.5.

Agrafos: Then how is that inductive reasoning if we don't have any cases for the claim?

Peismon: If induction means extrapolation, it is.

Kokinos: But with statistical inferences, are we not stating that we have a certain level of confidence in our conclusion? That's why we use confidence intervals or probabilities like *p*, rather than just a single value.

Peismon: I agree. For example, we take a sample of marbles from a jar and see that 80% are green, so we conclude that probably the next marble taken from the jar is green.

Socrates: There are two forms of your argument. In one, we know the proportion of green marbles and in the other we don't. The argument for the former would be this:

- Premise 1 (Fact): *p*% of the marbles in jar J are green.
- Premise 2 (Fact): The next marble will come from jar J.
- Conclusion: Therefore, the probability that the next marble from jar J is green is *p*%.

Mathitis: That argument is deductive.

Socrates: Not completely deductive.

Mathitis: Why not? That's the definition of probability. We can include the definition as a premise so it is clear and complete.

Socrates: What if the next marble selected from the jar is not randomly selected?

Mathitis: I see. This argument is inductive case 2. For inductive case 1 and to apply probabilities, we have to select the samples randomly. That means we need to adjust the second premise:

- Premise 2 (Fact): The next marble will be randomly selected from jar J.

Socrates: That's correct. Now it's a deductive argument.

Agrafos: But why do you use the premise that the next marble will come from the jar rather than the premise that this marble did come from the jar? In other words, why a premise about a future selected marble rather than one already randomly selected?

Socrates: Does the color of a specific marble change?

Agrafos: No, it is what it is.

Socrates: Then, once you select a specific marble, it is either green or it is not. The probability is either zero or one.

Agrafos: What if we don't know what it is? Isn't the probability something between?

Socrates: We typically think so. However, don't your probability and my probability have to be the same? If you can't see what color it is and I can, what is the probability?

Agrafos: If you can see it, it would be one, so it must be one for me too. That makes sense.

Mathitis: It's like confidence intervals. Before selecting a sample, there is a probability, say 95%, that an interval will contain the true parameter value. But once we select the sample and calculate the confidence interval, it either contains it or not. That's why it is not called a probability interval but a confidence interval.

Kokinos: Then, the case that Peismon gave would be inductive since we don't know what the proportion of green marbles in the jar is.

Peismon: This is the argument I would make:

- Premise 1 (Fact): A random sample of marbles from jar J has p% green marbles.
- Premise 2 (Fact): The next marble will randomly come from jar J.
- Premise 3 (Definition): A probability of an event is the frequency of its occurrence in the population.
- Conclusion: Therefore, the probability that the next random marble from jar J is green is p%.

Mathitis: But that conclusion is false.

Peismon: Why?

Mathitis: I should say it is not a valid conclusion—as it could be true but not necessarily true. You have a random sample, so the proportion of green marbles in the sample does not have to necessarily equal the proportion in the jar, despite the sample being random. We know that samples have characteristics that differ from the population because of sampling error.

Peismon: But when we use confidence intervals, we take into account the sampling error. Okay, even confidence intervals may not include the population parameter.

Kokinos: Then all statistical inferences from samples to populations are invalid.

Socrates: If they are inductive inferences.

Kokinos: But we said that all statistical inferences are inductive.

DEDUCTIVE STATISTICAL INFERENCES: HYPOTHESIS TESTING

Socrates: That depends on the structure of the argument. Many, if not all, statistical inferences can be converted to valid deductive arguments.

Peismon: How if we are making an inference from the sample to the population and we said that all four scenarios were types of inductive reasoning?

Socrates: First, recall that the four scenarios did not include any probability distributions, so they were not statistical inferences.

Peismon: But if we add the two premises about random sampling and assuming a probability distribution, then at least the first and third cases can be statistical inferences.

Socrates: True. Let's start with hypothesis testing. We can show that the calculation of p can be a valid deductive argument based on the assumption of the null hypothesis. A simple example is the calculation of getting three heads in a row *assuming a fair coin*. The calculation is a mathematical deduction based on the null hypothesis of a fair coin, which defines the assumed probability distribution model. We have these premises:

- Premise (Assumption) 1: The results of tossing a coin have the probability distribution with events {head, tail} and probabilities Prob(head) = Prob(tail) = 0.5.
- Premises (Assumption) 2: The next three results will be random.

Mathitis: Then, it deductively follows that the probability of {head, head, head} is $0.5^3 = 0.125$.

Peismon: But we make another conclusion about whether to accept or reject hypotheses.

Socrates: We do. But that is not a statistical inference since we are not calculating a probability of a hypothesis being true (Myth 10).

Mathitis: But we could for hypotheses about distribution parameters by assuming they are random. If we use Bayesian statistics, then the conclusion is the probability of the parameter being within an interval.

Peismon: That inference requires a prior probability distribution leading to a posterior one.

Socrates: We also can add to the deductive argument a premise stating a criterion for rejecting the null hypothesis H_o:

- Premise (Definition of rejectable): A hypothesis H_o is rejectable if, and only if, $p <$ alpha.

Mathitis: Then, once we have calculated p, using that and the additional premise of what alpha is, we can state a conclusion that either $p <$ alpha or it isn't and then the final conclusion that H_o is either rejectable or not.

Socrates: Or, we could replace rejectable by unacceptable in the definition premise. We could instead make it a requirement or an obligation, for example, H_o must or should be rejected or not accepted if $p <$ alpha. In any case, the argument is valid and deductive.

Kokinos: Then, the issue is whether each premise is true.

Socrates: Exactly. The arguments may not be sound. For example, the distributional assumption of the population or of the statistic may be false.

Kokinos: If we think all models are wrong, then whether inductive or deductive, premises about the assumed probability distribution will be false but approximately true to varying degrees.

Mathitis: But if we use inductive reasoning, we still require that assumption. Such reasoning would be unsound because of false premises and invalid conclusions. Then, hypothesis tests based on assumptions or premises of random sampling from a probability distribution are deductive.

DEDUCTIVE STATISTICAL INFERENCES: ESTIMATION

Peismon: But creating valid deductive arguments won't be possible for estimating. We calculate a 95% confidence interval to make an inductive inference about the population parameter value.

Socrates: We can create an analogous argument to hypothesis testing. We only need one premise that defines *acceptance* or *rejection* and make a conclusion on the basis of that definition. What conclusion do you want to make?

Peismon: If we have a $(1 - \alpha)100\%$ confidence interval, we want to conclude that we are $(1 - \alpha)100\%$ confident that the parameter value lies in that interval.

Socrates: Then, we can add the following premises and conclusion:

- Premise (Definition of confidence): If $(1 - \alpha)100\%$ of the intervals (L_n, U_n) based on samples of size n using the estimator $\hat{\theta}_n$ include the true parameter θ, then each (L_n, U_n) has $(1 - \alpha)100\%$ confidence it contains θ.
- Premise (Fact): (l_n, u_n) is one of the (L_n, U_n) intervals.
- Conclusion: Therefore, there is $(1 - \alpha)100\%$ confidence that (l_n, u_n) contains θ.

Peismon: That's just artificially creating a definition.

Kokinos: Isn't that exactly the definition of a confidence interval?

Socrates: This is exactly like Wheeler and Stauffer's probability question.

Peismon: That's not true. We don't know the population, that's why we are sampling.

Socrates: Not everything about the population, but we do know some relevant characteristics.

Mathitis: I had not thought about that. Yes, we know that if we are calculating a 95% confidence interval, 95% of them contain the parameter we are estimating and 5% will not. If we draw a random sample, then we know the probability of getting one of those intervals that will include the parameter. Rather than being a distribution with 95% confidence intervals, it could be a jar with 95% green marbles. The probability of getting one of those intervals that includes the population parameter is the same as the probability of getting a green marble.

Peismon: But the conclusion is not a probability.

Socrates: No, because once we select a confidence interval, it either contains the parameter or it doesn't. Similarly, once we select a marble, it is either green or it isn't.

Peismon: But when we select a marble without looking, we don't know what it is.

Socrates: When we calculate a confidence interval, we don't know whether it contains the parameter value or not. Whatever is true with respect to probabilities of the selected marble is true of the confidence interval. Either both reasonings are deductive or neither is.

Mathitis: But we only take one sample of size n, not several, to see what percent of the confidence intervals contain θ. As Kokinos said, the confidence level is a measure of the support for the conclusion that the interval contains θ. We can use that to judge its reasonableness.

Agrafos: Isn't that the purpose of using statistics: to provide some quantification of the evidence for a conclusion?

Kokinos: I think it is. But we still have the issue of soundness. Is each assumption true?

Agrafos: But that is the same issue for induction using statistics. In all cases, we assume a probability distribution model, and aren't they all wrong?

Kokinos: I think that's a key difference between nonstatistical inferences and statistical inferences using induction. When it's nonstatistical, we can use facts as our premises making them true. All the observations about swans or marbles are facts, for example. But the conclusion need not be true because the arguments are invalid. However, some are more reasonable than others. That's not the case with statistical inferences.

Mathitis: That's because whether inductive or deductive, statistical inferences have two premises that may not be true—perhaps, are never true: (1) samples are random and (2) the data come from an assumed probability distribution model. Then, we are better off using deductive arguments for statistical inferences—the arguments can be valid.

Socrates: Then, can inductive cases 3 or 4, an inference from one population to another, be a statistical inference?

Mathitis: No, because that inference violates the requirement of random sampling.

Kokinos: But it can be inductively reasonable or deductively valid if there are other premises that show a relevant relationship between the two distributions; for example, the marbles are produced by the same process.

Mathitis: Then, it is no longer just a statistical inference. We may not even need any assumptions about randomness and probability distributions.

REAL-WORLD CASES OF STATISTICAL INFERENCES

Socrates: Note that the third application, answering the Inference question, can be done theoretically or applied to the real world. For the real-world problems, we need to convert them into statistical questions or issues as Kokinos mentioned before.

Kokinos: Start with a practical problem, convert it to a statistical problem, solve the statistical problem to get a statistical solution, and then convert the statistical solution to a practical solution. The practical question might be "Can we sell this product?" or "Should I use this treatment?" or "Is this product or line equivalent to another?" Those questions might be translated statistically as "Is the average performance of this product equivalent to another in the market?" or "Is the cure rate better than a placebo?"

Agrafos: What if the translation is wrong?

Mathitis: That is an issue and perhaps why people disagree about statistical conclusions.

Socrates: Even for real-world problems, what is required to get a statistical solution?

Mathitis: You still have to randomly sample. For example, when doing clinical trials to see whether a treatment is better than a placebo, subjects are randomly assigned to each. When making a statistical inference about the defect rate of a lot, we randomly sample from the lot. Otherwise, the inference is invalid.

Peismon: Our inferences can be confidence intervals or p values for hypothesis tests.

Socrates: In these cases, we assume that sampling is random and we make statistical inferences and translate them to practical inferences. In these cases, are the universes known?

Peismon: No, that's what makes them applications of the third question.

Socrates: Consider the bowl of 4000 white beads and 1000 black beads. If I randomly sample with replacement 200 beads and get 50% white and 50% black, what would you conclude?

Peismon: That your sampling really wasn't random. The probability of getting 50% white when the universe has 80% white with a sample of 200 is too low.

Socrates: Why can we infer that or believe your conclusion?

Peismon: Because we know the universe. But this is not the third question of Inference.

Socrates: You are correct. So let me make it so. Let's assume we have a bowl of 5000 beads, but we do not know what proportion is white. We randomly sample with replacement 200 and get 50% white beads and 50% black beads. What is your conclusion?

Peismon: We can conclude that the proportion of white beads and black beads is about the same. If we had tested whether they were the same, the p value would be 1. Confidence intervals of each would be exactly the same.

Socrates: What if the bowl actually had 80% white beads and 20% black beads? How would you know the sampling process was fair or unbiased?

Mathitis: I don't think we can tell unless we look at the entire universe or a sufficient amount.

Socrates: Consider a coin. We know the universe is {head, tail}. But when we ask whether the coin is fair, are we asking what the distribution of heads and tails is or are we trying to estimate what proportion of the universe is heads?

Mathitis: We know the universe and proportion of heads. We are asking whether the coin is balanced.

Socrates: But why flip the coin rather than just determine physically whether it is weighted equally across the thickness?

Peismon: We may not be able to do that. That's why we use statistics. We toss the coin to get a sample of results and check, as before, through confidence intervals or hypothesis tests.

Socrates: Are you then testing the fairness of the coin or are you testing the sampling procedure?

Kokinos: I think it is the sampling procedure. We know a deck of cards has only four aces, but if when dealing the cards through several hands the dealer seems to be getting all the aces, we might suspect that the cards are not being dealt randomly.

Socrates: Doesn't that tell us that there are two cases of statistical inferences. In the first, with an unknown universe, we assume as fact that we can randomly sample to make statistical inferences about the characteristics of the universe, for example, proportions. But when the universe is known, we can check whether the sampling procedure is random by checking statistically whether the samples are close enough to the actual or expected

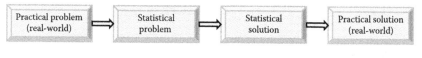

Two scenarios of statistical inferences

Scenario I: Inference to a universe

What is the distribution in the unknown universe, for example, lot, batch, deck of cards, bowl?	For example: Estimate the proportion π or test if proportion π is a specific value.	Confidence interval for π; or, test H_0: π = specified value	Approximate knowledge of the (real or theoretical) distribution

Scenario II: Inference to a process

| Is the sampling procedure from a known universe "fair" or random? | For example: Does flipping a coin result in the probability of heads $\pi = 0.5$? | Confidence interval for π includes 0.5; or, $p \geq$ alpha testing H_0: $\pi = 0.5$ | Yes, it is fair/random; use it. |
| | | Confidence interval for π excludes 0.5; or, $p <$ alpha testing H_0: $\pi = 0.5$ | No, it is not fair/random; don't use it. |

FIGURE 19.3

Practical–statistical approach applied to the general use of statistical control charts.

distribution. We can state these two cases using the practical-to-statistical format like in Figure 19.3.

Kokinos: Isn't that what casinos do? If someone starts winning "too much," then they try to determine whether it is chance—streaks that will occur periodically given the number of times people gamble—or whether someone is cheating or counting cards. That's the second scenario.

Mathitis: But both are variations of the Inference question.

20

---

*Myth 20: There Is One Universe
or Population if Data Are Homogeneous*

Scene: *Break room (continuation of the discussion on statistical inferences)*

BACKGROUND

Peismon: Socrates, regardless of whether statistical inferences are inductive or deductive, statistical reasoning requires the premise that the data come from a stable process.

Socrates: Why?

Peismon: According to Wheeler and Stauffer, if the process is unstable you do not have a homogeneous population (Wheeler and Stauffer 2014). Without homogeneity, all the analyses will be meaningless.

Socrates: What does homogeneous mean?

Peismon: In his *SPC* book, Wheeler says "Data collected, in the same way, from a process which is *known* to *display statistical control* can be said to be homogeneous. The homogeneity of all real-world data is suspect" (Wheeler and Chambers 1992, 25, footnote).

Mathitis: Then homogeneous means stability?

Socrates: He doesn't say must be stable, only that it displays stability. But why is that necessary for analysis of data?

Peismon: Recall Wheeler and Stauffer's questions. "'Are these data homogeneous?' must be the first question of any analysis. Process behavior charts [e.g., control charts] provide the easiest way to address this question. Hence, any analysis that doesn't begin by organizing the data in some rational manner and placing them on a process behavior chart is inherently flawed" (Wheeler 2009a).

Socrates: *Any* analysis?

Peismon: Yes.

Socrates: We talked about different applications of statistics (Myth 19), Description, Probability, and Inference. Wheeler's example of descriptive statistics was the bowl of beads that answer questions about summarily describing a population.

Peismon: That's right. We can use descriptive statistics to describe the contents of the bowl and the characteristics of the beads.

Socrates: Do we need to plot all data used for descriptive statistics on a control chart and display stability to avoid inherently flawed analyses? Did Wheeler and Stauffer do that with their beads?

Peismon: But you don't need that for a known, static population—a single universe. They say that the issue about homogeneity occurs when "[g]iven a collection of observations, is it reasonable to assume that they came from one universe, or do they show evidence of having come from multiple universes, processes or populations?" (Wheeler and Stauffer 2014). Since you have the known universe, you already know it is a single universe.

Mathitis: Then, question of homogeneity does not apply to the first (Description) and second (Probability) applications (Myth 19) since we know the universe and are either describing it statistically or calculating probabilities.

Peismon: That's right. The question applies in the third application using inferential statistics.

DEFINITION OF HOMOGENEOUS

Socrates: I'm confused. Don't we mean the same thing by universe and population?

Peismon: Yes.

Socrates: Is either the same as a process?

Peismon: No. But we could have multiple processes or universes. A control chart can tell us whether there are one or more of whichever it is.

Socrates: What does *homogeneous* mean?

Agrafos: Can't we use the dictionary definition?

Mathitis: Not if they use it in a technical sense. In statistics, homogeneous can mean equal. For example, a test of equal variances is also called a test of homogeneity of variance.

Kokinos: In science, homogeneous is defined as the same or having the same characteristic, parts, type, or nature. Having a common property. For example, the Japanese population is more homogeneous than the U.S. population. We mean more similar with respect to a characteristic, for example, physical characteristics, genetics, or where they are born or language they speak.

Socrates: Also note that *homogeneous* is an adjective while *population, universe, process,* and *data* are nouns. In the phrase "red car," *red* is the adjective that describes some characteristic of the noun *car.* Tell us what noun and which characteristic of that noun homogeneous describes. Is it of the universe/population, the process, the data, or something else?

Peismon: It is a characteristic of the data. The data can be homogeneous or not. If the data come from the same universe, then they are homogeneous.

Kokinos: That's not what I expected. A heterogeneous population also has members from the same population but the members would not be homogeneous by definition.

Socrates: Then, the word has a technical meaning referring to the characteristic that the data have a universe or population in common.

Peismon: That's right.

Mathitis: What do stability and control charts have to do with it then?

Peismon: We check for homogeneity of data using a process behavior or control chart. If it displays stability, the data are homogeneous; if not, then the data are not homogeneous.

Agrafos: Are you saying we must check whether the data came from one universe, population, or process by plotting the data on a control chart? Aren't there simpler ways to check that? Can't we tell whether the data came from one physical production process without plotting them? Can't we tell whether they come from one universe, e.g., the bowl of 5000 beads, without plotting the data?

Peismon: The purpose of a control chart is to identify changes in a process. If an assignable cause occurs, the process has changed. It means that there is more than one distribution or population. Remember that Wheeler's simulations showed that a single probability distribution would have a false alarm rate of no more than approximately 1% using the 3 standard error control chart limits. So, a point beyond those limits would indicate that the distribution changed.

Kokinos: So, two or more probability distributions.

Peismon: Exactly.

Agrafos: But it will still be the same process.

Peismon: If the process changed, how can it be the same?

Agrafos: If I injure my leg and now limp, I am still the same person.

Peismon: That's you, a person. We're talking about processes.

Socrates: Consider two production lines producing the same thing. How many processes?

Mathitis: Clearly two since there are two lines.

Socrates: On a control chart, we plot the results in the sequence produced but alternating from the lines. According to the authors, if the chart displays stability, how many processes?

Mathitis: One—and one universe.

Kokinos: We could do it in reverse also. We have only one production line and if the control chart shows instability, there is still only one process but he would conclude there are two.

Peismon: Wheeler and Stauffer quote Deming: "In applying statistical theory, the main consideration is not what the shape of the universe is, but whether there is any universe at all. No universe can be assumed, nor the statistical theory of this book applied, unless the observations show statistical control.... In a state of control, the observations may be regarded as a sample from the universe of whatever shape it is" (Deming 1950, 502–503).

Mathitis: But Deming doesn't use the word *homogeneous* nor does he refer to a process, whether it's zero, one, or more. Yet, he seems to be saying the same thing, otherwise why quote him?

Socrates: Let's separate process from universe and population and focus on the latter two, which are the same. Is the gist that homogeneity of data means coming from a single universe and that process behavior or control charts determine there is a single universe if they display stability?

Peismon: Yes, that captures it.

Kokinos: Then, we don't really need the word *homogeneous* as we can say it as Deming did.

Socrates: But there seems to be a contradiction between Deming and Wheeler-Stauffer.

Peismon: No, they agree; otherwise, why quote him?

Socrates: Wheeler and Stauffer say that if the data do not display stability, then there are multiple universes while Deming says without stability there is no universe. Which is it?

Peismon: I hadn't noticed that.

Socrates: In either case, there are three claims: (1) if appropriately plotted data display instability, there is no universe; (2) if appropriately plotted data display instability, there are multiple universes; and (3) if appropriately plotted data display stability, there is only one universe. The first is Deming's; the second, Wheeler and Stauffer's; and the third, all three.

IS DISPLAYING STABILITY REQUIRED FOR UNIVERSES TO EXIST?

Mathitis: I think the first claim is false, even when applied to statistical inferences.

Socrates: With 100% sampling, we can calculate the defect rate of a lot, even if the lot was produced by an unstable process. We can also estimate the defect rate with a random sample less than 100% using point or confidence interval estimates or test of hypotheses.

Mathitis: Deming himself says that sampling theory can be applied to enumerative studies and gives the example of a lot.

Kokinos: The key to applying sampling theory is that one can enumerate the universe. Any finite collection can be enumerated, which makes it a universe.

Agrafos: Stauffer seems to agree as he replied to that question as follows: "When we use the term 'universe' we are using it to mean the whole that we are studying. When you have a population (about which it is, at least in principle, possible to enumerate every member), then the universe is the population" (Wheeler and Stauffer 2014).

Kokinos: Does Deming give examples of populations and samples?

Mathitis: In the earlier chapter on the distinction between enumerative and analytical studies, he describes various populations of people in cities, items in a bowl, crop yield from a field. First, he said "In the analytic problem, on the other hand, something is to be done to regulate and predict results of the cause system that has produced the universe (city, market, lot of industrial product, crop of wheat in the past) and will continue to produce it in the future" (Deming 1950, 247). Ten years later, he wrote: "If it appears that a survey or experiment may be worth discussion, the next step is to carefully define the universe. The universe is all the people or firms or material, conditions, concentrations, models, levels, etc. that one wishes to learn about, whether accessible or not" (Deming 1960, 8).

Socrates: These seem to contradict the Deming quote Wheeler and Stauffer used. But it does match the discussion we just had where the possible outcomes is the universe.

Kokinos: To me, it's clear that it does not matter the order or how the population of the city, the products in a market, the items in a lot, the crop from a field, or any other thing that is in the universe came to be. They are definable, enumerable, so they are universes. Besides, if it's not accessible, then how can you determine whether it came from a stable process?

Mathitis: When the universe is inaccessible, Deming says to use a frame (Deming 1960, 39).

Kokinos: That also appears to contradict the quote Wheeler and Stauffer cite.

Peismon: But that's for descriptive statistics. The discussion is about inferential statistics.

Mathitis: The same thing applies. Statistical inferences apply to unknown universes knowing things about the sample. The key was having a random sample.

Kokinos: Deming says that every sampling unit must be uniquely identified so that there is a way to know which sampling unit is selected when making random drawings. The collection of sampling units is a universe, or more precisely, the frame to which inferences are valid.

Mathitis: In addition, Deming specifically states that there is one requirement for a frame and it is not stability. "The one basic requirement of a frame, without which there is really no frame at all, is that it must show a definite location, address, boundary, or set of rules by which to delineate and to find any sampling unit that the random numbers draw" (Deming 1960, 40).

Socrates: Remember that probability distributions apply to the theoretical world while the list-universes apply to the real world. We use the former to model the latter. Is Wheeler and Stauffer's bowl of 4000 white beads and 1000 black beads one universe?

Peismon: Yes, we've said that and used it to illustrate descriptive statistics and probabilities.

Socrates: If the beads were produced by an unstable process, is it no longer a universe? If we sample nonrandomly from the bowl, does the universe disappear or cease to be a universe?

Mathitis: No, of course not. The list of possible results or collection of real-world things are universes. That proves that real-world universes exist independently of their assigned probabilities as real-world relative frequencies or probability distribution models. That makes the first claim that displaying instability means there is no universe false.

ARE THERE ALWAYS MULTIPLE UNIVERSES IF DATA DISPLAY INSTABILITY?

Agrafos: Then isn't the second claim also false?

Peismon: No, the Empirical Rule shows it's true. The likelihood of being the same or only one probability distribution is too remote when a point exceeds the control limits.

Mathitis: But remote does not mean never. Besides, don't we have the same situation as with Deming's claim regarding the bowl of beads? If we nonrandomly sample from the bowl, does the single universe multiply to become two or more?

Socrates: Also, recall the example of a die with the numbers {1, 2, 3, 4, 5, 6} versus one with the numbers {1, 2, 3, 4, 5, 60}. Each list is one universe. But random results from the first would show stability, while those from the latter would show instability. You should confirm this.

Agrafos: Then, we would incorrectly conclude that there are two or more universes if using the second die, which is false.

Socrates: In fact, there is only one list and one probability assignment during the entire time.

Kokinos: Probability distributions consist of the possible results and their probabilities. If the list of possible results is the universe, then it can stay the same but the probabilities change.

Agrafos: Then the probability distribution changes, but the universe does not. So, the second claim is not always true. What about the third claim?

IS THERE ONLY ONE UNIVERSE IF DATA APPROPRIATELY PLOTTED DISPLAY STABILITY?

Peismon: The third claim is proven true by the Empirical Rule. In fact, Wheeler and Stauffer say, "This assumption of a single universe is equivalent to the assumption that the behavior of the outcomes in the sample is described by one probability model" (Wheeler and Stauffer 2014).

Socrates: That's not always true. Probability distributions are models applied to the real world. The people in a city, crop from a field, and other examples are real-world populations to which we apply statistics. Universes or populations are not probability distributions, although they are a necessary component of them. They are the collection of possible results.

Mathitis: Just as a car has an engine does not mean that all engines are cars.

Socrates: Wheeler's Empirical Rule is based on analyses of many different probability distributions. He used six in his *SPC* book. Are they different universes?

Peismon: Yes, that's why the limits work universally.

Socrates: Then, consider the alarm of points beyond the 3 standard error limits. Take a sample of four consecutive points from the normal distribution, then four from the chi-square, and so on from each of the six distributions. Put the 24 values on a control chart. What do you expect to occur?

Peismon: With six different distributions, I expect lack of stability.

Socrates: Yet, doesn't the Empirical Rule state that regardless of the probability distribution, we should expect no points out of 24 to fall beyond the 3 standard error control limits?

Mathitis: Yes, that's right. But we would in fact have six universes. So, displaying stability does not mean a single universe.

Socrates: All we know is that we have six *probability distributions*. How many *universes* given that a universe is just one component of a probability distribution?

Peismon: See. It can still show there is one universe.

Mathitis: The universe for the normal distribution is $\{-\infty, +\infty\}$; for the chi-square, Burr, and exponential, it is $\{0, +\infty\}$; and for the uniform and triangle, it is finite $\{a, b\}$. They are different.

Socrates: In fact, the Empirical Rule disproves the third claim. Displaying stability does not mean we have only one universe.

Peismon: Your examples are about sampling from a single, known universe. But we're talking about a process that produces only a sample and making inferences to the process.

Socrates: That's true. So let's change the order. Consider the possible process sequences for producing the bowl of beads we've discussed. Suppose the sequence consisted of 20 white beads followed by 5 black beads until 5000 beads were produced. Label the white beads 1 and the black beads 0. What would a control chart show?

Mathitis: Using an *I* chart, then $\bar{x} = 0.8$, the percent of white beads. We can estimate the limits since there are approximately 400 changes from 0 to 1 or vice versa out of 5000. The moving range $\bar{R} \approx 0.08$. The control chart factor is 2.66, so the LCL = $0.8 - 2.66 \times 0.08 = 0.59$. All the black beads (0's) would be below this, displaying instability. Yet, there is only one universe and one process.

CONTROL CHART FRAMEWORK:
VALID AND INVALID CONCLUSIONS

Peismon: But that's how the control chart works and successfully for decades.

Socrates: I think you are confusing the statistical solution with the practical solution. What is the practical, real-world problem?

Peismon: To see whether the process is consistently behaving during a period *T* covered by the data. That's why Wheeler prefers the term *process behavior charts.* The purpose of the control chart is to check whether process performance changed during that period.

Socrates: What does *consistent* mean in the real world?

Peismon: We know the results will not be exactly the same, so we allow some variation.

Mathitis: The advantage is that we can apply the statistical model of variation, random variables, or probability distributions, to describe that variation.

Socrates: How much variation or difference is allowed to be consistent?

Peismon: That's what the control chart limits are for.

Socrates: Are not the control chart limits determined statistically using the standard error?

Mathitis: Yes, so that means they are not the real-world definition of *consistent.*

Kokinos: That has been the case for decades in the pharmaceutical and medical diagnostic industries. They have defined *good* or *acceptable* statistically rather than with respect to the purpose of the product.

Kokinos: But we can define how much variation is acceptable probabilistically.

Agrafos: Even though we originally said that the limits were not probability limits?

Socrates: That adds to the evidence against that belief. Then, what is the statistical problem?

Mathitis: Do the plotted points occur within the acceptable statistical limits?

Socrates: If we use other alarms or special cause variation, how much variation is allowed?

Peismon: Each rule or special cause indicates nonrandomness, so that must be the general definition we use rather than just a certain amount of variation.

Mathitis: Then, the statistical question is, "Do the plotted points exhibit randomness during T?"

Socrates: How do we get the statistical solution or answer? How do we check for randomness?

Mathitis: Since randomness is a statistical term, we use probability theory to determine which events are unlikely. But that means we must assume a probability distribution.

Peismon: That is why the control chart is a test for a single universe. The special cause variation patterns or alarms are the unlikely events. If we get alarms, we conclude that the process is not consistent. That means there is more than one universe.

Socrates: Are you sure? Note that the statistical solution is whether the data display randomness. We already saw that results can come from a single universe without being random. The bowl of beads is a single universe, regardless of the order of production or selecting from the bowl.

Mathitis: I agree. The conclusion is invalid.

Kokinos: Also, can't nonrandomness in the statistical context mean that only the probabilities changed, only the universe changed, or both changed?

Mathitis: That's right. So if only the probabilities changed, then there is still a single universe.

Socrates: And the practical solution?

Mathitis: The process is in the real world. So, the practical solution is that the process behaved consistently or didn't change.

Peismon: But in this case, *consistent* is defined as having the same probability distribution during the period *T*. It's the premise we used in our argument and to check for randomness.

Socrates: The assumed probability distribution is used to establish the control chart limits and is not the real-world definition as probability distributions are theoretical models used to describe real-world phenomena. Isn't this the logical argument for the statistical solution?

1. (Premise) If the list of possible results and the probabilities assigned to each stay the same during period *T*, then the data will (almost certainly) display randomness.
2. (Observation) The data display nonrandomness.
3. (Conclusion) Hence (almost certainly), the list of possible results, the probabilities assigned to each, or both changed during *T*.

Peismon: But it's highly reasonable that the real-world conclusion is the process changed.

Mathitis: Yes, if changed means inconsistent and that means nonrandom.

Kokinos: Then, we can describe the problem and solution like in Figure 20.1.

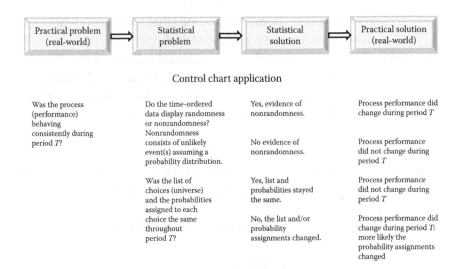

FIGURE 20.1
Practical–statistical approach applied to the general use of statistical control charts.

Socrates: That's good. However, the question we are discussing is this: "What conclusion is valid if the process displays stability (randomness)?"

Mathitis: Now, we can't conclude anything, although we do in practice. The invalid argument we use is: if A then B, we observe B, and conclude A. The conclusion about a single universe is invalid when the data display stability or randomness.

Socrates: Do you now see why we were confused about the meanings of stability, universe, and homogeneity? We sometimes talk about the statistical model and other times about the real world. Our interest is in the process consistency *in the real world*. We have been making conclusions about the statistical model and using those terms. Even then, we make the mistake of thinking that every change in probability distribution is a change in the universe.

Mathitis: I see. For example, if a die changes from being fair to biased, the possible results from tossing the die don't change, but the probabilities do. That's a clear example of the universe not changing. If we get six 6s in a row, then the probability of a six apparently changed.

Socrates: If you toss a die with numbers 1–6 and got a 7, what would you conclude?

Mathitis: That someone switched the die. Clearly, the list changed.

Agrafos: What about Mathitis' example of six 6s in a row? Does that change the list also?

Socrates: Possibly. We don't know whether now the list contains only sixes or whether the probabilities changed so that a six is substantially more likely. But when the list changes to produce results not in the original assumed list, do you need a control chart to tell you that?

Mathitis: No. So, if we assume the list of possible results is all possible numbers, then the true and valid conclusion is that the probabilities changed.

Socrates: Yes. Consider a test of the equality of two population means. If the list is all possible values from negative infinity to positive infinity whether the means are equal or not, all we are doing is testing whether the probabilities of each value is the same or not.

Peismon: But now you are contradicting yourself by saying you have two populations when the lists are the same.

Socrates: You're right, I was ambiguous. See how easy it is to use statistical terms and think we are talking about the real world? I was talking about the real world. When I said two populations, I meant two physical lists with the same content. For example, you buy two decks of cards. The content of the lists is the same but you have two decks. But when we switch to the statistical world, the lists are the same although the probability distributions may not be. Consider two manufacturing lines. We have two processes. We assume the lists are the same but maybe not the probabilities— that's the null hypothesis. We test the means. If they are statistically different, we are simply saying that probabilities assigned to those identical lists are different. If either the lists or the probabilities or both are different, then the probability distributions are different. But if the probability distributions are different, it doesn't mean the lists are different. That's the mistake.

Mathitis: Then, a graph of two normal distributions with different means have the same list but different assignment of probabilities. Control charts are really testing whether the assignment of probabilities has changed from one time to another because we do not expect to get a value we never thought possible, like a 7 when the die has numbers 1 through 6.

Socrates: We already saw that to have one universe, randomness does not have to occur in either the production of the universe or in selecting from the universe. Recall the example of the two dice, with universes {1, 2, 3, 4, 5, 6} and {2, 3, 4, 5, 6, 7}. When we randomly select one die and randomly select from that die, for example, by rolling the die, what do we get?

Mathitis: An I chart and an $\bar{X}$ chart with subgroup size $n = 2$ showed stability. Displaying stability does not necessarily mean one universe. We could generate numerous counterexamples to the third claim for each type of control chart.

Socrates: In the real world, the process is the mechanism of assigning probabilities, but it is not the list, the universe. If we are measuring diameters of a pipe, we know the list before we start. If we know that the maximum possible diameter is M, then the list is all numbers between zero and M. If the behavior of the process changes, what changes in the probability distribution model?

Mathitis: Only the assigned probabilities. We keep switching back and forth between the real world and the model.

Socrates: I think so. When we say we have evidence of two universes, we don't always mean physically when there is only one process, for example, one die or one manufacturing line. In the real world, we often want to show that two actual processes are equivalent and a control chart can provide evidence of that. But it does not show that there is only one physical, actual process.

Kokinos: That's true. We often have to show that two products or lines or methods are equivalent. That's why the TOST (two one-sided test) was developed.

Socrates: When you plot the results from two products, lines, or methods on a control chart and it displays stability, does that mean that one of them disappears?

Kokinos: Of course not. We are just showing that they are equivalent in performance—not the same physical thing but statistically equivalent in performance.

Mathitis: That's why the control chart doesn't tell us how many real-world processes or universes we have; not even how many theoretical probability models we have. In many cases, we know before checking how many physical processes and actual universes there are.

Socrates: Recall the two types of scenarios to answer Wheeler-Stauffer's third question of inference (Figure 19.3). Are we making an inference to the process or to the universe?

Peismon: Both, using the process inference to conclude something about the universe.

Socrates: But for the process inference, we need to know the universe. If we know the universe, then why are we making an inference to it? What is it that we do not know?

Mathitis: The probability distribution for that known universe. Specifically, we don't know the probabilities of each member of the universe. So, when it changes, then we say there is evidence that the process changed.

21

Myth 21: Control Charts Are Analytic Studies

Scene: *Project team meeting (day after the control chart discussion)*

BACKGROUND

Peismon: I think the reason for our disagreements about control charts the other day is the failure to distinguish between enumerative and analytic studies, as Deming said.

Socrates: Which disagreements?

Peismon: Disagreements about control chart limits being probability or empirical limits or the best economic limits.

Mathitis: I agree with Peismon. There is much debate and disagreement about these issues.

Socrates: What is the connection between these disagreements and the distinction between enumerative and analytic studies?

Peismon: The fact is that control charts are analytic studies. Yet, some people want to use them as enumerative studies.

Socrates: What is the difference between an enumerative study and an analytic one?

Peismon: Enumerative statistics is an analysis of data from a study to describe populations. It's descriptive statistics but an inference is still required. It's answering Wheeler–Stauffer's third question (Myth 20).

Mathitis: For example, if you count 4 red marbles in a box of 10 marbles, there are 40% red marbles. If the jar has 1000 marbles, then you would take a sample to estimate the percent. In both cases, you are describing the population by determining the percent of red marbles.

Peismon: Another way to look at it is that enumerative statistics answer the question "How much or many?" while analytic statistics answer the question "Why?" Analytic statistics would get at why there are 40% red marbles to change that percentage. It gets at understanding the causes so that action can be taken on the process that produced the characteristics of the population. Enumerative analysis isn't interested in why or making any changes to the process.

Agrafos: What does this have to do with control charts?

Mathitis: Deming specifically said control charts are analytic studies: "The control-chart is a splendid example, the purpose being to control the production process and the quality of lots yet to be made" (Deming 1953).

Socrates: Is the difference, then, what population each addresses? "Quality of lots yet to be made" sounds like the possible results from coin tossing.

Mathitis: That's the analytic population. Enumerative studies investigate finite, real populations.

Peismon: That's right. Control charts are analytic studies, using analytic statistics. We want to find the reasons for special or assignable causes and take action on the process, the cause system. Enumerative statistics would simply state the percent of good product. If any action is taken, it would simply be on the units produced, not on the process.

Mathitis: The actions include sorting good from bad and then replacing or reworking the bad. In lot acceptance, average quality level sampling plans are enumerative analyses.

Kokinos: The purpose of an analytic study is to make a prediction about what will happen because you have changed the process. It's to make improvements.

Peismon: But because you are making predictions about future lots, statistical theory won't help. In addition, in the analytic study, the population is dynamic. It is going in and out of control, shifting. We can't take random samples from this infinite, hypothetical, dynamic population to make useful statistical inferences.

Agrafos: But then why is their use called *statistical* process control?

ENUMERATIVE VERSUS ANALYTIC DISTINGUISHING CHARACTERISTICS

Socrates: It seems that there are several issues that distinguish enumerative and analytic. Perhaps we should identify them all before continuing.

Mathitis: What do you mean?

Socrates: The three of you have mentioned different questions to answer, different actions to take, whether predictive or not, and other differences. Did Deming specifically define the terms?

Peismon: He did provide one clear way of distinguishing between the two: whether 100% sampling is conclusive or not. Wheeler and others also agree with these differences.

Socrates: Let's see if from the references we can list the distinguishing features of enumerative and analytic, respectively.

1. 100% sampling: conclusiveness versus inconclusive

There is a simple criterion by which to distinguish between enumerative and analytic studies. A 100 percent sample of the former provides the complete answer to the question posed for an enumerative problem, subject of course to the limitations of the method of investigation. In contrast, a 100 percent sample of a group of patients, or of a section of land, or of last week's product, industrial or agricultural, is still inconclusive in an analytic problem. (Deming 1975)

2. Question answered: How much/many? versus Why?

On the uses of data: Briefly, the enumerative question is how many? The analytic question is why? Is there a difference between the two classes, and if so, how big are the differences? (Deming 1953, 245)

Such problems are enumerative because they depend purely on a determination of the number of people in an area... [t]hey do not involve the analytic question of why all these people are there... (Deming 1953, 245)

3. Action taken: on frame versus on cause system to predict

The problem was enumerative because the action (viz., allocation of food and materials) depended on how many people were there and not why they were there. (Deming 1953, 245–246)

Two questions arise: B (analytic). Shall we leave the machine alone, or shall we adjust it? Shall we make it run slower or faster, or shall we change the type of chemical bath? A (enumerative). What shall we do with the batch of product just made? (Deming 1953, 246–247)

In the analytic problem, the action is to be directed at the underlying causes that have made the frequencies of the various classes of the population what they are, in order to govern the frequencies of these classes in time to come. (Deming 1953, 246)

In the enumerative problem, some action is to be taken (on the frame) because the frequency of some particular characteristic of the universe is found to exceed some critical value. (Deming 1953, 245)

4. Population: real, static, finite population versus hypothetical, dynamic, infinite population

The frame is an aggregate of identifiable tangible physical units of some kind, any or all of which may be selected and investigated. Enumerative: in which action will be taken on the material in the frame studied. Analytic: in which action will be taken on the process or cause-system that produced the frame studied... (Deming 1953, 246–247)

5. Prediction: past/present versus future

Suppose that the problem is to predict the proportion that would be cured in a succession of lots of patients. This is an analytic question... (Deming 1953, 251)

6. Statistics: applicable versus not applicable

But every inference (conclusion) is conditional, no matter how efficient be the design of the experiment. ...we cannot assert, on the basis of a conditional statistical inference alone that other patients, other hospitals, other pupils, other locations, would show similar differences nor greater differences. We fill in the gap by knowledge of the subject-matter. (Deming 1976)

Mathitis: I think we have them all.
Kokinos: It seems that Deming wasn't clear or was only providing features rather than defining the term.
Mathitis: But what if this isn't what Deming meant?
Socrates: That's okay. The purpose is to learn. We can try each of these and see where it takes us. We can be as accommodating as we want.

If we still have difficulty understanding or finding something well defined, we can consider other definitions. This is one way new ideas are developed, for example, heliocentric models of the solar system versus geocentric ones, relativity versus Newtonian physics, and so on. There are numerous examples in statistics also, for example, the controversy over hypothesis testing leading to Bayesian statistics.

Mathitis: So, we assume Deming meant that they all applied?

Socrates: Good question. Are each necessary? For example, is answering "How much/many?" either necessary or sufficient for a study to be enumerative?

Peismon: I would think it would be necessary. If it doesn't answer "How much/many?" then it can't be enumerative.

Mathitis: Would all six conditions have to hold before something is enumerative or analytic?

Kokinos: If that were the case, then there were other kinds of studies. Studies that had half the enumerative features and half the analytic would be neither.

Peismon: Maybe having one or two implies having the others. For example, maybe meeting the conclusiveness condition necessitates that you meet the other five conditions.

Mathitis: Can we check that?

Socrates: There appear to be three themes among five of the six characteristics that describe stages: problem, study, and solution. Let me organize them that way (Table 21.1).

Mathitis: What about statistics? They are either enumerative or analytic.

Socrates: Statistics might be usable in any of the stages.

Peismon: Here is another passage from Deming about statistics:

> Enumerative: A statistical study in which action will be taken on the material in the frame being studied.
> Analytic: A statistical study in which action will be taken on the process or cause-system that produced the frame being studied, the aim being to improve practice in the future. (Deming 1975)

Mathitis: He is saying that statistics can be used for both.

Socrates: Let's leave it open as to which statistics and which stage they can be used. We can clarify and change as we learn more.

Kokinos: Where do we start?

TABLE 21.1

Characteristics Differentiating Enumerative and Analytic Stages

Characteristic	Enumerative	Analytic	Stage
Question	How much/many?	Why?	Problem
Population	Real, finite, static	Hypothetical, infinite, dynamic	Study
100% sampling	Conclusive answer to question	Inconclusive answer to question	Study
Answer	Enumeration	Causes	Solution
Prediction	Past/present	Future	Solution
Action	Population	Cause of population	Solution
Statistics	Applicable?	Applicable?	Problem (?), study (?), solution (?)

ENUMERATIVE PROBLEM, STUDY, AND SOLUTION

Socrates: Before we start, let me ask why?

Kokinos: Why what?

Socrates: Exactly. It seems that we can't answer "Why?" unless we know *what* we are seeking an explanation for. In this case, *what* is how much or how many. So let's start with the enumerative stages and look at each characteristic. Then, do the same for the analytic stages.

Kokinos: I agree. I need to know how many first.

Question: "How Much/Many?"

Socrates: You gave an example of counting the number of red marbles in a jar and concluding that 40% of the marbles were red. Then, the question "How much/many?" can be answered either by an absolute number or relative number, for example, a proportion or percent.

Mathitis: That's right. We are just counting, which is what enumeration means. I suppose they could also be other descriptive statistics, such as a mean or range.

Socrates: What information do you need to determine that count? For example, before you can count how many people buy product A, what do you need to know?

Peismon: You need to know for each buyer whether he or she bought the product.

Socrates: You have to classify each as buyer of product A or not.

Mathitis: Yes. To enumerate, we first must determine the classes.

Socrates: Recall our discussion of measurement. Counting is a measurement. All measurements, whether quantitative or qualitative, are merely classifications (Myth 1). It's clearer when we have qualitative data. The classes are good or bad, for example, or five colors, or buyer of product A or not. Measuring the length of the table, we are also classifying. It is in the class "6 feet" and not the class "6.1 feet" or "5.9 feet."

Kokinos: After determining the classes, we assign each unit to a class and then count the number of units in each class. So, we must first measure to determine what class a unit is in and then measure by counting to determine the enumeration or how many.

Socrates: Then, we need criteria for classifying. For example, to classify invoices as paid or not paid, what do we do with those charged to a credit card that have not been paid yet?

Kokinos: That makes sense to have classification criteria.

Socrates: We also have to determine which units we classify. Do we classify everyone in the world as buyers of product A or not? Are we classifying as good or bad, the product from this lot or everything in sight, such as the people, the office furniture, unrelated materials?

Kokinos: Using the invoice example, what do we do with invoices of product returned?

Socrates: Exactly why we need criteria or operational definitions.

Peismon: That's the finite, real, static population.

Mathitis: But if we can't sample or enumerate all the population, we use a subset, a frame.

Socrates: Deming speaks often of surveying. First, decide who or what to survey. Then, as Kokinos said, determine the classes, then classify each, and then count how many in each class.

Peismon: That only tells us what. It's not an analytic study. It's not what a control chart does.

Socrates: Let's understand the procedure first. We're only in the first stage, determining what an enumerative problem is. I think we have what is necessary: defining the population, identifying the frame as an accessible portion of the population and to which the inference will be made, defining classes and criteria for classifying, and establishing the method for enumerating.

Mathitis: That sounds right. I think we have covered the next characteristic, the population.

Population: Real, Finite, Static

Socrates: Perhaps not completely. Let's first look at Deming's examples.

Mathitis: He mentions surveys for census, the number of people who buy product A, the crop yield using fertilizer A, and the number of production units that are good.

Peismon: All these enumerative studies' examples have real, finite, static populations.

Socrates: Recall the two scenarios for the Inference question (Figure 19.3). We can ask about selecting from the coin's two sides or about the mechanism of tossing the coin to land on one of the sides. Is there a known finite, static universe in both cases?

Mathitis: In the first, yes. It's {head, tail}.

Kokinos: But in the second case, aren't the results a sample from an infinite, hypothetical, even dynamic, population? Box thinks so: "The total aggregate of observations that conceptually *might* occur as the result of performing a particular operation in a particular way is referred to as the *population* of observations [italics in original]" (Box et al. 1978, 27). If so, we are applying enumerative analyses to estimate the probability of heads. That contradicts the claim that it only applies to real, finite, static populations.

Peismon: It's finite, static, and real in both cases. We can only get a head or a tail.

Socrates: Then, doesn't the same reasoning apply to processes? If we are producing beads with different diameters, don't we know the list of choices but not the probabilities?

Peismon: How can you know the list of choices?

Mathitis: We know that pipe diameters must be greater than zero and less than a finite amount. So, the universe and frame might be all real numbers from 5 centimeters to 15 centimeters. People's characteristics—age, height, weight—are all within a real range.

Peismon: Fine. But it's still a finite, real, static population for enumerative studies.

Kokinos: Then, we *can* define real, finite, static populations for real-world processes. The probabilities may waver. Even then, we apply enumerative statistics, for example, coin flipping.

100% Sampling: Conclusive Answer to the Question "How Much/Many?"

Socrates: Can we get 100% sampling for the real cases like the coin or normal distribution?

Kokinos: No.

Peismon: But if you could, it would conclusively answer the enumerative question "How much/many?"

Mathitis: Can you even do that in theory, do 100% sampling of an infinite population?

Socrates: Let's say for the sake of argument that we can and that, therefore, Peismon is right. We would know exactly what the probability of heads is or was or will be. Let's see if that is consistent with other characteristics of the enumerative stages.

Peismon: This is the advantage of statistics in the enumerative case. We can sample randomly and make an inference using point estimates, confidence intervals, and even test hypotheses. It's valid for enumerative studies but not analytic studies.

Socrates: Hold that thought when we get to the analytic stages.

Prediction: To the Frame (Past/Present)

Mathitis: Once we have the enumeration, if it comes from a sample, then we make a prediction, or better, an inference to the frame.

Peismon: And, only to the frame. That frame is what has occurred or currently exists. It's about the past or present.

Kokinos: But I thought that in the coin example, we are not talking about the present. All gambling houses depend on the prediction being about the future. A roulette wheel with a zero and double zero favors the house by 20/38 versus 18/36 without the zero and double zero. The house is predicting that these odds will continue in the future sufficiently the same to favor it. Otherwise, the houses wouldn't be gambling or making money.

Peismon: But predicting to the future makes it an analytic study. First, we determine the results we got, which is the purpose of analytic studies.

Socrates: I am not sure you want to make such studies analytic.

Peismon: Why not?

Socrates: If knowing only the enumeration, that is, the proportion of heads and tails, is sufficient for me to have an understanding of the cause system, then there is no difference between enumerative and analytic studies. If I sample a coin by flipping it 100 times and see that heads occur 50% or 53% or whatever percentage, do I know why that percentage occurred?

Mathitis: I don't think you do. I guess I would not want to say I did.

Kokinos: It could be because the coin is fair or not, or it could be because of the method of flipping the coin including the person who flips it. Or, some combination.

Mathitis: Perhaps a simple test of whether we know why is to see whether we can predict the probabilities.

Socrates: But most importantly, if just knowing the enumeration is sufficient to be analytic by providing us the answer to "Why?," then the distinction is not between enumerative and analytic. It is between studies that answer "How much/many?" or "Why?" or both or neither.

Peismon: But the prediction is based solely on statistics. It is a statistical inference.

Socrates: Didn't we say that an inference can only be made to the frame from which the sample came?

Mathitis: Yes, that's what Deming says is the purpose of the frame.

Peismon: But the future is part of the frame when the population is the hypothetical, infinite one.

Mathitis: So you do agree that the population for an enumerative study can be hypothetical and infinite?

Socrates: Unfortunately, that's not enough. We didn't randomly sample from hypothetical, future results. We also don't know whether the conditions will stay the same in the future. Deming also said this statistical inference requires more than just statistics. We also need the assumption of conditions being sufficiently the same and perhaps other assumptions about laws of nature.

Mathitis: For example, if we use an ice cube as a die, the probability of getting any specific number will change as the ice cube melts. And statistics won't tell us whether it will melt in a way to retain equally the probability of each number occurring.

Action: Taken on the Frame

Peismon: Perhaps you are right. But the action is still the same. When done with the enumerative analysis, we take action on the frame and not on the cause system.

Socrates: What are possible actions?

Peismon: We can dispose of the frame, such as a contaminated batch. We might accept a lot if it meets an acceptable quality level. We may do 100% inspection to identify bad product to either scrap individually or rework.

Socrates: It appears that we need criteria to determine what action to take. For example, criteria to determine whether the batch is contaminated, the lot is acceptable, 100% inspection is needed.

Mathitis: I agree. Or, as in Deming's example of number of people in the area, criteria are needed to decide what to do with them.

Peismon: We do have criteria. For example, if the count or percent good is less than 99.9% good, then we scrap it; otherwise, we ship it.

Socrates: Does this summarize what we have concluded for the enumerative case (Table 21.2)?

Mathitis: It looks good to me. So let's do the same for analytic stages.

TABLE 21.2

Clarification of Requirements and Statistics for the Enumerative Case

Characteristic	Requirements	Statistics
Problem question: How much/many?	• Definition of units to include • Classification criteria • Classified units	MSA
Study: Population	• Real, finite, static	
Study: 100% Sampling	Conclusive answer to question	
Solution: Answer	Enumeration	Sampling theory, estimation (point and confidence intervals), hypothesis testing, MSA
Solution: Prediction	• Primarily to past/present frame	Statistical inference (e.g., confidence and prediction interval estimation) to frame
	• Future results	Conditional inference to future performance
Solution: Action	Action criteria to determine action to take on frame	Statistical risk analysis, estimation, hypothesis testing

Note: MSA, measurement system analysis.

ANALYTIC PROBLEM, STUDY, AND SOLUTION

Socrates: Do we agree on the problem: we don't know the answer to "Why?"

Question: Why?

Kokinos: I agree. Do we know what the answer looks like and what is needed to answer it?

Peismon: The answer is the causes of the enumeration you got when you answered "How much/many?"

Socrates: How do we know we have causes?

Peismon: For example, a manufacturer produces several lots. An enumerative problem would be determining how many defective units in a lot. That answers "How much/many?" An analytic problem would be determining how to reduce the number of defective units in future lots because you know why they are caused.

Mathitis: Then, the answer would be the factors that caused the defective units.

Socrates: Is that enough? Do you also need to know how to reduce their occurrence?

Kokinos: Yes. We may learn the factors that cause defective units but they may not be controllable or we don't know how to control them.

Mathitis: I agree. But we still understand the causes. Maybe there are two parts: knowing the causes of the current enumeration and knowing the causes of future, desired enumerations.

Socrates: Is knowing the factors enough? You learn that temperature is a critical factor for baking. Is that enough to answer "Why?" In other words, someone asks "Why is the meal burnt?" Do you know why if you can only answer "It's the temperature?"

Kokinos: No, you have to be able to also know the levels of those factors.

Peismon: I think you have to be able to show that you can produce the defects and prevent the defects by controlling the causes.

Population: Hypothetical, Infinite, Dynamic

Kokinos: Does that mean the population consists of the causes?

Mathitis: I don't think so. It still has to be the frame for the units we enumerated.

Socrates: Peismon, what is the population in the example of a lot with too many defective units?

Peismon: It's all possible lots. That's why it is hypothetical, infinite, and dynamic.

Kokinos: It's no different from the example of the coin. The choices or list is finite.

Peismon: In the analytic case, the population from the process is ever changing. The past was not what the present shows, and if the current process is out of control, then it will be different from future performances. That's the dynamic characteristic of the population for the analytic case.

Mathitis: But we already saw that we must distinguish between the list of possible results and the probabilities. The list is static, real, finite; only the probabilities waver.

Peismon: But when you sample, it is only from one period, so it cannot represent that entire universe or population. It is no longer random.

Mathitis: It's like organisms, like people. If we view our entire life as a single universe, how do we answer the question "How old are you?" or "How tall are you?" There is no single answer.

Socrates: There is one answer for each specified time. Is process control forever?

Kokinos: No, of course not. That's why we monitor even after getting it stable.

Socrates: We have data from time t_0 to time t_k, from a finite production. Our conclusion is not about the entire hypothetical, dynamic population or for all time. The conclusion about control is only for units from the period t_0 to t_k, not about units before t_0 or after t_k.

Kokinos: You're saying that questions about control charts also depend on when during the "life" of the process the control chart is monitoring we are asking about.

Mathitis: That's it.

Peismon: But we do make a prediction when it is in control, about times after t_k.

Socrates: True, but on the basis of the control chart and application of special or assignable cause rules, we say it *is* in control and not it *will be* in control. We really should say it *was* in control or, better yet, it *appeared* to be in control during the time t_0 to t_k.

The prediction of "it will be in control" depends on more than just whether the process was in control during the time t_0 to t_k. It depends on the critical process conditions not varying significantly from what they were during that time. We can sit back and passively hope that happens or we can actually control the process, not statistically, but, for example, having specifications on raw materials, machine settings, operating conditions, and actively ensuring they are met.

Peismon: I'm not following the argument.

Socrates: Regardless of why the process transitions from in-control to out-of-control states, we are trying to detect which state it is in during the period t_0 to t_k and take action when it moves out of in-control states to move it back to in-control states. That's the control chart's purpose—to identify the state it is in. The specific question for control charts is "What state was the process in during the period t_0 to t_k?" Our interest, enumeratively, is first to determine which state it is in. That's all the control chart does—nothing more.

Peismon: But why would Deming say control charts are analytic studies?

Socrates: I don't know.

Mathitis: Maybe, he said that because to understand is the ultimate goal. So, we should never stop at just enumerative studies. But as Shakespeare said about roses still being roses no matter what they are called, the control chart itself never answers "Why?" no matter what it is called.

100% Sampling: Inconclusive

Socrates: How does that affect whether 100% sampling is conclusive or not?

Peismon: Now, we want to understand the causes, so we are looking at the cause system. In the coin example, it was simply to enumerate. The statistics are different. Deming gave an example of the cause system having a certain proportion of defective units.

Socrates: That's the hypothetical, infinite, dynamic population?

Peismon: Yes. It produces various lots. We randomly sample from a selected lot. The sample is units from the single lot for the purpose of enumerating that one lot. Using all the units, we would know conclusively the proportion defective. The same sample in the analytic case serves to enumerate all possible lots or cause system (Deming 1953, 248). That's why statistics does not apply.

Socrates: That seems contradictory in two ways. The purpose can't be to enumerate the cause system because then it is just answering "How much/many?" for the cause system.

Peismon: That's what he writes and he does so to show that the sampling error is different for each. And, to me, it shows why the population is always hypothetical and dynamic.

Socrates: That's the other contradiction. If he can compute a sampling error for both, then statistics do apply. We can in fact make statistical inferences, for example, test hypotheses or calculate confidence intervals. We also showed that we do it for many other processes like coin tossing.

Peismon: Not if the population is dynamic.

Socrates: If the population is changing, then how can there be a definable sampling error, which requires a single sampling distribution?

Mathitis: But what about determining and understanding the cause system? This issue of sampling error is to answer "How much/many?" not for answering "Why?"

Socrates: Do we need to do 100% sampling to conclusively understand the cause system? To answer the question "Why this enumeration?" or "How can I have this other enumeration?"

Mathitis: No, I don't think so. We find the causes of enumerations without looking at all the data.

Kokinos: Sometimes, we find them through trial-and-error. Other times, we find them by doing hypothesis testing or correlation analyses to identify possible candidates and follow with controlled studies. Design of experiments (DOE) is the most efficient way, as they tell us not only the factors but also the levels that will produce optimal conditions. They tell us why and how.

Peismon: Then, it is true that 100% sampling is not conclusive.

Mathitis: True but less than 100% sampling *can* be conclusive. So the comment doesn't make sense. We never intend to sample the entire population because we can use scientific laws and subject matter expertise. If the lightbulb is burned out and each replacement burns out, I can find the short or why the fuse is overloaded without intending to look at 100% of hypothetical lightbulbs or an infinite number of them. And, I can confirm without 100% sampling.

Socrates: Then, statistics do apply to find the causes as when analyzing data from a DOE. Is the issue of whether 100% sampling is conclusive or not, applicable to analytic studies?

Peismon: No, except when you want to find the enumeration for the cause system.

Kokinos: But even that is not quite right. If we can do it for the coin, then we can do it for the production system, and if it's only enumerating, then it's not analytic, as Socrates said.

Prediction: Future

Socrates: What about predictions? Are analytic predictions solely about the future?

Peismon: Yes. Analytic studies tell you what the causes are, so that you can change the cause system to get the enumeration you desire. Thus, we predict future enumerations.

Kokinos: Again, how is that different from what we did with the coin?

Mathitis: And with the coin, we don't find the causes but make a conditional prediction: if the coin and the flipping method stay sufficiently the same, the enumeration of heads and tails will be the same in the future as the past and present.

Peismon: There is one critical difference. With the coin, you can only predict that the future will be the same as the past because you do not know what the causes are. You did not do the analytic study to find the causes.

Mathitis: Before, when we said the coin's population was hypothetical, infinite, and dynamic, you were arguing that it was an analytic study. Now you are arguing that it is enumerative because of the prediction to the future.

Kokinos: Maybe this gets at the issue of whether all these conditions are necessary to define enumerative and analytic.

Socrates: Peismon, I don't think you finished your point, which I think is a valuable one to understand the difference between the two predictions.

Peismon: That's right. As I was saying, with the coin you can only predict that the future will be like the past in the enumerative case of the coin. That's a limited prediction. While in the analytic case of lots and defective units, we can predict any enumeration we want because we can cause it. We know the causes.

Mathitis: Any enumeration?

Peismon: Maybe that's too strong but we can cause various enumerations so we can predict them. Another way of seeing the difference is what Deming said about what is needed for the prediction or inference. For the coin, only the enumeration is needed while for analytic predictions, enumeration plus other knowledge.

Action: Taken on the Cause System

Socrates: Good. So, this leads us to the actions for the analytic case.

Peismon: Clearly, it is to change the cause system. We do nothing with the output or the frame we sampled to get the enumeration for the enumerative case.

Mathitis: I like that distinction. It reminds me of corrective and preventive actions. Corrective actions are actions taken on the output or what we have called the frame. But not on the process that produced it, the cause system.

Kokinos: For example, accepting or rejecting a lot or doing 100% inspection, and so on?

Mathitis: That's right. Preventive actions are the reverse. They are actions taken on the process or cause system but not on the output or frame. Thus, preventive actions do nothing to the lot that has too many defective units. But because it changes the cause system, it creates a different quantity of defective units in future lots.

Peismon: It makes future lots have different enumerations—the desired enumerations.

Socrates: Let's summarize for the analytic case (Table 21.3).

Kokinos: We made some changes to what we thought initially.

Mathitis: I like the details.

Peismon: I'm still not sure about some entries. I see the issue with coins and gambling that we do make predictions about the future with possibly dynamic populations and we do have hypothetical, infinite populations in theory. But I still feel uncomfortable since it is the opposite of what I've read. Also, I don't agree with the 100% sampling being not applicable in the analytic case. Many experts make a big deal about it. I think it's a critical difference.

Mathitis: Don't forget to distinguish between the list of possible results and their probabilities. The population for a coin is finite: {Head, Tail}. Only the probabilities maybe dynamic, changing over time.

TABLE 21.3

Clarification of Requirements and Statistics for the Analytic Case

Characteristic	Requirements	Statistics
Problem question: Why?	• Factors to study • Levels of factors to study	
Study: Population	• Hypothetical, dynamic	
Study: 100% Sampling	Not applicable	
Solution: Answer	• Causal relationship between factors, levels, and enumerations	Hypothesis testing (correlation analyses), DOE
Solution: Prediction	• Primarily to past/present frame • Future	Confidence and prediction interval estimation
Solution: Action	• Enumeration for various factors and levels and combinations	Statistical risk analysis, estimation, hypothesis testing

Socrates: As we said earlier, this may not match everyone's ideas, but it is the basis for discussing the original question. Maybe in that discussion, we learn more and make some adjustments.

Mathitis: I agree. I have learned a lot from this. It gives me a different and deeper understanding of the issues.

Kokinos: Do we now return to whether control charts are analytic studies?

PROCEDURES FOR ENUMERATIVE AND ANALYTIC STUDIES

Socrates: Before we do that, let's use the information we have to create procedures for enumerative and analytic cases, from problem to study to solution.

Kokinos: We start with determining the requirements for enumeration: what things are in the frame, what are the classes, what criteria or rule is used to classify. These tell us what we will enumerate and how to answer the question "How much/many?"

Mathitis: Then, we simply enumerate. Of course, that could be 100% classification from a count, or sampling and then make an inference to the frame.

Kokinos: Then, using the classification criterion or rule, determine what action to take.

Socrates: What about the analytic case?

Peismon: Once we know what the enumeration is, then we do the study to determine why. That will involve determining what factors and at what levels produced the enumeration and what factors and at what levels to produce the desired or better enumeration.

Mathitis: So, the action requires having a desired or goal enumeration to trigger making the change that will produce that enumeration.

Socrates: Does this capture what all of you said?

1. What things (units) are in the frame? For example, units from the most recent lot.
2. What are the classes? For example, defective or not defective.
3. In what class is each unit? For example, evaluate a random sample of n, determine for each unit whether defective or not using specifications for the unit.
4. How much/many? For example, construct a 95% confidence interval estimate of the proportion defective for the entire lot.
5. What action should be taken? For example, if the entire confidence interval is below the acceptable quality level, do 100% inspection, scrapping defective units.
6. Why (this enumeration)? For example, investigate (root cause analysis) to learn why (factors and levels) the defective rate was greater than the acceptable level, for example, the temperature was below the specification limits of 275–285°.
7. What action should be taken on the cause system to change the enumeration? For example, change the process per the understanding of what, how, how much, when, and who, such as, change the temperature to 280° and add an alarm when the temperature drops below 275°.

Kokinos: That looks good.

ARE CONTROL CHARTS ENUMERATIVE OR ANALYTIC STUDIES?

Mathitis: Now we can look at control charts.

Socrates: Which of the characteristics of enumerative and analytic problems, studies, and statistics describe control charts?

Peismon: The analytic ones have to describe control charts per Deming and others.

Mathitis: If control charts are analytic studies, then they must provide the answer to "Why?" What is the answer?

Peismon: The answer is the type of cause: special cause variation as indicated by a point beyond the control limits or common cause variation if that doesn't occur.

Socrates: Let's do this more systematically by using the tables we generated. Peismon, you say control charts are analytic, so let's compare the list and see where they match.

Kokinos: I like that approach.

Peismon: First is the problem question, which is "Why?" I said that the answer is the type of cause: special cause variation or common cause variation.

Socrates: Now, we agreed that the answer to the "Why?" question is a list of factors and their levels. Are you saying that special cause variation and common cause variation are factors? And if so, what are the levels?

Mathitis: Isn't it the other way around? Aren't special cause and common cause the levels and the factor is variation (type)?

Socrates: Let's look at the control chart. It consists of points representing a statistic of a subgroup. These statistics are plotted on a timeline. The center line is the average of those statistics. Two other lines are the limits lower control limit (LCL) and upper control limit (UCL). Now, we said that we cannot answer "Why?" unless we know "How much/many?" Let's return to the enumeration then.

Kokinos: Are you asking "What are we enumerating?"

Socrates: That's right. We have this graph with points and lines. What do we need to answer "How much/many?"

Peismon: The units that make up the frame. That would be the points.

Mathitis: We need classification criteria that define the classes. I see. The classes are special cause patterns, or better, special cause variation and common cause variation.

Kokinos: But we don't know if the alarm is a special cause or common cause until we investigate.

Socrates: Then, aren't the two classes {alarm, no alarm}?

Kokinos: Yes, I think so. If we only use Shewhart's rule of a point beyond the 3 standard error limits as an alarm, then the classifications of each plotted point are alarm or no alarm.

Socrates: Isn't that true for all rules? If a pattern occurs it's an alarm; otherwise, it isn't. So there are only two choices {Alarm, Nonalarm}.

Mathitis: Then, alarm, special or assignable cause variation are not factors. That makes sense because when we do process improvement, we look for process factors. One tool often used is the fishbone or cause–effect diagram, which is a list of factors that are possible causes. Traditionally, these factors are grouped into six categories: people, methods, machines, environment, materials, and measurements. Sometimes, these are referred to as the 6 Ms when People is replaced by Manpower and Environment is replaced by Mother Nature.

Socrates: Now that we have the requirements to answer "How much/many,?" we need to enumerate. What do we need to do to answer the question?

Kokinos: In this case, simply count. I guess we count the number of alarms and nonalarms there are. There is nothing complicated about it if we only use the one type of alarm. If we had eight types of alarms, we could count how many of each type or the grand total of alarms of all types.

Mathitis: To take action, we need another criterion. In this case, the action criterion is if one or more alarms occur, investigate.

Peismon: That's how Wheeler describes the use of control charts. Here, let me draw a flowchart of the steps (Figure 21.1). The diamond at Step 2 is the action criterion of one or more alarms.

Socrates: Then, we have at that point finished the enumerative study, right?

Mathitis: Yes, we completed the requirements for the enumerative problem. The question might be stated more specifically as "How many alarms are there?" We then classified plotted points using the classification criterion, enumerated how many there were, and now used the action criterion to determine what action to take.

Socrates: Are we ready for the analytic case?

Kokinos: Yes. That would be Step 3b in the flowchart.

Socrates: Do you agree, Peismon?

Peismon: Yes, that is the next step.

Socrates: Will we use the control chart to do Steps 3b and 4?

Peismon: What do you mean? We can't get to these steps without the control chart.

Socrates: That may be. But now we are asking "Why were there alarms?"

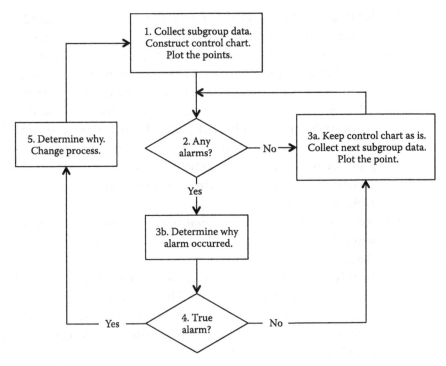

FIGURE 21.1
Decision flow for determining actions when using a control chart.

Peismon: And the control chart told us because of special cause variations.

Socrates: Does special cause variation mean something other than alarm?

Peismon: Special cause variations are the patterns, like a point beyond the 3 sigma limits or the other seven patterns we discussed.

Mathitis: I think I understand what Socrates is getting at. It's what Kokinos said. Those patterns are just alarms. Calling them special cause variations is simply saying they are alarms. Step 3b asks "True alarm?" Remember that there is always a probability of an alarm being false. Let's use Shewhart's term. Do the patterns indicate assignable causes?

Kokinos: No, because then he would not have had to do his studies, like the second part of the megohm study. The first part was seeing the alarms. Then, he investigated whether there were assignable causes. He found some, removed them, and constructed a new control chart. That one did not have alarms. But to show that lack of alarms indicated no assignable causes, he looked for them and found none.

Socrates: Thus, the classifications—to be clearer—should be as Kokinos stated: alarm versus no alarm. The next step is to check whether the alarm was a true alarm or a false alarm, i.e., were there assignable causes or not. This was the basis of Wheeler's simulations: to show that, for many distributions, the control charts had a low false alarm rate. It was not to—and critically, did not—show that the false alarm rate was zero.

Peismon: I agree but I don't see how that changes the fact that control charts are analytic studies.

Socrates: It changes in this way. If the alarms are false alarms, did the control chart tell us that?

Peismon: No.

Socrates: If the alarm was a true alarm, did the control chart tell us that?

Peismon: No, but the probability is very high that an alarm is a true alarm.

Mathitis: But not guaranteed. Therefore, the control chart doesn't tell us whether the alarm is true or false. We need to determine that.

Socrates: Once we know that it is a true alarm, what have we determined?

Peismon: That there are assignable causes. Hence, it's the analytic study.

Mathitis: Except the control chart did not tell us that.

Socrates: Recall that Shewhart said assignable causes were found for the megohm data that had eight alarms or points beyond the control chart limits. He did not say that he found the cause for each point. When we say that an alarm is a true alarm, what do we mean?

Kokinos: In general that the process is out of control. It changed.

Socrates: It was performing one way and then it performed a different way (true alarm), and may have returned to the first way if there are no more true alarms. Are there not two steps then in investigating alarms?

Mathitis: You mean first determining whether in fact there was a change in the process performance and then finding out what caused it?

Socrates: Exactly. A true alarm only means that the process performance changed. Finding an assignable cause is ambiguous. It could mean simply that we confirmed the process changed but don't know why. Or, it could mean that we found what caused the process to change. Therefore, it makes sense to separate the task into two stages: determining that the process changed and determining what made it change, irrespective of what these are called.

Mathitis: I accept that. But the flowchart does not make that distinction unless "True alarm?" means the process did change and Step 3 is finding why it changed.

Socrates: In Shewhart's case, we also have another issue that can occur. A control chart could have multiple alarms, only some are true alarms even if we find assignable causes. Do we need to confirm each and every point? When have we answered the analytic question "Why?": when we know that the process changed or when we know what caused the process to change?

Mathitis: Clearly, when we know what caused it to change.

Socrates: Then, only confirming that the process changed is not enough. After we have found the causes (assignable causes if that is what we mean), what do we do next?

Peismon: We determine how to change the process to reduce the true alarm rate for the future. We eliminate the assignable cause.

Socrates: That's one possible action. But in any case, to take the analytic action, we need to find the factors and their levels that caused the process change and alter the factors and levels to address the enumeration.

Mathitis: None of that is part of the control chart. As we said earlier, how we do that could be statistical tests or DOEs but not control charts.

Socrates: In other words, "How much/many?" is asked twice. Once for alarms and again for true alarms/nonalarms. The control chart only gets us to the first time the question is asked. In the flowchart, that is Step 2. Step 3b gets at the second question, whether it was a true alarm.

Mathitis: So, the analytic study only begins after Step 4. Control chart analysis cannot be an analytic study, according to our descriptions.

Peismon: This is a large study encompassing all the steps. That's why it is called an analytic study. It ends with an analytic analysis, conclusion, and action.

Mathitis: But it begins with an enumerative analysis, conclusion, and action, so why not call it an enumerative study? Even more reason to call it enumerative is that you need the enumerative study to do the analytic study but you don't need the analytic to do the enumerative study. And, the ultimate purpose is to get the desired enumeration.

Kokinos: But some say that a study cannot be both. Even Deming says that it is difficult to answer both questions with one study. In that case, there are two studies in the flowchart. The control chart only plays a role in the enumerative portion.

Socrates: It doesn't matter what it is called. The issue is what is done. That's why we went to defining the terms instead of debating what it is called. We've used the detailed descriptions and it seems that the flowchart of studies using control charts include both enumerative and analytic studies. Now that we have a procedure, we can see that the flowchart follows the procedure, confirming that the overall study has two or more parts depending on how many times "How much/many?" is answered.

Mathitis: And the part that includes the control chart is only the enumerative part.

Socrates: We can look at it another way. If analytic studies answer "Why,?" where is that answered in the flowchart?

Mathitis: At Step 5. Without control charts.

Socrates: There is one more thing about the flowchart. It is missing a crucial analysis.

Peismon: What is that?

Socrates: It is missing the question "True no alarm?" where no alarm means no process change.

Kokinos: Do we have to have two flowcharts then?

Socrates: Not if we modify it (Figure 21.2). We can see this point better if we changed the first decision to "Alarm or no alarm?" Just as we seek to confirm alarms, we now need to confirm no alarms or answer "True no alarm?" If the answer is no, then there really was a process change.

Kokinos: The next step is to answer "Why?" That's the same as with true alarms. So, false no alarms and true alarms result in analytic studies. I see now that the flowchart is incomplete.

Mathitis: Doesn't that lead to the discussion we had before? If the purpose of the control chart is to detect changes, we want to know when the silence (no alarms) is false (Myth 18). This flow chart misses the critical test for the purpose of detecting when there is lack of control.

Socrates: Exactly.

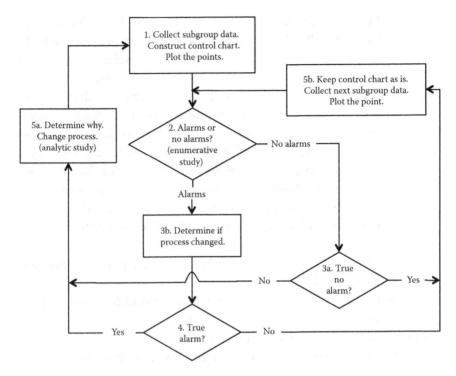

FIGURE 21.2
Modified decision flow for determining actions when using a control chart to include (1) check for both true alarms and true no alarms and (2) where enumerative and analytic studies are done.

CAUSE–EFFECT RELATIONSHIP

Mathitis: Wait! I see it now! "Why?" questions are addressing the cause part of a cause–effect relationship. The answer to "Why?" is a description of the cause. The answer to "How much/many?" describes the effect.

Socrates: Well done, Mathitis. What is the effect and what is the cause?

Mathitis: Any point beyond the control limits is the effect and what needs to be mitigated by the analytic action is the cause. Or, more generally, any alarm is an effect—all the rules of special cause patterns are effects. We seek to describe the effects first, which is the enumerative study where part of the description is enumerating the different classes of effects.

Kokinos: Then, the control chart cannot be an analytic study as it never tells us what needs to be mitigated or the cause. It only tells us about the effect.

Socrates: Even Shewhart in his 1939 book, makes this distinction:

> It is essential for an understanding of the operation of control that we distinguish three kinds of acts that are involved. These are (a) mental operations or judgments typical of which is the judgment that two or more observations are made under the same or different conditions, (b) mathematical operations such as are involved in constructing a criterion of control, and (c) physical operations such as looking for an assignable cause when an observed point fails to satisfy a criterion of control. (Shewhart, 25)

Kokinos: The third action tells me that the search for an assignable cause is different from determining whether a point falls beyond the control limits, which is a criterion of lack of control. The search is an analytic study and determining where the point falls or if it meets or fails a control criterion is an enumerative study.

AN ANALYTIC STUDY ANSWERS "WHEN?"

Peismon: But Deming also says "The methods of Shewhart control charts are essentially analytic, as they tell when to take action on the process" (Deming 1953, 247).

Socrates: Is he saying that it's analytic if it answers "Why?" and "When?" or either one?

Peismon: Both. To know when, you have to know why.

Socrates: A fire alarm goes off in this building. It tells you when to take action—and even what action to take: evacuate the building using specific routes. Do you know why?

Peismon: Because of a fire—it is a fire alarm.

Mathitis: But it could be a false alarm, so you wouldn't know why.

Kokinos: Or a fire drill.

Agrafos: Doesn't that mean that the alarm could signal three or four things and not just one? It could be a true alarm, a test of whether it works, a fire drill, or a false alarm.

Socrates: It may not matter what it is called. Was there a fire? But as Agrafos stated, when the alarm goes off, do we know which of these four options was the reason for the alarm?

Mathitis: No. Then, it's not true that you know why when you know when.

Peismon: But for control charts, you do know why when you know when. An out-of-control signal tells you to take action on the process. You know why—the process is out of control.

Mathitis: As with fire alarms, there are false out-of-control signals. If we don't know whether it is a true or false alarm, we can't know why. Yes, true alarms tell us that the process went out of control but we don't know why it went out of control. The alarm alone doesn't tell us if it's a true alarm.

Socrates: Remember the difference between corrective and preventive actions that Mathitis uses?

Peismon: Yes. The difference was *on what* was the action taken.

Socrates: If the information is used to act on the output of the process, it is enumerative; if the information is used to act on the process, it is analytic, right?

Mathitis: Yes, that's right.

Socrates: Does the information from a control chart alone tell you to act on the process or to act on the output?

Mathitis: It doesn't tell you either. The alarm tells you to act on the data. Investigate the points to reclassify them as either true alarms or false alarms.

Kokinos: Deming's statement is right only after an investigation that determines the alarm was a true alarm, followed by an investigation into why the process changed, then deciding on the enumeration desired, and finally a determination of what changes to make to the process to get the desired enumeration. After all that, we can now say we know when and why to take action on the process. So much more necessary information than what the control chart provides is needed to conclude that it is the control chart that gave you the answer to "Why?" or "When?"

22

Myth 22: Control Charts Are Not Tests of Hypotheses

Scene: *Project team meeting (day after control chart as analytic studies discussion)*

BACKGROUND

Socrates: Peismon, you had mentioned that control charts are not tests of hypotheses?

Peismon: That's right. I'm still thinking about the analytic study discussion we had.

Kokinos: Deming said "Rules for detection of special causes and for action on them are not tests of a hypothesis that a system is in a stable state" (Deming 1986, 335).

Mathitis: There is considerable controversy on the subject. Woodall provides a nice description of the controversy (Woodall 2000).

Socrates: Can you summarize some of the keys issues?

Peismon: The major issue for me is that when we are working with processes, we need to view a process as a cause system. Our objective is to understand that cause system so we can improve the quality of its output. That requires an analytic study. Statistical hypothesis testing doesn't apply to analytic studies.

Socrates: To what kind of studies do hypothesis tests apply?

Peismon: Enumerative studies. That distinction is one reason. The other is that a process is dynamic. As Deming said, there is no definable, finite population, and therefore, there is no sampling frame. No sampling frame, no hypothesis test.

Mathitis: But we still do statistical tests on the results of the process of flipping a coin. Also, we can define a finite list of possible results of real-world processes (Myth 21).

Peismon: But he and others also argue that using a statistical-based approach makes control charts less effective. Such an approach creates barriers to making the process better.

Socrates: How is it less effective? What barriers?

Kokinos: I'm not sure—I don't know that they explain.

Socrates: Is it how people respond to it or is it an inherent problem with treating it statistically?

Kokinos: It could be the first.

Mathitis: Yet, the other side of the controversy states that control charts have all the appearances and characteristics of hypothesis testing so why aren't they? In addition, some authors, including Juran, explicitly say they are. Others, like Woodall, and apparently Shewhart, argue that while the setup is like hypothesis testing, the requirement of randomness for statistical testing either does not occur or is highly suspect.

Peismon: I agree with that, which is just a different view of Deming's point.

Mathitis: However, Woodall thinks that the controversy is caused by a failure to distinguish between two phases of statistical process control. In Phase 1, control charts are used to determine whether the process is in control or stable. In Phase 2, control charts are used to determine whether an in-control process stays stable.

Kokinos: What difference does that make?

Mathitis: According to Woodall, in Phase 2, the stability of the process allows assumptions necessary for hypothesis testing but not in Phase 1.

DEFINITION AND STRUCTURE OF HYPOTHESIS TEST

Socrates: Even if Deming, Wheeler, and others are right about their reasons related to misuse or creating obstacles to using control charts, it doesn't necessarily mean that control charts are not tests of hypotheses.

Peismon: How could they be right and their conclusion wrong?

Socrates: Just because people misuse drugs or cars or tools, doesn't mean that those things are not drugs or cars or tools. The conclusion doesn't follow from their premises. But, it could be mere appearances that the conclusions are wrong and their premises right. We could mean different things. Let's start there. What is a hypothesis?

Kokinos: It could be an explanation of something.

Mathitis: It could be a guess or prediction. For example, that route A is faster than route B.

Peismon: It's a conjecture, a claim, or a theory.

Mathitis: Like your claim that control charts are not tests of hypotheses?

Peismon: (smiling) Yes—only that claim is true!

Socrates: Then, hypotheses can be false or true?

Mathitis: Yes. Hypotheses are statements about the world or reality. That is what a hypothesis test determines—whether the hypothesis is true or there is evidence for it.

Socrates: What kinds of tests are there?

Mathitis: All kinds. Medical tests, class tests, driving tests, tests of ability to use tools.

Socrates: Do they all test hypotheses?

Mathitis: A medical test can test the hypothesis that you have a disease or a broken bone. Class tests check the hypothesis that you know certain information or have certain skills.

Socrates: What about scientific hypotheses?

Kokinos: Those are conjectures about some physical-world phenomenon. About what causes a specific phenomenon.

Socrates: Are scientific hypotheses tested?

Kokinos: All the time. They are never proven, so they must be continually tested using the scientific method.

Socrates: What is the scientific method?

Kokinos: It's a procedure. Make an observation. Develop a hypothesis about why what you observed happened. Predict what the outcome would be if the conditions occurred again. Create or seek situations with those conditions. Check whether the prediction is true.

Socrates: It sounds exactly like the cause–effect relation we spoke of before. The hypothesis is a statement about relationship specifying both the cause and the effect.

Kokinos: That's true. But we don't prove that the scientific hypothesis is true.

Peismon: But with scientific hypotheses, you are only testing one hypothesis. When we do statistical hypothesis testing, the two hypotheses have to cover all possible options.

Kokinos: I was taught that one hypothesis is the opposite of the other, or its negation.

Socrates: They should be mutually exclusive and exhaustive. Do we have that with scientific testing of hypotheses?

Peismon: No. Kokinos mentioned testing Einstein's theory of relativity. When we test only that theory, either the results support it or they do not.

Kokinos: Right. There isn't necessarily a competing theory. We don't have to test it against something else, another theory. In science, theories are explanations of why things happen or how they happen. As Socrates said, they are about cause–effect relationships. We make a prediction assuming the theory is correct and check whether the prediction is correct.

Socrates: But we can state the hypotheses differently given that the evidence either supports the hypothesis or it doesn't. One hypothesis is the statement of the prediction and the competing hypothesis is that that statement is false, for example, gravity causes objects to fall versus it is not gravity that causes objects to fall, or simply, gravity does not cause objects to fall.

Mathitis: That's what we do with simple statistical tests, as Kokinos said. The mean equals 30 versus the mean does not equal 30, for example.

Kokinos: That's right. We have evidence either for it or against it. Even Einstein's theory of relativity is still tested even though, and as far as I know, it hasn't failed any tests. If I remember correctly, early tests of the general theory of relativity had to do with the Mercury's orbit and with the deflection of the sun's light. More recent ones have to do with black holes, I think.

Socrates: How are such tests conducted?

Kokinos: Einstein's theory was competing with Newtonian physics. Newton's theory predicted a certain amount of change called precession in Mercury's orbit that did not agree with the actual change. General theory of relativity predicted the correct amount as with the deflection of light. The then current theory predicted an incorrect amount.

Socrates: Do all tests involve collecting actual data and comparing to the predictions of the hypotheses? For example, when we debated the hypothesis that, for control charts, 3 standard error limits are the best economically, we looked at data from studies Shewhart did. We listed four tests for such studies, and he did two of the four on the megohm data.

Mathitis: No, you don't need actual data. You could do simulations, like Wheeler did to develop his Empirical Rule, which we might call a hypothesis.

Kokinos: Or, theoretically like the derivation of Tchebycheff's and Camp–Meidell's theorems.

Socrates: In all cases, is a test a procedure to verify that something is true?

Mathitis: Yes, it's a procedure with the purpose of checking or confirming or identifying.

Socrates: Do all hypothesis tests confirm?

Peismon: No. Medical tests don't always confirm that you are sick, have cancer for example. They have false-positive and false-negative rates.

Socrates: There are degrees of confirmation or verification.

Peismon: Definitely. Medical tests have different false-positive rates.

Kokinos: A test in a course doesn't always provide a good indication of what a person knows or can do. Some people have test phobias, or they may cheat.

Socrates: We can agree that a test is a procedure for ascertaining whether something is the case. The result of a test can range from proving a hypothesis to merely providing evidence for or against it to being inconclusive.

Mathitis: That sounds right. As we do in trials. We may have circumstantial evidence or we may have video showing the crime and confessions, or it may be a hung jury.

Socrates: Are these then not the elements of a hypothesis test: two hypotheses that exhaust all possibilities and are mutually exclusive, possible truths, a decision rule, and two types of errors?

Kokinos: We can create a table (Table 22.1) that's similar to the structure of statistical hypothesis testing (Table 10.1) we discussed before.

Peismon: Except there is one critical difference. The decision rule for statistical hypothesis testing requires calculating a probability from the data and comparing to the alpha risk or probability of a Type I error.

TABLE 22.1

Decision Structure for General Hypothesis Testing

	Decision	
Truth	**Accept Hypothesis**	**Reject Hypothesis**
Hypothesis true	Correct decision	Incorrect decision
Hypothesis false	Incorrect decision	Correct decision

Mathitis: That's true.

Socrates: But do all hypothesis tests require probabilities to make a decision?

Kokinos: Not with scientific hypotheses.

Mathitis: Nor other hypotheses. If I wanted to test the hypotheses about which route is faster, A or B, to get to an appointment, I wouldn't use probabilities but actual times.

CONTROL CHART AS A GENERAL HYPOTHESIS TEST

Socrates: Are control charts general tests of hypotheses? Let's compare their structure to that of general hypothesis tests.

Mathitis: We used this table (Table 22.2) when I was taught control charts. It's the decision table and what action we take as a result of our decision.

Socrates: Which features are common? Which are different?

Kokinos: They each have two rows listing the possible truths. The columns list the decisions, that is, which truth is accepted. The control chart table doesn't list errors.

TABLE 22.2

Control Chart Interpretation Table

	Interpretation of Variation	
True Variation	**In Control (Common Cause)**	**Not in Control (Special Cause)**
In control (common cause)	Stable—assess capability	Tampering— unnecessary action
Not in control (special cause)	Underreacting— failing to act	Unstable—find root causes

Mathitis: There are two errors: whenever the decision about what variation exists doesn't match True Variation. In the table, stable = in control and unstable = not in control. All the features of hypothesis testing are there.

Socrates: We are missing Shewhart's decision rule. Let's add that to the table in this manner (Table 22.3).

Mathitis: Now we have two possible truths, a decision rule, two possible decisions, two errors, and two correct decisions. It's exactly the same structure as general hypothesis testing. In fact, it is the same structure as statistical hypothesis testing we did before (Table 10.1).

Peismon: Except that the decision rule for statistical hypothesis testing requires a probability, the *p* value to compare to alpha. With control charts, we don't use probabilities.

Socrates: Before we get to that point, do the two choices cover all possibilities, as in general hypothesis testing?

Mathitis: Yes, as there are only two types of causes, common or special—or, assignable or not assignable. Each defines whether the process is stable or not. Whoever denies that control charts are not tests of hypotheses would have to show that either "process in control" or its negation are not hypotheses or that the control chart itself is not used to make a decision.

Kokinos: Or that errors cannot be made.

Socrates: Is that necessary? There could be a test with 100% accuracy for either hypothesis. For example, if we only have two choices of jars and one consists of all red marbles and the other of all green, a test of the color of only marble tells us which it came from.

Peismon: It still doesn't tell us that control charts are tests of statistical hypotheses.

TABLE 22.3

Control Chart with Errors Based on Decision Rule of Point beyond the Limits

True Cause	Decision (Interpretation of Points)	
	In Control (Common Cause)	**Not in Control (Special Cause)**
In control (common cause)	Correct decision (within limits)	Incorrect decision (beyond limits)
Not in control (special cause)	Incorrect decision (within limits)	Correct decision (beyond limits)

Socrates: That's true. But it does tell us that control charts are general tests of hypotheses.

Mathitis: I see your point. The controversy is whether control charts are tests of hypotheses. But even in Woodall's review, the description is "hypothesis test" or "test of hypotheses" not "statistical hypothesis test."

Socrates: Or, even "test of statistical hypotheses."

Agrafos: What is the difference between test of statistical hypotheses and statistical tests of hypotheses?

Kokinos: Recall the difference between practical problem and statistical problem. The latter would use a statistical test. But the former need not. For example, whether route A is faster than route B is not a statistical hypothesis or problem but may be tested using statistics.

Mathitis: Statistical hypotheses are on parameters of probability distributions. Means, variances, ranges, and so on are parameters or characteristics of probability distributions. We test whether the mean is a particular value or equal to one or more other means. Do these populations have the same standard deviation? Are the proportions the same?

Kokinos: What about a test for normality?

Mathitis: That's also a hypothesis about a probability distribution. So, yes, about probability distribution parameters, characteristics, or shape.

Kokinos: You are saying that it isn't clear which the controversy is about?

Mathitis: Although, on occasion, Deming and Juran have said "test of significance." I think that Juran views control charts as continual "tests of significance."

Peismon: But Deming said no test of significance applies to control charts. I think it's also clear that while he and others do not say tests of statistical hypotheses or statistical tests of hypotheses, they are referring at least to the latter.

Kokinos: But which one, if there is a difference?

Peismon: The one that uses probabilities to make decisions. They are saying that the control chart limits are not probability limits so the charts can't be statistical tests of hypotheses.

Socrates: Can we agree then that control charts are tests of hypotheses?

Mathitis: I think we have shown that control charts are tests of hypotheses. The hypotheses are the process was stable versus the process was not stable.

Peismon: Then, I say that Deming meant that control charts are not *statistical* hypothesis tests.

Socrates: While Deming wasn't specific, Box and Kramer clearly refer to *statistical* testing: "process monitoring resembles a system of continuous statistical hypothesis testing" (Box and Kramer 1992).

Peismon: I can accept that it is nonstatistical hypothesis testing. Shewhart developed the limits to be economically based. You can tell from the title of his book: *Economic Control of Quality of Manufactured Product.*

Mathitis: On the other hand, his follow-up book has the title *Statistical Method from the View of Quality Control.*

STATISTICAL HYPOTHESIS TESTING: ALPHA AND *p*

Socrates: Do you recall our discussion about statistical inferences?

Mathitis: Yes. We concluded that a statistical inference required random sampling and an assumed probability distribution so that the conclusion was either a probability or probability-based, like a confidence level. Those are statistical tests—even on statistical hypotheses.

Peismon: Control charts can be tests on statistical hypotheses, for example, without being statistical tests.

Agrafos: How is that possible?

Socrates: What do we plot on a control chart? What kind of control charts are there?

Kokinos: $\bar{X}$ chart for means, S chart for standard deviations, c chart for counts, and so on.

Mathitis: Even the I chart on individual values could be viewed as a chart on the distribution.

Socrates: When we say that the process is in control using an $\bar{X}$ chart, what do we mean?

Mathitis: That the mean is constant. There is one distribution. But we are looking at the means of multiple samples or subgroups.

Socrates: When we test whether the process is stable, we test whether there is evidence that the plotted statistics came from the same distribution. We have at least tests on statistical hypotheses.

Peismon: But, to be a statistical test of hypotheses, or statistical inference, we need to have probabilities. Where is alpha and p using a control chart? They have to come from an assumed probability distribution. What is that probability distribution?

Socrates: What statistical hypothesis test tests the equality of means from different populations?

Mathitis: The analysis of variance.

Kokinos: Or various nonparametric tests like Wilcoxon Rank or Mood's Median tests.

Socrates: Are these statistical hypothesis tests?

Kokinos: Of course.

Socrates: Why are we saying that they are statistical hypothesis tests?

Peismon: We use probabilities and statistics. As I said, we compare a probability p to a predetermined alpha. The difference is that, in statistical hypothesis testing, we decide on alpha and then determine p based on the null hypothesis. Control charts seem to be set up the opposite way, in reverse and incomplete. We determine the limits at ±3 standard errors. Is this p or alpha? Whichever it is, where is the other?

Socrates: In statistical hypothesis testing, we decide on alpha before testing. When constructing control charts, what do we decide before collecting data?

Peismon: We construct the chart after we collect data. Everything is decided after that.

Socrates: Is everything decided after that? How do we know what calculations to make?

Mathitis: We choose the default limits at ±3 standard errors. Those are the formulas we use, so we are choosing the limits before collecting data. So they must represent alpha.

Kokinos: Of course! They are the false alarm rates when we assume the control chart is stable—which is the null hypothesis. That's what the Empirical Rule defines: alpha.

Socrates: In statistical hypothesis testing, p comes from the data assuming the null hypothesis is true. In control charts, the null hypothesis is that there is one distribution. What represents p?

Mathitis: It must be the points because that is what we compare to the control limits.

Peismon: Only for the alarm of a point beyond the control limits. What about other alarms?

Mathitis: That's easier. Recall that you said the other alarms were designed to have approximately the same false alarm rate. Hence, if the pattern exists—we get an alarm—then it is equivalent to $p <$ alpha, whatever those values are.

Socrates: We showed you before that we can reverse engineer to determine what probability limits the 3 standard error limits equal assuming a probability distribution. Just as we did with the Poisson distribution for the c chart (Myth 17). Let's assume a normal distribution, say for the $\bar{X}$ chart. We could work backward to determine what alpha corresponds to 3 standard errors based on the sampling distribution of $\bar{X}$ to determine the sampling error.

Peismon: But many people have shown that a normal distribution is not required for control charts. They were wrong to use the normal distribution to explain control charts and to use those probabilities.

Agrafos: But even Wheeler said it was applicable to the $\bar{X}$ chart because of the central limit theorem although false about other charts.

Socrates: Exactly. Such graphs are used to explain the general concept of control charts but they give the impression that it applies to all control charts. Continuous probability distributions do not strictly apply to discrete measures, so they may not be good models to use for, say, the c chart on counts. Yet, we can assume other probability distributions to determine alpha and p. For example, what distribution might we use for testing a mean?

Kokinos: The t distribution, and we showed that we can determine the probability for 3 standard error limits or create probability limits, as Wheeler did for the c chart.

Mathitis: It just occurred to me that the analysis you did assumes there is only one sample. With the $\bar{X}$ chart, we are checking the means of the subgroups.

Socrates: Yes, you're right.

Kokinos: How do we test multiple means?

ANALYSIS OF MEANS

Socrates: Analysis of means (ANOM) (Ott 1967).
Kokinos: We didn't cover that in any of my classes.

Mathitis: The ANOM is a graphical test that looks like an $\bar{X}$ control chart. It has the means of the samples plotted as points with the center line the average of the means, and two limits that represent the decision limits.

Kokinos: You mean like alpha?

Mathitis: Yes, they are the distribution's critical values for the confidence level chosen. These limits represent the ends of a 100(1 – alpha)% confidence interval assuming the null hypothesis that each population mean equals the grand mean.

Socrates: Let's take a look at an example. Kokinos, can you do a search for ANOM and Ott?

Kokinos: Here is a list.

Socrates: Okay, here is one. It mentions Ott who developed the method. This example consists of six samples or subgroups, each of size 7. Can you calculate the center line and control limits?

Kokinos: Yes. The grand mean = 15.952, the control limits are upper confidence limit (UCL) = 16.854, lower confidence limit (LCL) = 15.051.

Socrates: Is there a point outside the limits?

Kokinos: Yes, the sixth subgroup is below the LCL.

Socrates: Using ANOM, what is the lower 95% confidence limit?

Kokinos: It's 15.147.

Mathitis: That's slightly more than the LCL, so the limits are slightly narrower, but not by much.

Socrates: Is the conclusion the same? Does that sixth point fall below the LCL?

Kokinos: Yes, it does.

Socrates: With a full table of critical values for an ANOM test, we could determine the confidence level for the UCL to LCL range. It appears that it is slightly more than 95%.

Kokinos: This takes into account both the subgroup size and the number of subgroups, getting at Mathitis' concerns.

Mathitis: I think so.

Kokinos: Does that mean when we find the confidence level for this number of subgroups and subgroup size, we can redraw the chart so it is indistinguishable from the control chart?

Socrates: Yes.

Peismon: But that doesn't make it a control chart.

Socrates: Consider this, Peismon. I show you a chart that looks just like this and ask you "Is this control chart a statistical test of hypotheses?"

Peismon: I would say no.

Socrates: But if I then said, "Oh, my mistake. This chart is actually an ANOM chart. Is it now a statistical test of hypotheses?"

Peismon: If it's an ANOM chart, then yes. But not if it is a control chart.

Kokinos: How can the same graph be a statistical test and not be a statistical test simultaneously?

Mathitis: Well, you do get to the limits in different ways. But in the end, it is the same graph.

Socrates: Are the hypotheses the same, regardless of which way I determine the limits?

Mathitis: Generically, they aren't the same. The control chart is testing for stability or control.

Socrates: But what is the specific null hypothesis for ANOM?

Mathitis: All subgroup means are equal to the grand mean. In other words, all the data came from the same population or distribution.

Socrates: Is it the same hypothesis for control charts? Does the construction of the $\bar{X}$ chart assume that all subgroup means have the same mean as the mean of all the subgroup means?

Peismon: The control chart test for means is done by checking whether there is only common cause variation or that there is no special cause variation. That's not testing for equal means.

Socrates: How are those control limits determined?

Kokinos: It assumes that there is one distribution, like ANOM, calculated from the grand mean. Both charts use the grand mean for the center line.

Mathitis: And the limits are set at some delta from the center line that represents within subgroup or sample variation.

Mathitis: So, yes, the null hypothesis to construct the limits for both charts is the same.

Socrates: With one exception.

Peismon: What's that?

Mathitis: My point that the control chart limits use a predetermined multiple of a standard error that doesn't take into account sampling error.

Socrates: But by assuming a probability distribution for that common mean, we can determine what confidence level the control limits represent.

Mathitis: I'm inclined to believe you. In practice, every $\bar{X}$ chart is an ANOM chart with an unknown but determinable alpha level when a probability distribution is assumed. All you do is change what you call it.

Socrates: I guess people should read more Shakespeare. "What's in a name? A rose by any other name is just...."

Kokinos: The chart is the same regardless of the name.

Socrates: What is one special cause variation?

Kokinos: At least one point beyond the control limits.

Socrates: What does that mean in each case, for each chart?

Kokinos: That the assumption of equal means or sampling from the same distribution is false.

Agrafos: But what about the other charts? Can we still say that an S chart, for example, is not a statistical test of hypotheses?

Socrates: What do you think, Mathitis?

Mathitis: If all we are doing is finding a sampling distribution for the statistic that we chart, then it seems that they would all be statistical tests. We may just not know at what confidence level each is being tested.

Socrates: For the S chart, we can use several tests for equal variances from two or more populations. For example, we could use Bartlett's test and reverse engineer to determine alpha.

Mathitis: That test is sensitive to the assumption of normality. Maybe Levene's test or the Brown–Forsythe test.

Kokinos: What about proportions?

Mathitis: I suppose, to test two or more proportions, you can use the concept of contingency tables. That uses the chi-square distribution using the observed counts versus the expected counts, where the expected counts are determined assuming that all the proportions are equal.

Mathitis: We saw that Wheeler used a Poisson distribution to create probability limits and therefore we can do hypothesis testing.

Socrates: Or, ANOM, since proportions are means.

Mathitis: That's true and the p chart uses a normal distribution approximation.

Socrates: The point is that the only thing lacking for control charts to be statistical tests of hypotheses is the assumption of a probability distribution. Nothing prevents me from making that assumption and calculating the corresponding confidence level for the control limits. Whether people don't want to do it is a separate issue.

Peismon: But you won't be able to do that for other special cause variation patterns.

Socrates: Let's see. Pick one of the other ones.

Peismon: What about 2 out of 3 consecutive points beyond the 2 standard error limits?

Socrates: Do you agree that we can use the ANOM table to determine the alpha that corresponds to those limits—just as we did for 3 standard error limits?

Peismon: Yes.

Mathitis: Then it can be done. Because now, it's like the probability of getting 2 out of 3 heads with that calculated alpha being the probability of a head. That determines your p, which can then be compared to any alpha you select before collecting the data.

Kokinos: It looks like they all have calculable p values if we assume a probability distribution.

Peismon: That's the whole point. You can assume a probability distribution but how likely is it to be correct?

Socrates: When Wheeler recommended using probability limits for c and u charts, was that an issue? When we do statistical tests on means or variances, is that an issue?

Peismon: Yes. That's why we test assumptions.

Mathitis: So we can do the same here.

———

SHEWHART'S VIEW ON CONTROL CHARTS AS TESTS OF HYPOTHESES

Socrates: We concluded that control limits could be probability limits. Doesn't Shewhart view them as such in the sense that there is a calculable probability associated with the limits? Did he think of control charts as statistical tests of hypotheses?

Mathitis: Woodall puts him in the middle. He quotes Shewhart as distinguishing between formal testing of statistical hypotheses versus empirical testing of hypotheses.

Socrates: Shewhart does recognize different kinds of tests of hypotheses.

Kokinos: I don't see in the index of his 1931 book anything on hypothesis testing but I do see the topic of detection of lack of control from standard quality. This is at the end of Chapter XIX "Detection of Lack of Control"; he states, "For such an engineer the statistical tests described in this chapter constitute a powerful tool in testing his hypotheses..." (Shewhart 1980, 299). He ends the chapter with "...statistical tests often indicate the presence of trouble. Of course, these advantages are attained with a knowledge that we shall not look for trouble when it does exist more than a certain known fraction $(1 - P)$ of the total number of times that a sample of size n is observed" (Shewhart 1980, 300).

Mathitis: That seems to support both that he viewed control limits as probability limits and control charts as tests of hypotheses.

Kokinos: In his 1939 book, he is even more explicit. First, he states that there are two kinds of errors in scientific testing: wrongly rejecting a hypothesis and wrongly accepting a hypothesis. He then states that the Neyman–Pearson (Neyman and Pearson 1928) approach to hypothesis testing is a special case of the general two-error framework. Third, he then gives examples of what the corresponding errors might be for control chart decisions:

e_{13} We may reject the hypothesis that the production process or repetitive operation is in a state of statistical control when this hypothesis is nevertheless true.

e_{23} We may accept the hypothesis that the production process or repetitive operation is in a state of statistical control when this hypothesis is nevertheless false. (Shewhart, 40)

Mathitis: It sounds like he viewed control charts as tests of hypotheses.

Kokinos: Yes and no. He ends this section with the statement "it would seem that *we must continually keep in mind the fundamental differences between the formal theory of testing statistical hypotheses and the empirical testing of hypotheses employed in the operation of statistical control* [italics in the original]" (Shewhart, 40). That difference is that for the general errors for scientific testing "we can no longer compute with mathematical exactness the probabilities associated with any pair of errors for a given hypothesis" (Shewhart, 40).

Peismon: That is what I have been saying.

Socrates: Not being able to compute an exact p value doesn't prevent the test from being a statistical test of hypotheses. Sometimes, the equation for calculating the standard error is intractable, so approximations are used. If statistical tests can and do use approximations, for example, tests on proportions or tests on means with unequal variances, why not control charts? Since few if any real distributions match the theoretical models, isn't it true that for most if not all statistical tests of hypotheses, we can't calculate the exact or true probability?

Kokinos: I don't understand the difference.

Socrates: Even if I can calculate the exact probability when the distribution is normal, that doesn't make the calculation the exact probability for that specific application. Real distributions are not normal. They are all discrete and therefore can't be normal. Using a normal distribution to calculate the probability can only be an approximation.

Mathitis: I see what you are saying. All real distributions are discrete. Any calculation based on a continuous probability distribution can be an exact calculation assuming that continuous distribution but must be an approximation for the real probability.

Socrates: These assumed probability distributions are models and probability is a theory applied to real-world phenomena. One might question whether probabilities actually exist, or are they theoretical constructs. Where calculation is impossible, there are conservative approximations. Wheeler did this with his analyses of different distributions. He's shown that the worst-case scenario might be a probability of 1% that a stable process will exceed a 3 standard error limit.

Peismon: But Woodall thinks that Shewhart is on the fence on this issue because of the statement Kokinos read.

Kokinos: Perhaps Peismon is right. While Shewhart started with general error statements related to scientific hypotheses, he said that the Neyman–Pearson approach was a special case using statistics.

Peismon: He also said that there was this difference between the statistical test of Neyman–Pearson and the test of a control chart.

Socrates: That difference was only that we needed to remember that the condition of randomness must be scrutinized. However, Shewhart also continues using the term *testing hypotheses*, stating that that is what is done in the use of control charts.

Peismon: How does he say they are used?

Socrates: He states that the operation of control has three steps: specification, production, and judgment of quality. After arguing that these steps are cyclical, he states, "It may be helpful to think of the three steps in the mass production process as steps in the scientific method. In this sense, specification, production, and

336 • *Understanding Statistics and Statistical Myths*

inspection correspond respectively to making a hypothesis, carrying out an experiment, and testing the hypothesis. The three steps constitute a dynamic scientific process of acquiring knowledge" (Shewhart, 44–45).

Peismon: He isn't saying he is doing statistical testing. He seems to be saying that it is very similar to statistical testing.

Kokinos: Could Shewhart think they are tests of hypotheses but not statistical tests of hypotheses?

Peismon: I accept that, as I think did Deming and others.

DEMING'S ARGUMENT: NO DEFINABLE, FINITE, STATIC POPULATION

Socrates: Earlier, you said that Deming and others said a control chart isn't a statistical hypothesis test because, for that, you need a definable population. Since all real-world processes waver, there is no definable frame or universe.

Peismon: No probability distribution, no probabilities, no *p* values.

Agrafos: But if processes always waver, doesn't that contradict that a process can ever be stable?

Mathitis: That's what Woodall noted also. Why test whether they are stable? We know the answer.

Kokinos: If all processes are unstable, then the truth-decision table (Table 10.1) only has one row.

Peismon: The process is wavering except when in control, stable.

Socrates: We already showed (Myth 21) that for all real-world processes, we can define a finite, static population and frame—the list of possible results. It is their probabilities that may waver.

Peismon: Even if that is the case, it's enough to prevent hypothesis testing.

Kokinos: Then, why doesn't it prevent us from doing it with the coin or other processes?

Mathitis: Or, prevent Wheeler from using a Poisson distribution to construct the limits? Or, prevent Shewhart from using the normal distribution to construct limits for $\bar{X}$ and S charts?

Socrates: Recall that we had two scenarios for statistical inferences (Figure 19.3). In one, the interest was solely in the universe, while in the other, the interest was in the process of randomization.

Peismon: Yes. We used a coin and the bowl of beads to illustrate these scenarios.

Socrates: The process of tossing the coin is the real-world mechanism that generates results, which can be modeled using theoretical probability distributions. With the coin, we know the universe {head, tail}, but what don't we know?

Mathitis: We don't know the probabilities. So we know one component of a probability distribution. Even if they waver, we still estimate the probabilities by tossing the coin to get a sample by assuming they are fixed and making predictions. As with all gambling processes.

Socrates: Which case is this: assuming a random selection process to learn about an unknown universe or checking whether the selection process from a known universe is random?

Peismon: The former.

Socrates: What is required for the inference to be valid?

Mathitis: The sample has to be random.

Socrates: Are not randomization processes also processes? Do they not also waver?

Mathitis: Yes. So, our calculations, estimates, tests of the probability of heads using the randomization process of flipping a coin are suspect. We don't even check if it is wavering.

Kokinos: Are you saying that whenever we assume we have randomly selected samples, we must check that the process of randomization is stable?

Socrates: If we assume all processes waver—until confirmed to be stable— we should.

Kokinos: But then, we should continually check that it stays stable.

Agrafos: That's a lot of work. No one does that, do they?

Mathitis: Not that I know of. Especially for doing Design of Experiments (DOEs) where even a simple three-factor study with two levels each has eight combinations. That would mean checking that the processes for each combination are stable during the entire study and that the randomization process is stable.

Peismon: But there are other reasons against control charts being statistical hypothesis tests.

Mathitis: Woodall raises concerns about independence in his article.

Socrates: Do you know that the results of flipping a coin are independent or do you assume that?

Peismon: We assume they are.

Socrates: And if they are false?

Mathitis: The test may produce the wrong decision. We saw this before with statistical tests when we discussed whether to use parametric or nonparametric tests.

Socrates: We assumed that the results were random and independent although they may not be. We assumed a particular distribution even though it may not be exactly that. But we don't say "Stop! Never do statistical tests because of these reasons." What do we say instead?

Mathitis: Know the assumptions, check that they are tenable, know how robust the test is to the assumptions, look for other ways to accomplish the same thing if the assumptions are violated.

Socrates: Can we do the same with control charts?

Mathitis: Yes.

Kokinos: Not just for the coin but in many other cases. We do it for reliability of products and products that are themselves processes. For example, your company produces washing machines. You test one washing machine by washing several loads but are making an inference to a hypothetical population of possible washes. There are many products that are intended to be used over and over again. They all have hypothetical, wavering populations.

Socrates: Perhaps not the population since that is the list of possible results. But certainly the probabilities, and therefore, the probability distribution, may be changing and could be dynamic.

Mathitis: These products are themselves processes and that's why they have this characteristic. The process of washing, the processes of medical testing, the process of making a call, and so on.

Socrates: Yet, for each, we make an inference to a hypothetical, wavering population. That's what reliability testing is about. We accept or reject and quantify probabilistically washability, durability, reliability, accuracy, and so on using statistical hypothesis tests. Do people say "I'm sorry but statistical testing is not applicable because you have a wavering, hypothetical distribution. What is your frame? Do you have a random sample from that frame?"

Kokinos: Doesn't Box's comment about models apply? They are all wrong, but some are useful. Applying probability models helps us understand the coin and the process of flipping.

Mathitis: Also, Deming calculated sampling errors for analytical studies. If we have sampling errors, we have sampling distributions. That means we can calculate probabilities.

Socrates: If Deming and others are right, we are in a strange situation where we can apply statistical hypothesis tests to hypothetical, infinite populations if we don't put the results on a control chart but not if we do and they show instability. Remember these are models, tools.

Agrafos: You mean like why is it a statistical test using ANOM but not using a control chart if the exact same calculations, chart, and interpretation apply to both.

Kokinos: I agree. Why are we asking more of control charts than of other applications of probability theory? Why can't we do the same with control charts?

Mathitis: I think we can.

WOODALL'S TWO PHASES OF CONTROL CHART USE

Peismon: I think the mistake you are making is the one Woodall discusses in his review of the controversy.

Socrates: What mistake is that?

Peismon: There are two phases in the use of control charts: first, to check whether the process in the past was in control; second, to check whether it continues to be in control once control has been established.

Kokinos: Why does that make a difference?

Peismon: In the first phase, we do not know what the probability distribution is because we have not established or confirmed control. This is a retrospective study. It uses historical data to set the limits and see whether the process is or is not in control.

Mathitis: But construction of the control chart is based on the null hypothesis that the process is stable, that all the points plotted come from a single distribution.

Kokinos: What about the second phase?

Peismon: In the second phase, because the process is in control, we have evidence of what that probability distribution is and so we can calculate probabilities. This is a prospective study and use of the control chart. Each new point tells us whether the process continues to be in control.

Socrates: In either phase, are any points plotted on a control chart from the future?

Peismon: No, that's impossible.

Socrates: Then, all points on a control chart—all data for that matter—describe the past. Even in Phase 2, no future data are plotted on a control chart. All control charts use historical data and all decisions are about whether it *was* in control. We make a subtle distinction that, in Phase 1, we are asking if the process was in control and, in Phase 2, was the process still in control, but both amount to answering "*Was* the process in control from t_{start} to t_{end}?" Both characterize the past.

Kokinos: Then, what is the difference?

Mathitis: In what t_{start} and t_{end} represent. In Phase 1, $t_{start} = t_1$, the time the first datum occurs and $t_{end} = t_k$, the time the last datum occurs. In Phase 2, $t_{start} = t_k$ is the last point on the chart showing stability and $t_{end} = t_{k+1}$ is the next point plotted. Phase 1 answers "Was the process in control from t_1 to t_k?" Phase 2 answers "Was the process still in control at t_{k+1}?"

Peismon: But that is the distinction that Woodall wants to make. While subtle, it makes a difference as to what you can do statistically.

Socrates: But Phase 2 really answers "Was the process in control from t_k to t_{k+1}?" Now, you see there is no substantive difference. There are other reasons why probability distributions can be applied. Did Wheeler not show that a *c* chart could be constructed with probability limits assuming a probability distribution and that those limits could be constructed to equal the same 3 standard error limits without the assumption?

Mathitis: Yes he did. And we showed that we could do it for any control chart.

Kokinos: That completely counters the argument that it can't be done.

Mathitis: What is the second reason?

Socrates: I agree with Woodall's distinction. He just didn't take it far enough.

Peismon: Why?

Socrates: As Woodall said, the transition from his Phase 1 to Phase 2 requires much work. What is that work?

Peismon: If the process is unstable, then we need to find out why and make a change to the process based on that answer.

Socrates: Then that work *is* the analytic study as it answers "Why?" Look at the modified flowchart (Figure 21.2) to see that Woodall's first phase represents exactly what we talked about before: an enumerative phase followed by an analytic phase. Let's call the enumerative study Phase 1a and the analytic study Phase 1b. His Phase 2 occurs in two ways.

1. Phase 1a: At Step 2, answer yes to "Any alarms?" with the data from t_1 to t_k. Phase 1b: At Step 3b, determine whether the process has changed based on an analytic study and answer no at Step 4 "True alarm?" Phase 2: At Step 5b, we keep the control chart as is, collect the data for the next point t_{k+1}, and check for alarms again at Step 2.

2. Phase 1a: At Step 2, answer no to "Any alarms?" using the data from t_1 to t_k. Phase 1b: At Step 3a, answer yes to "True no alarm?" presumably based on an analytic study (although, in the original flowchart, verifying no alarms was not done). Phase 2: At Step 5b, keep the control chart as is, collect data for the point t_{k+1}, and check for alarms again at Step 2.

There are two other cases resulting in true instability or true alarms followed by modifying the process until the process is under control during time t_1 to t_k.

Mathitis: Then, the purpose of Woodall's Phases 1a and 2 are to check stability: in Phase 1a whether stable and in Phase 2 if still stable. The purpose of the transition phase, Phase 1b, is to determine and remove the causes of instability or to identify and include the causes of stability. That transition can occur at Step 5a or 5b depending on whether there are assignable causes.

Kokinos: To understand the causes is the purpose of analytic studies.

Socrates: Exactly. One transition phase is the analytic study at Step 5a. The other two studies are not. They are enumerative studies—or neither.

Peismon: Woodall says that, in Phase 1, we can't make distributional assumptions: "In practical applications of control charts in Phase 1, however, no such assumptions are or can be made initially and the control chart more closely resembles a tool of exploratory data analysis" (Woodall 2000, 343).

Socrates: Woodall is talking about Phase 1a. Have we not shown that we can and do make assumptions all the time—nothing prevents anyone from making assumptions? He can't mean that it is physically impossible.

Peismon: I think he meant *should* rather *can* not. But also he says the assumptions are untenable.

Mathitis: Why is that?

Peismon: Because later in the article, he adds, "In fact, distributional assumptions cannot even be checked before a control chart is initially applied in a Phase 1 situation because one may not have process stability" (Woodall 2000, 344). I interpret his earlier statement as saying one shouldn't make those assumptions because even if you do, you can't check whether they're reasonable or viable.

Kokinos: Why can't we check then?

Peismon: Because the distribution is not stable.

Kokinos: But if it is stable, isn't it like Phase 2? So, we can assume a probability distribution.

Peismon: But not when it isn't. And, we don't know which it is.

Socrates: So, in Phase 2, it appears we can do statistical hypothesis testing, and in Phase 1a, we can also when the process is stable. Then, the only case still to debate is whether in Phase 1a, when the process is not in control, the control chart can be a statistical hypothesis test.

Peismon: I'm not sure you can under any conditions in Phase 1a.

Kokinos: And that's because it is not in control, it is unstable.

Peismon: Exactly.

Socrates: However, Phases 1a and 2 have two conditions that are exactly the same.

Peismon: What are those?

Socrates: He said that, in Phase 1a, the process could be in control or not, stable or unstable. And we don't know which it is.

Peismon: True.

Socrates: The same occurs in Phase 2.

Peismon: In Phase 2, the process *is* in control. That's why he said that statistical hypothesis testing might be applicable.

Socrates: *Was* in control. But that is not when statistical hypothesis testing would be done.

Kokinos: Why not? We want to know the probability of getting a point beyond the control limits when the process is/was in control.

Socrates: Not exactly. In Phase 2, the process was in control from time t_1 to t_k. When we plot the point from t_k, time has passed. We no longer know whether the process is in control between t_k and t_{k+1}. We plot the point to check for stability during the period after t_k up to and including t_{k+1}.

Kokinos: Then, the next point to be plotted in Phase 2 at time t_{k+1} has the same conditions as all the points plotted in Phase 1a.

Mathitis: Except for one difference. We don't know whether the process was in control before our first subgroup at t_1 in Phase 1a, while we do know in Phase 2 it was in control during t_1 to t_k.

Socrates: That's true. Yet, regardless of how long the period and whether it is 1 point or 25, 1 subgroup or 25, it is that period covered by the data that we are checking whether the process was—not will be or is—but was in control.

Peismon: How does that change whether the control chart is a statistical hypothesis test?

Socrates: If Woodall claims that, in Phase 2, we can do statistical hypothesis testing, then we can also in Phase 1a. Or, I should say it the other way around. If Woodall claims that a control chart cannot be a statistical hypothesis test in Phase 1a solely because we don't know whether the process is stable, then the same applies in Phase 2. At that moment, we are checking whether the process was stable, in control, we don't know whether it was.

Agrafos: If it is in statistical control from t_k to t_{k+1}, then why check? Is that the point?

Peismon: Yes, but it is not statistical hypothesis testing.

Socrates: But it can be, since we can assume probability distributions, like Wheeler did.

Agrafos: Why, if we don't know whether the process is stable?

Kokinos: But in Phase 2, we have more information. We can use the average and standard deviation from the control charts to determine a probability distribution.

Mathitis: We can use all the data from the stable process to check the viability of the assumed probability distribution.

Socrates: That statistical test is testing the hypothesis that the subgroup at time t_{k+1} came from the same distribution that the subgroups during the alleged stable period t_1 to t_k came from.

Mathitis: That's right.

Socrates: In Phase 1a, the hypothesis tested is different. It is whether all the subgroups during the period t_1 to t_k came from the same probability distribution.

Kokinos: I don't see a difference.

Socrates: The difference is between the null hypotheses. Do we assume that there is a common, unknown probability distribution that is estimated by the average of the subgroups, or do we assume that we know the probability distribution?

Kokinos: Isn't it the same distribution?

Mathitis: No, because in the second case, the information from the t_{k+1} subgroup is not used to estimate that common distribution.

Peismon: But that difference is what prevents Phase 1a control charts from being statistical hypothesis tests.

Socrates: When you calculated probabilities for flipping a coin, did you have evidence that the process was stable?

Peismon: No.

Socrates: Yet, you are able to statistically test hypotheses about the probability of heads. How? By making assumptions and testing whether the predictions from those assumptions are accurate.

Peismon: But we collect the data randomly. That's a critical difference. The other critical difference is that there is no probability distribution.

Socrates: A statistical test in Phase 1a would use the grand average of all the subgroups and see whether any individual subgroup average is different from that average statistically. In Phase 2, we would use the average of the stable process and compare the single, new average against it.

Peismon: The results are not necessarily the same. That's why Woodall distinguishes between the two phases.

Kokinos: Why not?

Socrates: The amount of data is different and so might the standard error. But in both cases, we are testing. That's what ANOM does, and what control charts can do. There are also other tests. We can use one subgroup as a control to test whether the others are different or the same as that control. For example, we might be comparing several treatments or new products against a placebo or old product. We want to know if they are the same and whether the treatments are different from one in particular, for example, the control or placebo. That would be like Phase 2 where all the subgroups from t_1 to t_k are the control and t_{k+1} is the treatment.

Mathitis: It seems that Deming's reasons for a control chart not being a statistical hypothesis test do not apply since the control chart cannot be an analytic study. It is enumerative as it only answers the question "How much/many?" Woodall's separation of Phase 1 and Phase 2 doesn't help since, in both phases, we don't know whether the process is stable at the time. We plot the data and check—otherwise, why ask as Agrafos said. If we can do statistical hypothesis testing in one phase, we can do it in the other.

Socrates: There is also one more piece of evidence that we do not know whether the process is in control from t_k to t_j. Do you recall that we said the flowchart (Figure 21.1) Peismon drew first was missing the check on no alarms so we amended it?

Mathitis: Yes. I see. Since we do not check whether the no alarms are true or false, even when we believe the process is stable, we could be wrong. That's another reason why we don't know.

FINITE, STATIC UNIVERSE

Socrates: Perhaps there is another way around the whole issue and avoiding the controversy about a static, finite population, and whether it is reasonable or tenable to assume a single probability distribution.

Peismon: How, if like the coin, must we consider what could have been produced by the process as a cause system? We want to understand the cause system and adjust it to predict what future performance will be.

Kokinos: Don't you have to show that production processes are different from flipping a coin?

Mathitis: The point is that when we make statistical inferences, we use a probability distribution. It has two components: the list of possible results and their probabilities. We know what the list or population is. We don't know the probabilities, just as we don't know what they are with the coin-tossing process. It seems that, in both cases, we can make an assumption about what it is, and then test that assumption statistically.

Peismon: What would be the hypothesis for a production process?

Mathitis: The same as with ANOM and what we just discussed: all the subgroups are samples from universes with the same value for whatever statistic is plotted.

Socrates: I agree we can do that. But remember we are only talking about the control chart. We concluded that it was an enumerative study, not an analytic study. That means we can take another approach. Remember Woodall defined Phase 1—or 1a—as looking at historical data to see whether the process was in control. We take data from t_1 to time t_k. Is our conclusion about whether the process is in control applicable to times before t_1?

Mathitis: No, it can't be because we don't have data before t_1. Who knows what happened?

Peismon: I agree. It's a wavering system—constantly changing.

Socrates: Is our conclusion in Phase 1a about times after t_k?

Mathitis: It can't be for the same reason.

Peismon: But if the process is in control, we can make predictions.

Socrates: That would be Phase 2, correct?

Peismon: Yes, that's true.

Mathitis: In Phase 1a, control is only during the period from t_1 to t_k, then.

Socrates: Those who believe that in Phase 1a we cannot be doing hypothesis testing, statistical or otherwise, say it is because the universe is a wavering, possibly infinite, hypothetical population, from which we cannot sample.

Peismon: That's right.

Socrates: In the enumerative part of Phase 1, Phase 1a, all we want to know is whether the process was stable during the period covered by the data: from t_1 to t_k. Let's only assume we are sampling from the actual production during that time. My inference is to whether the actual production came from a single distribution or from more than one.

Kokinos: If you do that, you have a static, finite universe.

Socrates: Yes, I do. And it's real. Therefore, I can determine the actual probability distribution.

Mathitis: How?

Peismon: That's not what others say you are doing.

Socrates: What stops me from actually, as opposed to saying, my sample is from the actual production rather than some hypothetical, dynamic distribution? Isn't it from that actual production? If I can flip a coin 1000 times and state the frequency of heads and tails for that population of 1000, I can do the same for the population of the actual production from t_1 to t_k.

Kokinos: Does that change the use of control charts?

Peismon: I think it does.

Socrates: How? What is the purpose of Phase 1a?

Mathitis: To see if the process was in control during the period covered by the data.

Socrates: Do I change that by using the actual, finite production during the period t_1 to t_k?

Mathitis: No, it doesn't change.

Socrates: Then what has changed?

Mathitis: I want to say the investigation and understanding of the cause system, but you'll just say that is not Phase 1a, which is determining whether the process is, was, in control.

Socrates: That's right. Remember that probability distributions are models applied to real-world phenomena to understand them better. I can apply this model to a hypothetical, dynamic universe or to the real-world, finite, static universe. Which do you want to do?

Mathitis: I would rather work with reality.

Socrates: In the former case, I can calculate an exact hypothetical probability by assuming a hypothetical probability distribution— might as well make everything hypothetical—and conclude hypothetically that the process was in control. In the latter case, I apply my model to an actual population, which I can verify.

Kokinos: By making the actual, finite production the population, then you also overcome the objection of not knowing whether it is stable and of not being able to confirm its distribution.

Socrates: Exactly. I am only answering the question "How much/many?" That is why I think others are mistaken. Deming was right in distinguishing between enumerative and analytic studies. However, he failed to see that a control chart doesn't answer any "Why?" question, only "How much/many?" questions.

Kokinos: And Woodall?

Socrates: Woodall was right in distinguishing between the two phases. He didn't go far enough in two ways. He didn't identify in which phase enumerative studies and in which analytic studies are done. He didn't include the cycle of enumerative–analytic–enumerative phases.

Kokinos: By separating Phase 1 into the enumerative study of Phase 1a and the analytic study of Phase 1b. And Phase 2 is also enumerative.

Socrates: Exactly.

Kokinos: We can do statistical hypothesis testing in Phase 1a.

Socrates: Not only can we, we do statistical hypothesis testing when processes are not in control.

Peismon: How? When?

Socrates: In the enumerative cases. We produce a batch of 1000 units. We have a release test that says if the quality level is at least 99%, then we release the batch; otherwise, we do 100% inspection and replace or rework the defective units.

Mathitis: We do such tests. We are testing the proportion of good units against a standard and it is a statistical hypothesis test. We have two hypotheses: the batch has an acceptable quality level versus it does not. Deming gives clear examples of this being an enumerative study.

Socrates: We talked about this before with the bowl of beads to answer the descriptive question and the probability question. The process that produced them can be unstable.

Mathitis: I forgot that the bowl of beads was a universe even if the process of producing the beads was unstable or the process of selecting from the bowl is not random.

Peismon: But you lose something by restricting yourself to the finite, actual population.

Socrates: What is lost?

Peismon: You are not getting at the cause system.

Mathitis: That occurs in Woodall's Phase 1b, not Phase 1a or Phase 2. The analytic study.

Socrates: The three phases make it clearer when one is getting at the cause system. The question is "What is gained by assuming this hypothetical, infinite, wavering universe in Phase 1a?"

Kokinos: When does this hypothetical, infinite, wavering universe come into play?

Mathitis: Good question. Perhaps it applies in an analytic study to investigate the causes of a process not being in control or to look for ways to improve the process. The cause system can be viewed as possibly generating an infinite number of results. The answer to the "Why?" question leads us to take action to alter the characteristics of that hypothetical universe.

Socrates: That would be Phase 1b. No hypothetical population of possible results is required to answer the "How much/many?" question. We could assume it to answer the analytic question, but even then it may not be necessary.

Peismon: Why not?

Socrates: When we described the analytic case, we listed characteristics of the problem, the study, and the solution. For the study, we identified such statistical tools as hypothesis testing and design of experiments used to identify causes. Response surface studies help us determine the critical factors and their optimal levels. No hypothetical, wavering, infinite population is needed.

Mathitis: Or, it's false that hypothesis testing can't be done on such populations.

Peismon: But Deming said that those analyses are not applicable in the analytic case.

Socrates: That puzzles me since we do in fact use DOEs to find root causes.

Kokinos: I think Deming was inconsistent in his view of control charts. He distinguishes the frame from the population. The frame is what we sample from and the inference is only valid to that frame. If he is saying that there is this hypothetical, infinite, wavering population from which we cannot sample, then how can he claim that any inferences made to it are valid?

Socrates: We only need to assume that the sample comes from the actual output of the process. That actual output is the population for that period.

Mathitis: In that case, we have all the features for viewing control charts as statistical hypothesis tests. With a finite, static population that can be enumerated, we can apply probability models.

Peismon: But another issue is that the data for a control chart are not randomly sampled.

Socrates: That seems to be another reason for using the real, finite, static population rather than the hypothetical, infinite, wavering one. Are subgroups randomly sampled for control charts?

Kokinos: No. We pick a time interval and collect data at each interval.

Mathitis: Except for the *I* chart for individual values. Those are often collected regardless of the interval, in my experience.

Socrates: Then why not randomly sample from the entire production during the period t_1 to t_k?

Peismon: But then you have to wait until all the production is done.

Socrates: No. You could select k random times from t_1 to t_k before the process starts and at each time chosen, select n consecutive units.

Peismon: What's the advantage of that?

Socrates: With your approach, if your intervals coincide with a systematic event or fail to capture a systematic event, then your control chart will deceive you. However, by randomly selecting the times you will collect data, you increase the chance of identifying those systematic events. Also, your inference from the data to the entire period is valid. The traditional way of sampling subgroups at fixed intervals does not provide a valid inference to the entire period in question.

Mathitis: And that's simply because it is not a random sample and therefore may not represent the entire output for that period. This could be one reason why Shewhart said it was a conditional prediction.

CONTROL CHARTS AS NONPARAMETRIC TESTS OF HYPOTHESES

Socrates: There is one more reason why control charts can be considered tests of hypotheses.

Peismon: Another reason?

Socrates: Your claim and that of other experts was that there is no specific probability distribution so we cannot calculate a p value.

Peismon: That's correct.

Socrates: Why do we sometimes use nonparametric tests?

Peismon: When the assumptions of a parametric test are not tenable or are violated.

Mathitis: I see your point, Socrates. We even call certain tests distribution-free, although we always assume a probability distribution. We can calculate p values for these tests.

Socrates: How do we calculate those probabilities?

Mathitis: We make as few assumptions as possible to include several distributions. For example, we might only assume a symmetrical distribution and use the median or ranks.

Socrates: Recall the run tests, which we studied before discussing control charts.

Mathitis: Yes. They are statistical tests producing p values on whether too few or too many runs occurred. No specific or parametric distributional assumptions, like normality, are used.

Kokinos: Then we could do the same here.

Socrates: We don't need to since it has already been done.

Peismon: Who did it? When?

Socrates: Wheeler and Chambers. Their analyses of a wide range of distributions provide the evidence that nonparametric tests can be used. The Empirical Rule states the worst-case scenario for a variety of distributions. The Empirical Rule shows that the 3 standard error rule cannot distinguish between different lists and different probabilities assigned to members of the list.

Peismon: Why not?

Mathitis: The Empirical Rule shows that a uniform distribution with a finite range is indistinguishable from a normal distribution with an infinite range with respect to the probability of getting a result beyond the 3 standard error limits.

Peismon: The probabilities are different. It's just that 1% is the worst case for each.

Socrates: When we compared the *p* values from nonparametric tests to parametric tests (Myth 12), what did we discover?

Mathitis: It could be more or less depending on the distributions assumed and whether that assumption was violated. The *p* for non-parametric was a "worst-case" value to accommodate various parametric distributions. That's what the Empirical Rule shows.

UTILITY OF VIEWING CONTROL CHARTS AS STATISTICAL HYPOTHESIS TESTS

Peismon: Maybe so, but Deming also said that even if control charts were statistical tests of hypotheses, it is misplaced to consider them. Control charts are practical applications and using them as hypothesis tests may undermine their use or prevent learning and understanding.

Socrates: Isn't it the reverse? Isn't an unwillingness to consider both options undermining learning and understanding? After all, it was Deming who said that there is no knowledge without theory. What we are providing is theory by saying that with a simple change in one's view of control charts, there is a considerable theoretical body than can be used—probability theory.

Those who argue that the limits are only empirical are without a theory. As Deming often asked rhetorically, "How could they know?"

Peismon: Woodall also suggested that viewing them as tests of hypotheses could "prevent the application of control charts in the initial part of Phase 1 because of the failure of independence and distributional assumptions to hold" (Woodall 2000, 343).

Socrates: Remember when I asked for studies confirming that 3 standard error limits are the most economical?

Kokinos: Yes, and we didn't know of any.

Socrates: When Deming and Woodall suggest that viewing control charts as statistical tests of hypotheses might prevent others—not them—from doing something, do they provide evidence of that? Do they have at least one example of someone not using control charts and when asked why not, they said they thought control charts were statistical tests of hypotheses?

Mathitis: Not that I know of. I think they were conjecturing.

Socrates: Also, this view seems to forget that the statistical application of probability theory is simply applying a model. If the model provides the information needed or completes the objective it was used for, is it useful?

Peismon: But are you not doing the same thing you are objecting about others? If the requirements of independence and of the distribution are violated, then why use it?

Socrates: You use a car to drive to work, don't you?

Peismon: Of course.

Socrates: If someone told you, cars can be abused. There are many drivers who don't understand the risks, resulting in thousands of accidents. Therefore, no one should use cars.

Peismon: I would not agree with that. Just because others don't know how to drive responsibly doesn't mean I don't know.

Socrates: Should practitioners who know how to use control charts and can modify them to serve their needs better be prevented from using them because others don't know how to use them? Should we advise everyone not to use them in a particular way because some people abuse or misuse them when used that way?

Mathitis: No, that isn't right.

Socrates: Don't those requirements of assumption and independence apply to all statistical hypothesis testing? Shall we then say don't view the two sample *t* test as a statistical test because these assumptions may not hold?

Mathitis: It seems that the teaching of control charts contains a mix of information that is inconsistent. We teach things without explanation and people learn them without understanding.

Socrates: There are many applications of hypothesis testing that don't strictly follow the assumptions on which the tests are based. Often, our samples are not truly random: we didn't define the frame, number the units in the frame, and then randomly sample from them. We might have auto-correlation and ignore it. Yet, we don't denounce the results as attempting to do something that cannot be done. These are tools—they help or not, are useful or not.

Kokinos: Isn't it the same as when we say a test is robust to an assumption? We caution in its use but since we don't know whether the assumption is strictly tenable, we may still apply the test.

Mathitis: I can see that. I don't see people refusing to use a control chart or fail to learn because of something of which they aren't even aware.

Kokinos: I wasn't aware of these issues. Yet I use control charts to decide whether the process is in control or not. When it isn't, we look for assignable causes and eliminate them when we find them. We've made improvements to processes and have extended their stability using them. But, we also use the original flowchart and not the modified one. We never check whether no alarms are true no alarms.

Peismon: Of course there are people, some experts, who lament exactly that. I think that is one reason why Wheeler has written often about the myths surrounding the use of control charts.

Mathitis: Yet, there are others who lament that some people, even experts, don't want any changes made to control charts, or any theory applied to them. Woodall commented on that a decade and a half ago. But control charts can be modified and used in different ways. Fine, don't call them control charts or Shewhart charts or process behavior charts but don't stop people from using them in other ways if they help. Wasn't the reason for their creation that they "work?"

Kokinos: On the one hand, use them because they work and forget about any theoretical rationale for using them, and on the other hand, don't modify them because we don't want to apply theory as the limits were supposedly empirically derived. Even when applying theory can't that result in improved applications or improvements of the tool?

Mathitis: Deming did say no one has improved on control charts.

Kokinos: Perhaps it's because we don't allow theory to be applied to them.

Socrates: If the "proof" for using a model is that it accomplishes or aids in meeting its purpose, then all one needs to do is show that. What's the risk of not recognizing that control charts are statistical hypothesis tests?

Kokinos: One thing is that when we may not recognize that there is a chance of error for either decision. I just realized something. When we talked about hypothesis testing, we initially said that we could only fail to reject when $p >$ alpha. Yet, when there are no special cause variations, we don't say we fail to reject the process is stable or in control. We say, quite strongly, that the process is stable. We act as if that were 100% conclusive.

Mathitis: I agree with Kokinos. Unless we determined the power of the test, we shouldn't conclude that the process is stable—only that we don't have enough evidence to reject stability.

Socrates: Unless you use the modified flow chart (Figure 21.2). What is the practical purpose and benefit of saying they are statistical tests of hypotheses?

Mathitis: Now, we can apply the same framework and lessons as we do for what we normally call hypothesis testing. We can talk about the null and alternative hypotheses, about Type I and II errors. For example, we can discuss power and what difference do you want to detect. It becomes a richer discussion and learning.

Kokinos: We avoid the mistake of accepting the hypothesis of stability when we shouldn't—when we don't have enough evidence to reject or enough power.

IS THE PROCESS IN CONTROL? VERSUS WHAT IS THE PROBABILITY THE PROCESS CHANGED?

Socrates: Why stop there?

Kokinos: What do you mean?

Socrates: Recall that we discussed that the question (Myths 10 and 14) we really want to answer when testing hypotheses is "What is the probability that null hypothesis is true?" Or, for a specific value of a parameter, "What is the probability that the parameter value is $X \pm \Delta$?"

Kokinos: Are you suggesting that we use Bayesian statistics for control charts or process behavior charts?

Mathitis: I'm intrigued by that. After all, we call it statistical process control and yet there are proponents including experts that claim probabilities do not exist or refuse to consider probabilities in evaluating different alarms. Why use statistics if probabilities cannot be applied?

Peismon: Because they don't work and don't serve a purpose.

Mathitis: But Wheeler's Empirical Rule shows that nonparametric statistics do apply and work.

Socrates: Peismon, you have said that Deming and others view the cause system as producing a dynamic population, correct?

Peismon: Yes, and that's why there is no single probability distribution.

Socrates: Those views do seem contradictory. Let's look at it differently. Let's accept that the process is dynamic and therefore no single distribution occurs. Doesn't that suggest that we view the probability of the statistic that is plotted on a control chart as a random variable?

Mathitis: Of course—that makes sense.

Agrafos: How does that apply to a control chart when we want to know whether the process is stable?

Socrates: What do we monitor with a control chart?

Agrafos: We can monitor several things. The mean, the variation using the range or standard deviation, a proportion, and so on.

Socrates: Consider the mean. Why not assume it is a random variable, especially if, as Peismon and others say, the process is dynamic?

Kokinos: If the value of parameter varies, then we can treat it as a random variable and model it with a probability distribution.

Agrafos: Yes, but specific distributions, like normality, are not required for control charts.

Mathitis: But we don't need a specific distribution. We just need the mean or any other parameter, to be random.

Peismon: How would you identify changes?

Socrates: Another point you made, Peismon, was that you don't want to identify every change but the significant ones. Then, we specify a range in which changes are insignificant and calculate the probability the mean is still within that range.

Peismon: But that would involve changing the entire approach to control charts.

Socrates: Look at it this way. What question do you want to answer and what question does the current approach to control charts answer?

Peismon: Control or process behavior charts answer the question "Is the process in control?"

Socrates: Does the control chart actually answer that question? Recall Wheeler's flowchart for using a control chart (Figure 21.1) and how we modified it (Figure 21.2).

Mathitis: No, the control chart only answers the question "Is there evidence that the process is in control?" If we get alarms, the answer is no, and if we don't get alarms, the answer is yes with respect to *evidence*. We then check whether the alarms are true alarms. We modified the flowchart to include checking whether no alarms were true no alarms.

Kokinos: We really want to answer "Is the process in control?"

Socrates: Or, given that we expect variation and, if some special cause variations are insignificant, don't we want to answer "Has the process changed significantly?" where *significantly* is defined ahead of time?

Kokinos: I would accept that.

Agrafos: As with hypothesis testing, it matches more of our thinking.

Peismon: How do we determine what is significant?

Mathitis: How do you determine what it is now? We don't. We just follow a procedure, using 3 standard error limits and several special cause patterns, whether it makes sense or is the best approach for our specific process and costs.

Peismon: But Deming has said that no one has improved on the control chart since its invention.

Mathitis: Woodall asks "Why shouldn't we?"

Socrates: Could we not combine this with capability, actual capability, or performance?

Mathitis: How?

Socrates: The point of controlling a process is not just to have it be stable but to have it be stable at a desired performance, that is, capable. If we know what range of centering and variation will produce that desired performance, we can assume that the parameters for centering and variation are random variables.

Mathitis: I see. Then we can monitor to see whether they are within the ranges that produce that desired performance.

Kokinos: Would that mean we didn't have to collect enough data to create a control chart?

Mathitis: It might be possible to monitor each subgroup without constructing the control chart.

Agrafos: It seems to eliminate the issues we discussed—are the limits probability limits, do we do hypothesis testing, small versus large changes—and gives us an answer to the right question.

Mathitis: I see several advantages. I like that rather than having confidence levels or nothing, we get actual probabilities.

23

Myth 23: Process Needs to Be Stable to Calculate Process Capability

Scene: *Project meeting room (second project charter meeting)*

BACKGROUND

Peismon: Now, you will need to show that the process is stable to calculate process capability.

Agrafos: In class, the instructor showed a flowchart of the steps for calculating process capability (Figure 23.1). The question before the calculation was "Is the process stable?" If yes, then the next steps lead to calculating process capability; if no, then the flowchart leads to the step of making the process stable.

Socrates: Did you understand her to mean that process capability cannot be calculated unless the process is stable?

Agrafos: Yes, that was my understanding.

Peismon: Stability is a requirement for all capability calculations.

Kokinos: All the class materials I've seen say that you must have a stable process. That's what my instructors said when I took my Black Belt (BB) class.

Socrates: Process capability is simply making a calculation. Clearly, I can enter data in software to get the calculation or manually calculate it.

Mathitis: But that is garbage in–garbage out.

Peismon: That's a good way of looking at this, Mathitis. When the process is unstable, then process capability is meaningless—it's garbage out.

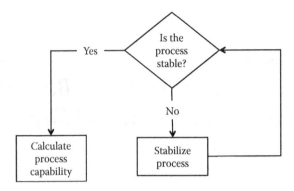

FIGURE 23.1
Decision flow for calculating process capability.

Socrates: I understand the concept of garbage in–garbage out. The decision chart that we saw in class said to check for and make sure there is process stability before calculating process capability.

Mathitis: What is confusing about that?

STABILITY AND CAPABILITY: DEPENDENT OR INDEPENDENT?

Socrates: Not about that. But in combination of another chart that was shown with respect to control charts—one about the four states of a process (Wheeler and Chambers 1992, 15) (Table 23.1).

Peismon: You're referring to the four possible combinations of process stability and capability.

Case 1: The process is unstable and not capable. That's called chaos.

Case 2: The process is unstable but capable. That's called the brink of chaos.

Case 3: The process is stable but not capable. That's the threshold state.

Case 4: The process is stable and capable. That's the ideal state.

TABLE 23.1

Four States of a Process

	Capable	Not Capable
Stable	Threshold	Ideal
Not stable	Brink of chaos	Chaos

Socrates: Yes, and it is the combination of those two views: the four states and the requirement of stability to determine capability that confuses me.

Peismon: How? They are referring to different things.

Socrates: Does capability mean the same thing in all four states? Does stability mean the same thing in all four states? And capability and stability mean the same thing in the four states as in the flow chart (Figure 23.1) Agrafos drew?

Peismon: Of course.

Socrates: Then, it appears stability is independent of capability with respect to the issue of the four states of a process.

Agrafos: They would have to be if all combinations are possible, wouldn't they?

Peismon: No, because we have to check for stability before we can determine capability.

Socrates: That is my confusion. They are independent for identifying the states but the procedure makes capability dependent on stability for the calculation. Let's combine the four states with the claim that stability is required to calculate capability; otherwise, it is meaningless. These are the situations (Table 23.2).

Peismon: That's right.

Mathitis: That shows that one or the other table is wrong since they can't both be true.

Kokinos: If it is meaningless, then we cannot know whether it is capable.

Peismon: That's right. Without stability, capability has no meaning, no sense.

Socrates: Does winning the lottery depend on buying a ticket?

Agrafos: Of course. No ticket, no winning.

Socrates: In the same way if being capable depends on being stable, then you can't be capable without being stable. Capability depends on stability.

Peismon: So, you agree.

TABLE 23.2

Four States of a Process with Respect to Ability to Calculate Capability

	Capable	Not Capable
Stable	Capability can be calculated	Capability can be calculated
Not stable	Capability cannot be calculated; meaningless	Capability cannot be calculated; meaningless

Socrates: Then, case 2 cannot occur. You can't have not stable but capable. You can't have a winner with no ticket.

Agrafos: I see what you are saying, Socrates.

Socrates: And, you can't have the first case either: not capable and not stable. Either they are independent or not. If dependent, then cases 1 and 2 never occur.

Peismon: We can know which of the four states we are in. Otherwise, why distinguish them?

Agrafos: I think I understand what Socrates is saying. If you have to have stability before determining capability, and the process is unstable, then how do you determine if it's capable?

Peismon: Just search the literature. You'll find that these four cases are taught in all Six Sigma and statistical process control classes and are the standard way of understanding the situation your process is in with respect to stability and capability.

Socrates: We can do a search. But this is logic: how can you know that a process is capable when it is unstable if you need to have a stable process to determine capability.

Peismon: You're just twisting my words, Socrates. Capability depends on stability.

Mathitis: For the calculation. Does capability depend on stability to identify which state it is in?

Peismon: That's it. For calculating yes but not for identifying the state it's in.

Mathitis: How would you know what state it's in if you can't calculate capability unless it is stable?

ACTUAL PERFORMANCE AND POTENTIAL CAPABILITY VERSUS STABILITY

Agrafos: Is stability required for all capability indices?

Peismon: P_{pk} and P_p are actual performance measures or indices while C_{pk} and C_p are potential capabilities. I interpret the "P" in P_{pk} and P_p as performance and the "C" in C_{pk} and C_p as capability in the sense of potential.

Mathitis: We saw that Equations 9.1 through 9.5 are the same whether calculating potentials are actual capability with only the standard deviation that is used changing.

Socrates: What does actual process performance mean?

Mathitis: Data are collected for a period and the actual, which we call total, variation of the process is used to determine the standard deviation for the formulas.

Peismon: For potential capability, we use the best estimate of process variation.

Socrates: How is that determined?

Peismon: Over the short term, it is expected that the process will be stable. That variation is assumed to be the best the process can achieve. That short-term variation is called within-group variation. We don't expect the variation for a small number of consecutive outputs to vary much. We use that to estimate the short-term variation.

Mathitis: We use long-term variation for P_{pk} and P_p but short-term variation for C_{pk} and C_p.

Socrates: You said that, over the long run, the process will be unstable. Then, long-term variation includes variation from both when the process is stable and unstable.

Mathitis: Yes, because total variation = short-term variation + long-term variation.

Peismon: Another way of looking at it is using the terms from control charts, which is more appropriate since we require stability as shown by a control chart. The short-term variation or within-subgroup variation is called common cause variation. Any additional variation is called special cause variation. Special cause variation occurs when the process is unstable. Then, long-term variation or total variation = common cause variation + special cause variation. It all depends on the definition.

Socrates: You use common cause variation to calculate C_p and C_{pk} and total variation to calculate P_p and P_{pk}.

Peismon: That's right.

Socrates: Then, you can calculate process performance when the process is unstable—you can calculate P_{pk} and P_p, and also P_{pU} and P_{pL}.

Peismon: Yes, those indices do not require stability. But estimates of C_p, C_{pL}, C_{pU}, and C_{pk} do require stability. We then use that to predict future performance.

Socrates: Let's postpone the discussion about prediction unless we must necessarily include it.

Peismon: That's a critical aspect of the stability.

PROCESS CAPABILITY: RELIABILITY OF ESTIMATES

Socrates: Then, it will be worth spending more time on it. In the formulas, what does the caret or "hat" over the σ mean?

Peismon: It means that we are estimating the standard deviation.

Socrates: Does requiring process stability imply that there is a correct way of estimating?

Peismon: I meant that it provided better results, not correct results.

Kokinos: We do not get a "true" process capability result when the process is unstable.

Mathitis: No, because I don't think we can know what the true process capability is.

Kokinos: Also, we know from previous discussions that estimators cannot be correct; there is no correct way of estimating.

Mathitis: But they can have other characteristics like not being biased or having small variation.

Socrates: As a result, we are discussing whether estimating with stability is better than without stability *with respect to some characteristics of estimators.*

Mathitis: Couldn't we say that estimating using control charts when there is evidence of stability is more reliable than these other ways? Can't we say that a desirable feature of estimating is the reliability of the estimate?

Kokinos: More reliable, for example, in that without stability, the estimate is biased.

Socrates: Has anyone shown that? Do you have any evidence that the estimate of C_p or C_{pk} is biased when the process is unstable and it is unbiased when the process is stable? Or, both are biased but less biased when unstable?

Mathitis: Not that I know.

Peismon: It couldn't be biased since we use the c_4 factor to remove the bias of the sample standard deviation formula.

Socrates: That would apply whether it was stable or not.

Peismon: Yes.

Socrates: What about other desirable characteristics of estimators?

Peismon: When it is unstable, then we do not have a reliable estimate. That's a different kind of bias. It's like taking a nonrandom sample.

Socrates: How do you determine stability and get estimates?

Peismon: We collect data in subgroups, a few consecutive outputs, so that each subgroup has this short-term variation. We calculate the standard deviation and mean of each subgroup. The standard deviations are plotted on an S chart and the averages are plotted on the $\bar{X}$ chart. If both charts are in control, stable, then we use the average of all the subgroup standard deviations adjusted by c_4 to remove the bias to calculate C_{pk} and C_p.

Mathitis: That will occur when no points fall outside the control limits. The control limits define the amount of common cause variation when the process is stable.

Socrates: For C_p, only the common cause variation, the standard deviation from the S chart, is used, correct?

Peismon: That's right.

Socrates: Then, the $\bar{X}$ chart does not need to be stable as it does not affect the calculation of C_p.

Mathitis: I hadn't thought of that. Regardless of the stability or instability of the $\bar{X}$ chart, the calculation of C_p will be exactly the same. But, the process is stable if both charts are stable and not stable if at least one chart is unstable. So, we can say that stability is not required for C_p since the $\bar{X}$ chart can be unstable.

Kokinos: So, of the eight indices, the four performance measures and the capability index C_p do not require stability.

Peismon: But the other three—C_{pk}, C_{pU}, and C_{pL}—do because they include both the standard deviation and the mean. So, the two charts that monitor the average and variation must be stable.

Mathitis: That's true since C_{pk} is the minimum of C_{pU} and C_{pL}.

Socrates: Otherwise, the estimate of the short-term variation won't be reliable.

Peismon: That's right.

CONTROL CHARTS ARE FALLIBLE

Socrates: So, it isn't that the process capability calculations from unstable processes cannot be calculated or are meaningless, they just aren't reliable or accurate enough.

Agrafos: Then, all four states are possible.

Peismon: That's what I have been saying. Meaningless means unreliable or inaccurate. Look, Socrates, I've done this for years and it has worked. When the process is not stable, you don't calculate process capability; when it is, you do.

Socrates: When you say it "works," do you mean that every time you calculate process capability, it is a reliable estimate?

Peismon: Yes, every time.

Socrates: And how did you determine the estimate was reliable? Did you compare it to truth?

Peismon: If I had known what it was, I wouldn't need to calculate it.

Kokinos: Besides, we said we didn't know what the true process capability is.

Socrates: How did you determine it was reliable?

Peismon: Because the process was stable.

Socrates: How did you determine it wasn't reliable when the process was unstable?

Peismon: That's the definition of unreliable—the process is unstable. This is getting us nowhere.

Kokinos: That seems like a circular argument.

Mathitis: I agree with Peismon but not his argument. If reliable means stable and stable means reliable, then just use one word. But I think he is misapplying the words. Reliable refers to the estimate and stable refers to the process. So, they can't mean the same thing. It's true that one depends on the other but they are not the same. Otherwise, there would only be two states.

Socrates: We need to separate the two. Can we assess the reliability of the estimate even if we don't know what truth is?

Kokinos: How can we do that?

Socrates: If the data on which the calculation depends were flawed, would that indicate the estimate was not reliable?

Kokinos: Sure. If the data are incorrect, then likely the calculation is incorrect. It's what Peismon said earlier: garbage in–garbage out.

Socrates: Then, let's see when we have garbage in. How did we determine stability?

Mathitis: We used a control chart. If the control chart showed no assignable or special cause patterns, then the process is said to be stable and we can calculate process capability.

Socrates: If at least one special or assignable cause pattern occurs, then the process is not stable and you claim you cannot calculate process capability—at least not C_{pU}, C_{pL}, and C_{pk}.

Peismon: That's right. The process is unstable. There is common cause and special cause variation. To calculate these indices, we need only common cause variation for the two charts monitoring centering and variation.

Socrates: When you say special cause variation pattern, you mean it is a signal?

Peismon: That's right, an alarm.

Socrates: Is the control chart infallible in detecting special cause variation? Is the alarm infallible? Is the lack of alarm infallible?

Peismon: Well, no, neither. But the false alarm rate is very low. That doesn't contradict what I am saying. If the alarm is false, then you could have calculated those capability indices.

Socrates: Is not our concern with the false nonalarm rate? When there are no special cause patterns, signals, or alarms, we conclude the process is stable. How often could we be wrong?

Mathitis: Yes, our concern for this issue is the false nonalarm rate.

Peismon: We don't know what that rate is.

Socrates: Do you calculate C_{pU}, C_{pL}, and C_{pk} even when no alarm is false?

Peismon: You shouldn't because the process is unstable.

Socrates: Do you verify that lack of assignable or special cause alarms or indicators or patterns are true signals?

Kokinos: We don't. The decision flowchart (Figure 23.1) says if the answer is yes to the question "Is the process stable,?" then we calculate process capability.

Socrates: However, this flowchart is misleading or inaccurately describes what you said.

Kokinos: How is that?

Socrates: If the question is really asking whether the process is stable, then you do confirm each answer. But if the question is only asking "Are there no alarms,?" then you are correct. Remember that Wheeler's decision flowchart (Figure 21.1) only asked to determine why an alarm occurred and never asked to determine why an alarm did not occur.

Mathitis: You're right. We only check for the signals or patterns.

Kokinos: We do the same and that is what our instructor said.

Socrates: Then, even in those cases when you mistakenly believe the process is stable, you use exactly the same procedure to calculate process capability and you believe exactly the same that it is a reliable and valid calculation.

Mathitis: Yes, because I don't know when I am mistaken, when I failed to detect a change.

Kokinos: And we are also saying that it has "worked."

Socrates: So, in fact you can and *do* calculate process capability when the process is not stable. Is the calculation of process capability reliable only if you believe the process is stable?

Mathitis: I see what you are saying. I don't know which case I have when the control chart shows no alarms. I do exactly the same thing whether it is a true or false nonalarm. Only when there is a special cause pattern do I say I can't do it. But you are saying nothing is preventing me from doing it because I do it in other cases when it is unstable and I didn't know it. And, as Peismon said, we believe it was reliable and worked.

Socrates: Are you saying that if you know it is unstable then you cannot calculate process capability, but if you don't know it is unstable you can? So, the ignorant person can do more than the knowledgeable person?

Mathitis: No, that doesn't make sense.

Peismon: But, when there are no special cause patterns, I have reason to believe that the process is stable.

Socrates: Are you saying that if you correctly believe it is unstable, then you cannot calculate process capability, but if you falsely believe it is stable you can? So, having false belief about stability allows you to do something that the person who has true belief cannot?

Mathitis: That doesn't make sense either.

Peismon: But it would be unreliable. It's garbage when the process is unstable.

Socrates: Now, when the process was actually unstable but you incorrectly thought it was stable and proceeded to calculate process capability, you said that that worked also and was reliable.

Mathitis: I see what you are saying. If when we were wrong about stability it "worked," then we are contradicting ourselves when we say that when we are right about instability it won't work.

Socrates: Have you done further investigation to show that when you didn't know if it was stable the calculation was less reliable only when it was unstable but not when it was stable?

Mathitis: I wouldn't even know how to do that.

Socrates: When you check for stability, do you take into account the power of the tests for special causes?

Peismon: I'm not sure I understand.

Socrates: Since you say that an unstable process will result in an unreliable estimate, how much instability is necessary to make the estimate practically or statistically unreliable?

Peismon: I don't know.

Socrates: If I increase the subgroup size, what happens to my ability to detect special causes?

Peismon: It should increase.

Mathitis: Now I get it. If I don't have a good likelihood of detecting special causes, then I will have a higher chance of failing to detect instability.

Socrates: Yet, will you say in those instances that you can calculate process capability reliably and validly?

Mathitis: Yes. I will say my estimate is reliable because the process is stable.

Peismon: It works for me. When there are no special cause patterns on the control chart, I conclude that the process is stable. I then calculate process capability and I get a reliable and valid estimate.

Socrates: Do you look at every output from the process to construct the control chart?

Peismon: No, we take a sample of 20–25 subgroups. Typically, each has the same sample size.

Socrates: Is it possible that, between the subgroups you collected, the process was not stable? For example, is it possible that, between the fifth and sixth subgroups, the process went out of control and it would have shown as a point beyond the control limits if you had collected data then?

Peismon: I suppose so.

Socrates: But you wouldn't know that from the control chart.

Peismon: Not from the control chart.

Socrates: Recall that other special cause patterns depend on more than one point.

Peismon: That's right.

Socrates: What if the out-of-control process only lasted through one less than the number of data points required for a special cause pattern to appear? For example, if you needed three points but the out-of-control condition only lasted through the time spanned by two points or for the pattern that requires nine points if out-of-control only lasted through seven or eight points, would you know it?

Kokinos: No, we wouldn't know it. It wouldn't show on the control chart.

Socrates: Yet, the process would be out of control, unstable.

Kokinos: That's true.

Socrates: Are those possibilities?

Peismon: I suppose so.

Kokinos: They are real possibilities.

Socrates: Do you always use all special cause patterns?

Kokinos: No, often we use less than that.

Socrates: In each case, when those occurred, you erroneously concluded that the process is stable. Peismon, in these cases, you would still calculate process capability?

Peismon: Not if it wasn't stable.

Kokinos: That's the point, Peismon. Since there would be no special cause variation patterns or alarms on the control chart, how would you know?

Agrafos: You convinced me, Socrates.

Kokinos: I think I believe it also.

CAPABLE: 100% OR LESS THAN 100% MEETING SPECIFICATIONS

Peismon: But I am not convinced. In addressing capability, if the process is unstable and I have a defect, then I know I have case 1: Chaos, unstable, and not capable.

Agrafos: That's true.

Socrates: Then, capability means no defects at all?

Peismon: Capability means that the process output meets customer needs as defined by specifications.

Socrates: All outputs?

Peismon: Yes, all, not a single output fails specifications. But that is for the Ideal state. In the Threshold state, you can have some defective, nonconforming units.

Socrates: Then, there are degrees of capability.

Kokinos: Yes. Capability indices range from 0 to 2 or more.

Socrates: And capability does not mean the same thing in all four states. Do we mean potential or actual capability?

Peismon: Potential.

Socrates: What value of these indices indicates the process is capable?

Kokinos: That's what the experts say. Wheeler defines capable, or no trouble (two kinds of trouble) as "When your process is operating in the Ideal State you will usually find that the centered capability ratio (C_{pk}) will be close to, or greater than, 1.00" (Wheeler 2009b).

Mathitis: In product development, we start with $C_{pk} = 2.0$, so we will get 1.33 in production.

Socrates: What value of C_{pk} means there are no defective outputs?

Peismon: Well, infinity.

Socrates: Then, these values of 1, 1.33, and 2.0 mean that some defective outputs occur.

Peismon: Yes, but a small amount. Less and less as C_{pk} increases.

Socrates: Then, by capability, we don't really mean zero defective outputs.

Mathitis: That's true.

Socrates: If you accept some defects to mean capable, then when you said having one bad output meant you knew it was not capable is false.

Peismon: Okay, but C_{pk} and C_p are the process potential. They are potential capability indices. Other capability indices, like P_{pk} and P_p, are actual performance indices.

Mathitis: But when the capability index $C_{pk} = 1$, we do have defective product. It only produces 99.73% good product, if we assume a normal distribution. For a business producing a million units, that would be 2700 nonconforming units.

Socrates: Why is 99.73% capable but not 99% or 85.3% or 99.99%?

Mathitis: Tradition.

Kokinos: Although with other initiatives like Six Sigma, capable means 99.99966% conforming, so less than 4 nonconforming units in a million.

Socrates: In any case, it is not 100%. In other words, to claim a process is capable, it has to be at least a certain prespecified value.

Peismon: Okay, that's true, but that doesn't change what I said.

Socrates: These two statements are contradictory: (a) we can't calculate process capability to compare to the amount required unless the process is stable and (b) the process can be unstable and either capable or not.

PROCESS CAPABILITY: "BEST" PERFORMANCE VERSUS SUSTAINABILITY

Peismon: But the process capability indices C_p, C_{pk}, C_{pU}, and C_{pL} are the best possible performance of the process. They represent the process potential. While we don't know what it is, we want to find that inherent process capability not just any estimate. The purpose of determining capability on the basis of a stable process is to be able to predict future performance and sustain that level of performance.

Socrates: That's fine. I accept that definition. But the definition does not include that the best process performance be predictable or sustainable. Those are additional desirable characteristics of a process but not a requirement to calculate process capability. There is nothing in the definition that says if a calculated capability to meet specifications is not sustainable, then that calculation is wrong or unreliable.

Mathitis: Even if it did, the definition doesn't say for how long it needs to be sustained. Is it for one more point or one more shift or one more year?

Kokinos: Then, even a prediction from an unstable process could be accurate and sustainable.

Socrates: Then, why can't we base the process capability simply on the smallest within-subgroup variation?

Peismon: That doesn't guarantee it is sustainable.

Socrates: Does being stable guarantee it is sustainable?

Peismon: No, but we are more confident.

Socrates: Again, that is a different issue from whether it can be done and is meaningful and reliable. Sustainability and predictability are separate issues. Once a process performs at one level, it has the capability of doing that since it actually did it. Now, we check for sustainability.

Peismon: I think I know why we want consistency, as shown by stability on a control chart. We are using samples, and samples have sampling error. If we toss a die with the numbers 1 through 6, and get an average of 3.1, it doesn't mean that is the true average. We are trying to get an estimate of that, remember? One way of addressing the sampling error is to have numerous subgroups displaying stability, to reduce that sampling error.

Agrafos: Then, the same would be true for the average estimate. I see now why we want both charts to be stable.

Socrates: Except for C_p, which uses only the best variation. The other capability indices use the variation observed and the average observed *when the process is stable*. How does stability by itself guarantee that either are the best?

Mathitis: It doesn't.

Peismon: But that is why C_p is based on common cause variation. Common cause variation is the minimal inherent variation of the process.

Socrates: But only for the period during which the subgroups were collected. We already agreed it tells us nothing about the past and you've said that unless it is stable it can't tell us about the future or how far into the future.

Mathitis: But C_{pk}, C_{pU}, and C_{pL} also include the average. Is that the best average?

C_p VERSUS *P/T*

Socrates: Recall our discussion on measurement system analysis?

Mathitis: Yes.

Socrates: Did we say that the measurement system had to be stable to evaluate it?

Mathitis: No. In fact, we use two control charts, an $\bar{X}$ chart for the parts to show part-to-part variation by operator and R chart to show repeatability or within operator-part variation. We expect the $\bar{X}$ chart to be out of control. Even if the R chart is not in control, we still evaluate using the four criteria.

Peismon: I don't understand the connection.

Socrates: What is the equation for *P/T*?

Peismon: It's six times the measurement system standard deviation divided by the tolerance:

$$P/T = 6\sigma_{\text{measurement system}}/(\text{USL} - \text{USL}) \qquad (23.1)$$

Socrates: What is the equation for C_p?

Peismon: It's the tolerance divided by six times the process standard deviation.

Mathitis: It's just the reciprocal but using different standard deviations. I see. Why would we have a requirement for one but not for the other? Shouldn't both require stability to be meaningful or neither?

Peismon: But assessing the *P/T* ratio is not assessing capability of the measurement system. There are no defects so we don't have the four states.

Mathitis: I disagree. I think it is measuring its capability to distinguish between good and bad units.

RANDOM SAMPLING

Socrates: What if there was another way of getting that estimate that didn't require stability?

Kokinos: I'd be interested in that.

Peismon: It's not possible since control charts are the only way to check for process stability and they are required.

Socrates: Consider a large jar of 5000 red and blue marbles. If I don't know the proportion of either and don't want to count them all, how can I estimate them?

Kokinos: By taking a random sample. You then make the inference to the population of the jar.

Socrates: Is random sampling a sufficient condition to make a valid statistical inference from the sample to the population?

Mathitis: Yes it is. And it would be reliable within the margin of the sampling error.

Peismon: You're not suggesting we do random sampling?

Socrates: I am simply wondering whether we can apply anything from that way of making valid and reliable estimates to the issue of making valid and reliable estimates of process capability. If random sampling from a fixed population provides reliable estimates, then can we apply that principle to getting reliable estimates of the within-subgroup variation?

Mathitis: I don't see how to apply it.

Peismon: I don't think it is valid.

Socrates: How are data for control charts collected?

Peismon: In subgroups of size greater than one typically, preferably.

Socrates: How are data for a single subgroup collected?

Peismon: The values should come from outputs that occur relatively close in time. Consecutive units is preferred or some other way to make them rational.

Socrates: Why do we want the outputs used for a subgroup to be consecutive?

Peismon: To make sure we have the best estimate of short-term or common cause variation.

Mathitis: We assume that consecutive outputs will have the least amount of variation.

Peismon: One reason we do this several times is to get that reliable estimate.

Socrates: In other words, the purpose is to see how alike, homogeneous, the outputs can be?

Mathitis: Yes, that's another way of looking at it.

Socrates: Then the same can be done using random sampling. Rather than using simple random sampling, one could use stratified random sampling. That is what control chart subgroup sampling is: a combination of stratified and systematic sampling.

Agrafos: How is it stratified?

Socrates: As Peismon and Mathitis stated, the purpose of stratified sampling is to find groups that are homogeneous to reduce not only the overall sample size but also the sampling error, which makes the estimates more reliable.

Mathitis: Polling uses stratified random sampling. That is why a sample of approximately 1000 can estimate a proportion from a population of 60 million with a margin of error of perhaps ±3%. The idea is to sample from homogeneous groups so you only need theoretically a sample of one from each group—at least for polls asking about two choices.

Kokinos: I see. If we collect consecutive units from a production line, we think that they will be as homogeneous as possible.

Socrates: That's the stratified part.

Mathitis: That makes sense. It matches exactly the purpose of subgroups.

Agrafos: And the systematic sampling?

Socrates: The systematic part is sampling periodically, for example, every 15 min or every hour or some other period.

Kokinos: How do you do this with random sampling?

Socrates: Suppose you plan to collect $k = 25$ subgroups each of size $n = 4$ for a total sample size of $nk = 25 \times 4 = 100$. You randomly choose times from the production period you are sampling. For example, if you were planning to collect data from 8 AM to 4 PM, then randomly select 25 times. Then, select four consecutive units starting at each period.

Peismon: You still have to check for stability.

Socrates: It depends on what you are estimating. The random sampling will give you a more accurate estimate of what happened during the time from 8 AM to 4 PM.

Kokinos: How is that?

Socrates: Suppose instead of randomly sampling the 25 subgroups you take them every 20 minutes starting at 8 AM. But something happens each half hour lasting 1–5 minutes and this occurs throughout the 8 hours of data collection. These events would change the variation or stability. But you will completely miss that.

Peismon: You could also miss it taking a random sample.

Socrates: Yes. The difference is that you will categorically miss it with the control chart sampling but have a chance of catching it with random sampling.

Kokinos: I don't know. Maybe both are equally reliable.

Socrates: If both are equally reliable, why shuffle a deck of cards instead of just taking every fourth card when playing bridge?

Mathitis: We want to make sure the deal is random. The original deck is ordered, and as hands are played, cards of the same suit are grouped together. I think a random sample is better.

Socrates: A random sample only requires one inference: from the sample to the population. Control charting requires two inferences: from the times when you collected data to the times you didn't and then from that sample to the population. In addition, you ought to confirm whether alarms are true alarms and non-alarms are true nonalarms.

Peismon: I still don't see how that gets us more reliable estimates if we don't check for stability.

Mathitis: I think I see how. We already said that random sampling provides reliable estimates and valid ones. The stratified samples, or subgroups, will tell us reliably and validly what the within-subgroup estimates are even if they are not in statistical control.

Socrates: Even if one or two subgroups are beyond the control limits, we can always exclude them to get an estimate of what the best, within-subgroup, or short-term variation is. Remember that we are estimating and estimators are not correct.

Kokinos: Couldn't we also calculate confidence intervals to understand the range of process capability? I don't see that done when control charts are used.

Peismon: The control chart limits provide the range.

Kokinos: Yes, but I don't see the range of C_{pk} calculated or provided. If the point estimate exceeds whatever goal, 1 or 1.33 or 2, then we say it is capable.

Peismon: I'm still not sure about this.

Mathitis: Why not?

Peismon: Everything I've read and have been taught says you need stability.

Socrates: The stratified random samples do get us reliable estimates of common cause variation. Certainly, it applies to C_p where that is the only estimate needed.

Peismon: That's why I don't think it works. We already addressed that issue. But for C_{pk}, C_{pU}, and C_{pL}, we need also an estimate of the average. We can't get that estimate without stability.

Mathitis: That isn't true. If randomly sampling from a jar provides valid, reliable estimates of the proportions of red and blue marbles, then random sampling of times will provide valid, reliable estimates of the average—proportions are averages.

Peismon: But it isn't for the process—the inherent variation of the process or the process mean.

Socrates: Recall that we talked about stability only applying to the period during which we sampled. Yes, we predict what might occur, but the assessment is only for that period.

RESPONSE SURFACE STUDIES

Socrates: Perhaps there is another way. What about DOE and response surface studies?

Mathitis: Yes. They are designed to find the optimal conditions for process performance.

Socrates: Is the "best" process capability achieved at the optimal conditions?

Mathitis: Of course.

Socrates: But sometimes, that optimal setting is not robust. It is not a "flat" surface, so it is difficult to maintain the best performance like this (Figure 23.2). Suppose there are two critical factors, X_1 and X_2, and more is better using the performance metric Y, on the vertical axis. If we have the result on the left (Figure 23.2a), the peak or optimal performance may not be stable. But if we have the result on the right (Figure 23.2b), it would be stable.

Mathitis: We would like the highest peak performance to be flat not a peak.

Socrates: If the optimal is a peak, is it likely we will find the best or the potential process capability, for example, C_p or C_{pk}?

Mathitis: No, because it's not very stable. It would be like me trying to walk across a balance beam. I'm sure I would lose my balance and fall.

Kokinos: Then, performance at the peak won't be stable.

Peismon: We're looking for the potentially best performance of the process.

Socrates: Now, we have a dilemma. Which do we use to determine process capability: the performance at the peak, which is best but unstable or something less than best but is stable? Is the best at the peak or flat area?

Mathitis: If it depends on stability, then it can never be the maximum potential performance if that is not stable.

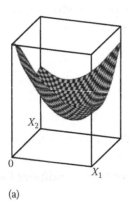

(a)

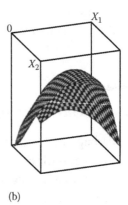

(b)

FIGURE 23.2

Response surface result of two variables X_1 and X_2 when (a) optimal performance is a peak and unstable or (b) optimal performance is flat and stable.

Kokinos: If we insist that process capability can only be determined reliably when the process is stable, and the optimal condition is unstable, we may never find it using control charts.

Socrates: And that is another issue. We know that there are global optimals and local optimals.

Mathitis: That's true.

Socrates: Unless we do a study that finds the global optimal settings and run the process under those conditions, even a control chart showing stability may only produce estimates of a local optimal—or not even that. If, by inherent process capability, we mean the best short-term variation at the global optimal, then any estimate not under those conditions—whether stable or not—will not yield an estimate of the inherent process capability, let alone a reliable estimate.

Mathitis: Neither will it be the inherent process capability at local optimals even if the process is or can be stable.

Peismon: That just means that the inherent process capability is dependent on the conditions.

Mathitis: That is the first time I have heard of that requirement.

Kokinos: So, we are saying that there are many process capabilities for the same process. One for each condition or setting?

Socrates: Exactly. Let's say we collect data at three different and independent periods to construct control charts. The charts all indicate that the process is stable during each period. But we see that the within-subgroup standard deviations and the grand means are different. The calculations of C_p and C_{pk} are different. Which reflects the inherent process variation and the "correct" process capability?

Kokinos: Or, worse. If we only have data for one condition and the process is stable but not optimal, we will mistakenly assume that is the best.

Mathitis: We either have a contradiction or we should always state the conditions when we state a process capability result.

Socrates: Do we always know the conditions?

Kokinos: Unfortunately, no. At least we don't always know which ones are critical.

Socrates: In any case, it seems that response surface studies are better suited than control charts for finding that inherent process capability globally, locally, or under any conditions.

Mathitis: I can see that. The control chart requirement only tells us whether a process was stable. Calculations of process capability when the process is stable only tell us the potential for the conditions under which the process was run, which may not be optimal. This contradicts the notion that it is the "best." The stability requirement merely gets at the reliability of that estimate.

Kokinos: But also a response surface experiment tells us what the optimal conditions are. So, it appears there are other ways to reliably and validly determine process capability than requiring stability and other ways to determine stability than using a control chart.

Mathitis: You mean, because an optimal point on a flat area of the response surface would be stable while that on a peak would not be stable.

Socrates: Exactly.

24

Myth 24: Specifications Don't Belong on Control Charts

Scene: *Break room (continuation of discussion on capability)*

BACKGROUND

Agrafos: There was a warning the instructor said about specifications and control charts.

Mathitis: I hope she said that specifications don't belong on control charts.

Agrafos: She did, but why not?

Peismon: Control chart limits represent the voice of the process. They are meant to distinguish between a stable process and an unstable one. The control limits, as we saw, are calculated from the data that come from the process. Specifications represent the voice of the customer. So, they represent different things. We would be comparing apples to oranges.

Kokinos: I was taught that control charts should not have specification limits because of the distinction between process stability and process capability. They are independent of each other, and as Peismon said, they represent different things.

Peismon: Control charts cannot tell you about process capability, only stability. An average can't tell you whether you are meeting specifications.

Mathitis: Putting specification limits on control charts will only confuse people. They will then use those limits rather than the control limits and maybe put them on all charts.

Socrates: The control chart has two axes. What is on the horizontal axis and what is on the vertical axis? Are they the same thing?

Mathitis: The horizontal axis has time and the vertical axis has the measure. Time and the value of the metric can be on the same chart because we are tracking the metric's stability over time. They are not the same thing so they are on different axes. The specification limits and control limits could be put on the same axis: the measurement axis or vertical axis.

Socrates: Are specification and control limits on the same scale or in the same units?

Mathitis: Yes, but specification limits define what is good while control limits define sampling variation, common cause variation, or the limits of process stability. Since stability is only an indication of consistency within some bounds, then stability does not mean consistently good. A stable process could produce only defects with respect to customer requirements.

Peismon: I just wouldn't put them on the same chart for another reason.

Socrates: What's that?

Peismon: Specifications apply to single units, while control charts apply to statistics on multiple units. The $\bar{X}$ chart applies to means of subgroups, the R chart applies to ranges of subgroups, the P chart applies to proportions of subgroups, and so on.

Socrates: So, specifications apply to single units and not to statistics?

Peismon: Yes, that's why they shouldn't be on the same chart.

Socrates: It seems all of you agree that there is no control chart that should have specification limits because it would be applying them to the wrong thing and it would confuse people.

Peismon: Yes, that's right.

RUN CHARTS

Socrates: Can I put specifications on a time series chart of individual units?

Mathitis: Sure. We do that often. But a time series chart doesn't test for stability.

Socrates: What are the axes on a time series charts?

Kokinos: They are the same as on a control chart with time on the horizontal axis and the measurement on the vertical axis. But that doesn't mean that specifications make sense on the chart. We discussed this before when we talked about the four states of a process, which were combinations of stability versus capability. Control limits answer whether stable or not and specification limits answer whether capable or not.

Socrates: How about a run chart, which is a time series chart?

Mathitis: Run chart, time series chart—they're the same thing.

Socrates: So, I can put specification limits on a run chart or time chart.

Mathitis: Yes. But note that run charts use the median for the center line not the mean, while a time series chart may not have anything else on the chart but the points.

Socrates: I can use the run chart to test for patterns of runs and trends, can't I?

Mathitis: Yes, various software programs do that.

Socrates: Aren't these tests of stability?

Mathitis: I don't think so.

Peismon: I agree. Besides Wheeler, and apparently Deming and others also say only a control chart can check for stability: "To determine if a process or system is in a stable state, a Shewhart (control) chart is needed" (Perla et al. 2011).

Socrates: What does stability mean?

Peismon: Stability means the process is consistent. There are two kinds of variation: common cause and special cause. Common cause will show on a chart as random variation because there is only one distribution. Special cause will show on the chart as nonrandom variation.

Socrates: Then, we are simply checking if the patterns are random or not (Myth 19).

Peismon: That's right.

Socrates: Can we not use some of the rules from a control chart on a run chart?

Kokinos: Yes. The trend pattern of six or more consecutive points increasing or decreasing can be applied to both a control chart and a run chart.

Mathitis: Also, eight or more consecutive points on the same side of the center line and 14 consecutive points alternating up and down.

Socrates: Are there any other patterns checked for on a run chart?

Kokinos: Yes, you can check for too many runs or too few runs.

Socrates: What defines too many or too few?

Kokinos: Given the number of points, we can calculate how a range of many runs are expected at a certain confidence level using combinatorial calculations.

Socrates: Then, are these not patterns of nonrandomness?

Mathitis: I've used run charts for that purpose. Runs and trends are patterns that shouldn't occur if there is a single, common distribution.

Kokinos: I hadn't thought of that.

Socrates: If such patterns occur, doesn't that signal that the process may be unstable? If no patterns occur, doesn't that signal the process may be stable? Are not these patterns alarms also? Doesn't a run chart test for process stability?

Mathitis: Yes, to all your questions. Now, I would say yes it does.

Peismon: I still say no. Run charts don't test for stability or at least are not very effectively. Using a run chart can't show stability because of failure to catch points beyond the control limits.

Socrates: Can't at all or can't using that pattern but can using other patterns? But you are right since there are no other lines but the center, median line. Then, isn't the reverse true also?

Peismon: What do you mean?

Socrates: The control chart doesn't test for too few or too many runs—although it could. In other words, since we now have more than one nonrandom pattern as an alarm, should we not conclude using your argument that Shewhart's charts showing no alarms does not show stability?

Mathitis: You're right. The nonrandom patterns are different, so perhaps we should use both types of charts or add the too few or too many runs as patterns to use on control charts.

Socrates: Are not each of these rules based on a low probability of occurring if the distribution stays the same through the period of the data?

Kokinos: Yes, you're right. We talked about the rules applied to control charts having the same false alarm rate and so they would have the same false alarm rate on the run chart.

Socrates: Do you find it odd that, for the "too few" or "too many" patterns, the false alarm rate is 5% but less than 1% using Wheeler's Empirical Rule for three standard errors?

Peismon: One more reason why run charts do not determine stability.

Kokinos: But the 5% would be equivalent to using narrow limits which is more consistent with the purpose of a control to detect lack of control, as we discussed before (Myth 18).

Socrates: Good point, Kokinos. Recall in the flowchart describing how to use a control chart that we said when there were no alarms we automatically assume the process was stable?

Mathitis: Yes, and we said that the step missing were checking whether nonalarms were true or false nonalarms. If too few or too many runs also indicate a change in the process and we are not checking for those changes, then even more of a reason why we should fail to reject that the process is stable rather than reject that it is when there are no alarms.

Socrates: Peismon's argument and one that others use is not a reason for rejecting run charts as checks on stability but an argument that the false nonalarm rate is too high. It misses some types of nonrandomness, which are indicators of process changes. But that is then inconsistent with those who use control charts and never check whether nonalarms are true or false.

Mathitis: If the argument is that the run chart says a process is stable when it isn't, that's a false nonalarm, not a false alarm.

Socrates: What about specifications on a time series chart or run chart?

Kokinos: We created histograms and put specifications on the graph; we created individual value plots and put specifications on the graph. In each case, it was to graphically show the capability of the process. How well we were meeting specifications.

Mathitis: I agree with those types of graphs. For that purpose, not for checking stability.

Socrates: On a run chart, you can check for stability and simultaneously see which units fail the specifications. That is, you can check for both stability and capability.

Mathitis: Okay, I see that.

Peismon: As others have shown, it would be a weak check for both. It won't detect extreme points and the other patterns based on one or two standard errors from the center line.

CHARTS OF INDIVIDUAL VALUES

Socrates: True. Let's see if we can address that issue. Isn't a control chart a time series chart?

Peismon: Yes.

Socrates: And isn't an *I* chart for individual values a time series chart?

Peismon: Yes, but a control chart including an *I* chart only checks for stability, which is not done by using specification limits.

Socrates: I'm not saying that the specification limits be used to check stability. I'm saying that both control and specification limits can be put on an *I* chart and then both stability and which units meet specifications can be evaluated simultaneously. And we can add the nonrandom patterns of too few or too many runs.

Mathitis: Yes, we can. That just seems to me so contrary to what I learned and was drilled into me. It also seems that it would be confusing.

Peismon: I don't think you can or should do that.

Socrates: Suppose I'm taking a sample at the end of production for product release. To release each unit, I measure and compare to the specifications. I can show that on a graph with the units on the horizontal axis and the measurement on the vertical axis and the specification limits parallel to the horizontal axis. Treat it as a run chart. Can I separately create a control chart for the individual values to check for stability?

Mathitis: Yes, one chart to compare to specifications and another to check for stability.

Socrates: Since both charts have the same axes, both horizontally and vertically, and the same points, can I overlay one graph on top of the other and use one chart instead of two?

Kokinos: Won't both horizontal axes cover exactly the same range on both charts and the same for both vertical axes?

CONFUSION HAVING BOTH CONTROL AND SPECIFICATION LIMITS ON CHARTS

Peismon: I see what you are saying but I still think it would be too confusing. People on the manufacturing line would find it confusing.

Socrates: Those people aren't trained on statistical process control?

Peismon: Yes, they are. Many are Green Belts.

Socrates: It sounds like you are saying they are capable enough to understand Six Sigma and various statistical tools but not intelligent

enough to understand the difference between specification limits to check whether units are good and control limits to check
whether a process is stable if they are on the same chart.

Mathitis: Well, not exactly. I agree with Peismon. From what I've heard,
I'd be hesitant to do it.

Socrates: Consider this—from Shewhart himself. He graphs a measurement that is outside the control limits but inside the specification limits, all on the same graph:

Thus we see that for reasons of economy and quality assurance it is necessary to go beyond the simple concept of go, no-go tolerance limits of the
customary specification and to include two actions limits *A* and *B* [sta-
tistical control limits] and an expected value *C*, as shown schematically
in fig. 6.... Likewise, it should be noted that, although the action limits
A and *B* may lie within the tolerance [specification] limits L_1 and L_2...
(Shewhart, 24)

Mathitis: It certainly seems as if he is plotting points on the same graph
that has both specifications and control limits. With the caveat
that the plotted points represent individual values. But, what he
said reminds me of why I think we shouldn't put specification
limits on control charts—on the *I* chart.

Socrates: Why is that?

Mathitis: The control limits, as the quote from Shewhart says, are action
limits. We are monitoring the process to see whether it will continue to produce the same quality of product as it has in the
past. So, special cause patterns are signals of process change.
That change could be in the centering or the spread. When we
see a change, we need to take action.

Agrafos: And if you have a defect, don't you also take action?

Socrates: Good point, Agrafos. Aren't control limits the voice of the process while specification limits are the voice of the customer?

Mathitis: Yes, that's right.

Socrates: So, control limits can only tell us about stability or the consistency of the process but nothing about whether that consistency
is good, mediocre, or bad, right?

Mathitis: Yes.

Socrates: From the customer's point of view, does it matter whether it is
consistently good, mediocre, or bad?

Mathitis: Of course. The customer wants consistently good.

Socrates: How does only responding to instability ensure consistently good product?

Mathitis: It doesn't. That's why you make sure the process is operating at the right level and then ensure consistency through control charts that monitor stability.

Socrates: Does that mean that both stability and capability must be checked?

Mathitis: Yes.

Socrates: Is it possible for one to change without the other changing or being evident that it changed? For example, is it possible for capability to worsen and there be no special cause variation on the control charts?

Peismon: How could that happen?

Socrates: Do all special cause patterns cover all possible ways the process could be become unstable? Does the run chart, for example, identify process changes that a control chart does not?

Peismon: Well, no, yes. Not all possible changes are identified by either chart.

Socrates: Isn't one special cause a point outside the control per Shewhart?

Peismon: Yes, that's the most commonly used pattern.

Socrates: How likely is that to occur with a consistent process?

Peismon: Much less than 1%—a fraction of a percent.

Socrates: If you have only 25–30 points on the chart, is that likely to occur?

Peismon: No, but I don't see your point.

Socrates: Can the process change but not enough to produce a point beyond the control limits? Can the process change so that the probability of being beyond the control limits is 1%?

Peismon: I don't know—that's what the control chart limits and patterns are supposed to detect.

Mathitis: I see what Socrates is saying. There could be a minor change and not show on a chart. A slight change in the mean might increase the probability of being beyond the control limits from say 0.27% to 1%. In that case, the odds are against us detecting it.

Kokinos: Also, if not all output is plotted on the chart, then the process change could occur between the times of plotted points. And we have been told that the 3 standard error limits are designed to detect mean shifts of at least 1.5 standard deviation.

STABILITY, PERFORMANCE, AND CAPABILITY

Socrates: Checking for stability and capability must be done separately because they use different limits.

Mathitis: That's correct.

Socrates: Does consistently good, or capable, mean 100% defect-free production?

Kokinos: We talked about this before. Wheeler, for example, views a $C_p = 1$ as capable. Shewhart always used 99.73% good as capable. But with modern quality initiatives like Six Sigma, we look for higher defect-free percentages.

Socrates: Do you need to know when a defective product is produced?

Mathitis: Of course. We should.

Socrates: Does the control chart tell you whether you are producing a defective product?

Mathitis: Not all. The charts on defects and defectives do. Continuous data charts—I, $\bar{X}$, S, and R charts—don't. They can't since the limits are control limits not specification limits. But if we have established the process capability for the stable process, then we know the defect rate.

Socrates: Do you know which units produced are defective from the control chart?

Mathitis: No.

Socrates: But you do need to know which ones they are to take action on them, such as scrap or rework the defectives ones, right?

Mathitis: That's correct.

Socrates: Let's summarize. There are three things you need to know: (1) the process capability, (2) whether the process is stable, and (3) which units are defective.

Mathitis: That's correct.

Socrates: Changes in the first two—capability or stability—require action and identified defective units require action.

Mathitis: Also true.

Socrates: Control charts, even individual charts, can only tell you about stability when they only have control limits.

Mathitis: I get it. If control charts also had specification limits, then I can begin to answer the other two. At least, I can identify those units I measured for the control chart that are defective if they are outside the specification limits. And, I can do a quick check on the process capability by calculating percent good from the points plotted on the chart. I suppose I could even determine a trigger point of no more than k points out of n outside the limits to ensure the process capability hasn't changed. But my sample size may be too small—maybe 15–25 points from a production lot of thousands.

Socrates: True. The question is, can the cost of having this early warning signal be worth the benefit of including the specifications and assuming you have people educated enough to know the difference between stability and capability?

Mathitis: I think it would. The cost of adding the specification limits is minimal. But I am worried about what we said earlier that may have a higher cost. Sometimes, people confuse specification limits and control limits. Won't that make it worse?

Socrates: If the people do not understand the difference between stability and capability or confuse the two, then adding them both on the same chart when appropriate may very well help clarify the difference. In addition, since the action or response plan to each is different—including the third issue of how to identify and what action to take on defective units—then that should also help clarify the difference.

Mathitis: I see. By only having one type of limit, I may confuse it with another. If I have both types and they are not the same, then it will make it clear that they are different.

Kokinos: In fact, sometimes people think that stable processes are always capable. So, putting both specifications and control limits on the I chart may help resolve that confusion.

Socrates: It sounds like you might be willing to try it.

Mathitis: Maybe. It's just that I've heard that people start putting specification limits on all types of control charts when it only applies to I charts.

Socrates: But recall that we discussed whether stability was required for assessing capability.

Mathitis: And we said that some capability indices do not require stability and there are other ways of assessing the other capabilities without relying on stability.

Kokinos: We also distinguished between performance indices and capability indices.

Socrates: Great. Now a p chart tells us what?

Kokinos: Typically the proportion of defective units.

Socrates: Did we agree that capability and performance indices are the proportion of good units?

Mathitis: We did.

Socrates: When unstable, does a *p* chart tell us about performance, for example, P_p and P_{pk}?

Mathitis: Yes, it does.

Socrates: When stable, would it tell us about C_p and C_{pk}?

Mathitis: It could if it were based on the within-subgroup variation. But that would occur when the process was stable.

Socrates: Don't *p* charts, *NP* charts, *c* charts, and *u* charts check for stability and capability simultaneously?

SPECIFICATIONS ON AVERAGES AND VARIATION

Peismon: Yes, but not with specifications on the charts. At best, specifications could be put on *I* charts for individual values. Shewhart recommended $\bar{X}$ charts with *n* = 4. The other charts he used were also not on individual values. It sounds like you are recommending only the *I* chart.

Kokinos: Wheeler recommends the *I* chart.

Mathitis: We are not saying not to use the other control charts. We are saying that *I* charts can have specification limits on them and the alarms of too many or too few runs. We also said that *p* charts or *u* charts, both proportions, tell us about stability and capability simultaneously.

Socrates: What if I told you that *legally* you can put specifications on control charts?

Mathitis: I don't understand. This isn't a legal issue.

Socrates: Did you know that there is at least one industry that does put specification limits on control charts other than *I* charts and is accepted (if not encouraged or required) by the federal government to do so?

Mathitis: I did not know that. What industry?

Socrates: The medical diagnostic industry including its customers, like clinical laboratories and peer-reviewed clinical journals. Despite some efforts to convince manufacturers and clinicians that specifications should be on single laboratory results and not on means and coefficients of variation (that's what they use rather than standard deviations), the industry continues to put specifications on control charts for means and variations.

Kokinos: Yes, that's an industry I worked in. It's accepted by the Food and Drug Administration (FDA) on product submissions. The FDA's latest *Guidelines* include an example of specifications on a mean chart (Global Harmonization Task Force 2004). Arguing whether it is right or wrong might be misplaced. The discussion should be on how inefficient and possibly ineffective it is to have specifications on statistics of multiple units (results) rather than on each unit.

Socrates: A question for debate is: Are there more medical decision errors because of specifications on statistics as opposed to specifications on individual results?

Mathitis: In that case, specifications can go on $\bar{X}$ control charts and S charts. Even R charts, by converting specifications on standard deviations to specifications on ranges, inverting the conversion of the average range to an average standard deviation.

Peismon: Even if they confuse people?

Socrates: Then, the issue is not that they cannot be done but whether it is useful. If one chart conveys the information needed, then maybe the issue is how to teach people so they are not confused. After all, if people are confused about which standard deviation to use or whether it is a standard deviation or standard error, should we stop using control charts?

Mathitis: No. The issue that some people are or would be confused is not a valid reason for not putting them on the appropriate control charts. If some people are confused by control charts, we wouldn't argue that no one should use them. At the very least, we should stop saying they can't or shouldn't be done, but provide a warning.

Kokinos: And, more effective teaching.

25

Myth 25: Identify and Eliminate Assignable or Assignable Causes of Variation

Scene: *Break room (continuation of project meeting discussion)*

BACKGROUND

Socrates: Can we review the flowchart or the steps for using control charts? The purpose of a control chart is to ensure process stability, correct?

Kokinos: That's one purpose.

Socrates: And we do that by monitoring a process to determine when we should take action and when we shouldn't?

Kokinos: Yes. The control limits are action limits to indicate when the process changed and the action is on the cause system, as we talked about before.

Socrates: That's instability.

Kokinos: Instability or out of control—they all mean the process has changed. If the process is stable, no action is needed. If the process is unstable, then find the causes and eliminate them.

Peismon: Shewhart called the variability inherent to the process common cause variation. It is variation that, no matter how we search for its cause, we are not likely to find it. That is why no action should be taken. The other kind of variation he called assignable cause variation. It was variation whose cause could be found, so he called it assignable cause of variation.

Socrates: And the action is to eliminate the assignable causes of variation.

Kokinos: To investigate, do the analytic study, and find the assignable cause.

Mathitis: And then, when you have found the assignable cause, eliminate it.

Peismon: Exactly. In his 1931 book, Shewhart provided three postulates in the section on the "Scientific Basis for Control." His third postulate states: *"Assignable causes of variation may be found and eliminated.* Hence, to secure control, the manufacturer must seek to find and eliminate assignable causes [italics in original]" (Shewhart 1980, 14).

Also, in his 1939 book, he repeatedly says, "as these assignable causes are found and eliminated, the variation in quality gradually approaches a state of statistical control" (Shewhart, 37).

Kokinos: Wheeler says the same thing when discussing the logic of control charts, which I used to create the flowchart: "Take action to identify and remove assignable causes" (Wheeler and Chambers 1992, 39). Later in the chapter on using control charts effectively, he says: "Whenever a point falls outside the limits, it is most likely due to the presence of some Assignable Cause, and it is appropriate to identify and remove this Assignable Cause" (Wheeler and Chambers 1992, 90).

Peismon: Even those who disagree about probability limits and hypothesis testing and other aspects of control charts agree that assignable causes should be eliminated.

Socrates: All assignable causes should be eliminated?

Peismon: Yes, all. What else would you do with them?

ASSIGNABLE CAUSES VERSUS PROCESS CHANGE

Socrates: Doesn't that assume that assignable causes of variation are always bad?

Mathitis: Yes. Why else would we remove them?

Kokinos: That's what I was taught. Everyone says to eliminate them.

Socrates: What does the occurrence of an assignable cause mean?

Peismon: That the process has changed.

Kokinos: Yes—that's what it means to have an out-of-control process. It changed because of the assignable cause. Out-of-control is what

Wheeler calls chaos, or brink of chaos if the process is capable (Myth 23).

Socrates: It certainly sounds like we don't want assignable causes if that means we have chaos. Then all assignable causes are bad—make the process worse?

Kokinos: Yes.

Mathitis: More defects are being produced as a result of the change.

Socrates: I understand that more defects is bad. Is fewer defects bad?

Peismon: No, and that's how we get fewer defects. By eliminating assignable cause variation.

Socrates: In your view, assignable cause variation always results in more defects.

Mathitis: Perhaps not all the time. The change could be minor.

Socrates: We said that an assignable cause means a process change. But the difficulty I am having is that we're saying assignable causes are bad, but is process change always bad? After all, to make an improvement in a process, we must change it. So that change is not bad.

Mathitis: Maybe not all process change is bad—certainly not process improvement.

IS INCREASE IN PROCESS VARIATION ALWAYS BAD?

Peismon: But more variation is bad. Having assignable cause variation is adding special cause variation to common cause variation, which increases the variation. That's bad.

Mathitis: When we look at Shewhart's work, we see that he used only the rule a point beyond the control limits. That clearly shows a change in variation, and it's an increase of variation.

Socrates: Is it always true that more variation is bad? Is more diverse products bad?

Peismon: Not products. We're talking about process performance. Increasing process variation is always bad.

Mathitis: If you have a stable process, which is good, and it becomes unstable, then it must change for the worse.

Kokinos: Instability is never good.

Socrates: I'm not asking whether instability is good—although that's another topic worth discussing. I am asking two questions: (1) Is all assignable cause variation bad? (2) Is all increases in variability bad?

Kokinos: What's the difference if the existence of an assignable cause means instability?

Socrates: In this technical sense, that's true. But instability doesn't mean that the addition of assignable cause is bad. Let's consider an example.

Kokinos: I think that would help.

Socrates: Let's say you are monitoring processing time, for example, how long it takes to go through the 10-item checkout line at the grocery store or some other queuing process. Is that an appropriate measure for a control chart?

Kokinos: Sure.

Mathitis: If you are measuring in minutes, you would have time in minutes on your vertical axis and you could have person or groups of people in the order in which they went through the line on your horizontal axis.

Socrates: Great! On the vertical axis, is any direction better than another?

Mathitis: Yes, longer times are worse. Lower times are better. I think I see what you mean.

Kokinos: I don't.

Socrates: Imagine we had an *I* control chart of individual times of every person going through the line. We construct the 3 standard error limits and for a while no individual time exceeds either limit. Is the process stable, in control?

Kokinos: Well, yes, all individual times are within the 3 standard error limits.

Socrates: Is it in control, stable, or can we just conclude that it appears to be in control? Or, that we have no evidence that it is not unstable?

Kokinos: If you want to be technical, yes, you are right. We have evidence that it is in control but we could be wrong. We already discussed that issue.

Socrates: Now, suppose that the next person's time occurs beyond the 3 standard error limits. Is there evidence that the process has changed?

Kokinos: Yes. Technically, we don't know if it's unstable but we have strong evidence of it being unstable. The 3 standard error limits provide a strong case for claiming it is unstable.

Peismon: I would say it's unstable because the evidence is so overwhelming.

Socrates: Is this an increase in variation?

Peismon: Yes. The variation between subgroups, individuals in this case, increased. It could be because there was an increase in within-subgroup variation or because there was a shift in the average, but there was an increase in variation.

Socrates: And that is why you say its cause should be found and eliminated.

Peismon: That's right.

Socrates: If lower times are better than higher times, should we treat the individual time that is beyond the 3 standard error limits the same regardless of whether it exceeds the upper limit or is less than the lower limit?

Mathitis: That's what I thought you were getting at. Clearly, the first case—greater than the upper control limit—is an indicator of performance getting worse while the second case is an indicator of performance getting better.

Socrates: Are both assignable causes?

Mathitis: Yes.

Peismon: Are you saying that you would not want to eliminate them both?

Socrates: I wouldn't. Would you?

Peismon: Yes, because they are both assignable causes—and that's what we do with them.

Socrates: I would eliminate the second because it is a bad assignable cause and incorporate the first into common cause variation because it is a good assignable cause.

Mathitis: I would too.

Kokinos: Before, I would have done what Peismon says because of how I was taught. But, the more I think about it, the more I agree with you, Socrates. I just accepted that all assignable causes are bad.

Socrates: In both cases, did the variation increase?

Mathitis: Yes, but in only one case was the increase bad.

Socrates: Do you now think that all instability is bad?

Kokinos: I'm not sure about that—that still seems right that all instability is bad.

Mathitis: I tend to agree with Kokinos. Yet, I see that in this example one kind of instability does provide us an opportunity for improving while the other kind made the process worse.

Socrates: That's why I hesitated to agree with Kokinos—it depends on what we mean by instability. If instability just means a change, then I would say it is not always bad. Switch the words so that its connotation doesn't influence your thinking.

Peismon: That's just one example.

GOOD ASSIGNABLE CAUSES

Socrates: However, we can create general cases that cover every possibility.

Peismon: How?

Socrates: Why were we able to determine which assignable causes were good and which were bad in this one example?

Mathitis: In this example, less time is better. Points above the upper control limit are bad and points below the lower control limit are good.

Socrates: Not just points below the lower control limit are good. Any change lowering time would be good.

Kokinos: Then, the opposite is true. If we have a measure where more is better, then changes increasing the result would be good while changes decreasing would be bad. For example, points greater than the upper control limit are good and points less than the lower control limit are bad.

Mathitis: But also increasing trends would be good in your case while decreasing trends would be good in the time example.

Agrafos: But we typically have the case where we want the measure to be a specific value, called the nominal or target. How do we know which direction is better then?

Socrates: You have identified three types of performance measures: more is better, less is better, and a specific value is better. The queuing time is an example of the second type. Sales per week might be an example of the first type.

Peismon: But when an exact value is best, both directions worsen performance. In manufacturing, I think this is the common situation and Shewhart was discussing manufactured product. That's what Shewhart focused on.

Socrates: Let's look at it more carefully. What do we mean by the third case? In the first case, we would have a lower specification limit to define good because more is better. In the second case, we would have an upper specification limit to define good. What about the third case?

Peismon: We need both specifications around the value that is best—the nominal or target. So, any change away from that is bad—especially if you believe in Taguchi's loss function.

Socrates: When you say "any change away from that is bad," to what does "that" refer?

Peismon: The nominal or target.

Socrates: Now, we can show that not all changes even in the third case are bad.

Peismon: How?

Socrates: Do we expect every production unit to have its measure equal the nominal?

Peismon: No. The nominal is where we set the process average.

Socrates: Do you believe that every process produces a product that has its average exactly equal to the nominal?

Mathitis: Of course not. We often find processes with centering problems. That's why for continuous, quantitative data, we have $\bar{X}$ charts for the mean and S or R charts for the variation.

Peismon: That's why we have C_{pk} besides C_p and P_{pk} besides P_p.

Socrates: If the process is not centered exactly on the nominal, then is there a change to that average that is an improvement?

Mathitis: I see what you are saying. As long as it's not perfectly centered, the average must be either greater than nominal or less than nominal. In either case, there is one direction that is good and another direction that is bad.

Kokinos: Okay. In the situation where the average does not equal nominal, we have a case similar to either the first or the second case we talked about before.

Socrates: What about variation?

Mathitis: That too. In general, less variation is better. If we use a variation control chart like an S chart or R chart, points beyond the upper control limit are bad assignable cause indicators and points below the lower control limit are good assignable cause indicators on these charts.

Kokinos: But we often don't have a lower control limit. Well, it is zero and no variation can be negative, not the range or the moving range or the standard deviation.

Socrates: What about the other assignable cause alarms or rules? Can they indicate good assignable causes?

Kokinos: I guess so. Sure. We can have two out of three consecutive points beyond the 2 sigma limits. These can occur either in the zone above the center line or in the zone below the center line. The first case would be a bad assignable cause indicator when less is better or the mean is above the center line. And the second would be a good assignable cause indicator for variation.

Mathitis: It's possible that each rule can indicate an opportunity for improvement—a good assignable cause indicator.

Kokinos: I can see that now and see its value.

Agrafos: I'm not sure how that connects to the discussions we had.

Socrates: One connection is that if we only include the cost of missed opportunities to eliminate bad assignable causes, we haven't accurately or completely assessed the economic value of control chart limits. We did not include the missed opportunities of improving—not just of stabilizing—a process. Our discussion on whether 3 standard error limits are the most economical did not include the benefit of making improvements by incorporating good assignable causes (Myth 18).

Mathitis: I get it. If we only use control charts to look for indicators of when things worsened, we miss all the indicators when things improved serendipitously.

Socrates: Another point is connected to the debates about control chart limits being probability limits and control charts being statistical tests of hypotheses. One view was that these ideas—even if true—are not useful, confuse people, and can prevent the use of control charts. This idea that all assignable causes should be eliminated is not true. But by stating and restating it confuses people and prevents them from taking advantage of opportunities to improve.

Kokinos: I was never taught that.

Socrates: What about the discussion on whether specifications belong on control charts?

Mathitis: This adds to that discussion. With specifications and the target on appropriate charts, it would be easier to identify which assignable cause signals are potentially good and which are potentially bad. Or, at least, know which direction is good and which is bad.

26

Myth 26: Process Needs to Be Stable before You Can Improve It

Scene: *Break room (continuation of control chart discussion on assignable causes)*

BACKGROUND

Peismon: We finished with the idea that not all assignable causes are an indication that the process has gotten worse, even when the assignable causes create increased variation. But when we have assignable causes, it means the process is unstable or not in control. The process is in one of the two states that Wheeler identified: chaos or brink of chaos (Wheeler and Chambers 1992, 17).

Mathitis: That was why identifying assignable causes that would improve process performance would help get us out of those states into the other ones.

Peismon: But we still need to get the process under control, before we can make any improvements. We can't just move directly from good assignable causes to an improved process.

Socrates: Are you saying that the process could not be improved unless it was stable?

Kokinos: I was taught the process needs to be in control before you can improve it.

Socrates: Why would you need to stabilize it first?

Kokinos: The instructor showed a flowchart for the procedure. You define the problem, determine a baseline, and then the next question in the flowchart was "Is the process stable?" If the answer is yes, then you continue, but if the answer is no, you have to find the causes of instability, make changes to address them, and check for stability by collecting more data.

Socrates: That seems like a lot of work. You're sure it is all needed?

Peismon: Yes, and stabilizing a process doesn't improve it. You then have to find causes of the current performance to make more changes to the process to improve the performance from the stable performance level.

Socrates: Are there two kinds of assignable causes and therefore two analytic studies must be done? One is to find the causes of instability and the other is to find the causes of improvement.

Kokinos: I hadn't thought about it that way, but the flowchart does seem to indicate that.

Socrates: Are there any exceptions? Do you always need a stable process before making improvements, no matter the performance measure or process or type of out-of-control alarm?

Peismon: Yes. Why would it be true in some cases and not in others?

HISTORY OF IMPROVEMENT BEFORE THE 1920s

Socrates: What are the ways to show stability?

Kokinos: With a control chart. Wheeler says that "[c]ontrol charts are the only way to break out of the Cycle of Despair" (Wheeler and Chambers 1992, 18), which is cycling between the two chaos states and brink of chaos.

Mathitis: We know that run charts also check for stability but not as rigorously as control charts.

Peismon: Others also say that control charts are the only way to check for stability.

Socrates: When were control charts invented?

Mathitis: Last century. Shewhart's book was published in 1931. He was using them in the 1920s.

Socrates: Before their invention, no one had a way of testing for stability, if that's the only way to check for stability.

Peismon: That's true.

Socrates: Before the 1920s, did anyone ever improve a process?

Mathitis: Of course.

Socrates: How did they know that the process was stable?

Mathitis: They couldn't have known.

Socrates: Then, if stability is required, every time before 1920 that anyone improved a process, it coincidentally happened that process was stable at the moment they decided to improve it.

Kokinos: I suppose that's a valid conclusion although it does seem too coincidental.

Socrates: Do you think that there were just a few improvements before 1920?

Mathitis: No, thousands if not millions.

Peismon: On the other hand, we don't know if people had tried various times to improve processes and it wasn't until the process became stable that they were then able to successfully improve it. They wouldn't know why it was possible since they didn't know about stability.

CONTROL CHART FALLIBILITY

Socrates: Today, you use a control chart to get the process under control, before you try to improve it.

Peismon: That's right. That's what I teach class participants and project teams.

Socrates: Recall that the flowchart for control charts shows that alarms aren't 100% accurate.

Peismon: But when they do occur, the chances of being wrong are very low.

Kokinos: We saw that through Wheeler's simulations, the chances are no more than approximately 2%.

Peismon: And they can be much less, even 0.1%.

Socrates: If such a pattern occurs, you conclude that the process is unstable with an error rate of 2% or less.

Kokinos: That's right.

Socrates: And that rate is for both cases: (a) being wrong when the process is stable and (b) being wrong when the process is unstable?

Kokinos: No. Only for false alarms.

Mathitis: We also saw that control charts don't check for too few or too many runs, which are assignable causes also.

Peismon: That's not a problem, because we can improve the process when it is stable, whether we know it or not. That's what I was saying probably happened before the 1920s.

Socrates: Peismon, when there are no patterns, you conclude that the process is stable and then—and only then—you can improve the process. Is that right?

Peismon: That's right.

Socrates: Do you do other checks or tests to confirm the process is stable?

Peismon: No, I just use control charts. But we extend the control chart limits to continue checking process stability.

Socrates: You don't know when the lack of out-of-control patterns is wrong, do you?

Peismon: No, but it works for me.

Kokinos: But we saw that the flowchart of the steps for using control charts didn't test whether nonalarms were true or false. We don't know how often we thought the chart was stable, in control, and it really wasn't.

Socrates: You have improved processes when they don't show these patterns?

Peismon: Yes, scores of times.

Socrates: Even when the lack of out-of-control alarms was wrong, you still concluded that the process was stable and you improved the process.

Mathitis: I see what you are getting at, Socrates. If he was able to improve when the process actually was unstable because the lack of indicators was wrong, then stability can't be required for improving processes.

Peismon: But not having any of these assignable cause patterns means that the process is stable.

Mathitis: Not 100% of the time. That means that you were able to improve unstable processes without first stabilizing them.

Agrafos: We agreed that control charts are fallible. In the cases we mistakenly think that the process is stable, change the process based on the false belief assumption and it "works," we are confirming that stability is not required.

Socrates: It may be that stabilizing makes it easier to improve or that there are other benefits to stabilizing first, but that is a different claim from stability being required. Determining when stability is useful or more useful is a different issue.

Peismon: But stabilizing doesn't improve the process. You have to eliminate assignable cause variation to improve.

STABILIZING A PROCESS AND IMPROVING IT

Socrates: Suppose a control chart shows that the process has been in control up until this past week. During this week, it went out of control, resulting in an increase in defects. Can we reduce the defect rate without first making the process stable?

Kokinos: Not based on what I was taught.

Socrates: Is making the process stable—like it was before it went out of control—improving the process? In this example, is stabilizing the process to the previous level improving the process?

Kokinos: You haven't improved the process; you've just made it the same as before. That's what you need to do before making improvements.

Socrates: What aspect of the process might you be monitoring with a control chart?

Kokinos: It could be the average and the variation. So, you would use an $\bar{X}$ chart and either an R chart or an S chart.

Socrates: Would these also be the measures of process performance you were trying to improve?

Kokinos: They could be.

Socrates: Did we not just discuss that assignable causes could be good (Myth 25)?

Kokinos: Yes, for all charts, an assignable cause might be an indication that performance improved, for example, less variation, fewer defects or defectives, more favorable average.

Socrates: Then, if we find the assignable cause in these cases and incorporate them into common cause variation, won't the process change in the direction of improved performance?

Kokinos: I suppose so.

Socrates: In those cases, you would then improve process performance before stabilizing.

Peismon: But, as I said, you are only bringing the process to its level of performance before the assignable cause variation.

Mathitis: He is talking about, say, a point below the lower control limit for an S chart. We find the cause of that less-variation signal and adjust the process accordingly. We don't remove the assignable cause; we incorporate it so the process performs more often like that one point. That improves process performance with respect to variation by reducing it.

Peismon: But adjusting the process according to the assignable cause doesn't necessarily bring it under control. That one point showed that the process was out of control then, so why would it become in control incorporating the assignable cause?

Socrates: If the setting for that factor had not been optimal before the assignable cause event and that is why it was out of control, and you then set it to a more optimal level, would you not have improved the performance of the process?

Mathitis: I see what you mean. It's applying the response surface approach. The process could be at a local optimal for a specific factor, it goes out of control, we find the cause is the setting for that factor, and we adjust it to a global optimal. Yes, that would improve the process, whether it applied to the variation or the centering.

Peismon: My question was about stabilizing the process at the new level of performance.

Socrates: With response surface, we can also see whether the process could be stable. The issue of stability after improvement is different from the issue of improvement without stability.

Mathitis: Also, the same question occurs when you stabilize first and then change the process to improve it. How do you know it will be, let alone stay, stable?

STABILITY REQUIRED VERSUS FOUR STATES OF A PROCESS

Socrates: Can lack of stability be the reason you have a low process capability? We did discuss the four states of a process and one state was unstable, not capable. If these are independent of each other, then we can move to the state of unstable but capable.

Kokinos: On the basis of the four states, moving from chaos to brink of chaos is an improvement.

Peismon: And it is an improvement in capability, which is what we are talking about.

Socrates: Do we improve the process as we move from unstable to stable to not capable to capable? Cannot the four states be ordered from worst to best?

Mathitis: Yes: chaos, brink of chaos, threshold, and ideal. So, you are right with respect to these states. Going from unstable to stable, whether capable or not capable, means you have improved the process. Stabilizing does improve the process.

Socrates: Recall our conversation about process capability (Myth 9) and proportions.

Mathitis: Yes, we realized that capability indices are equivalent to percentages or transformation of percentages to another scale.

Socrates: If we show that P_p or P_{pk} has increased, does that not show that performance has improved?

Mathitis: Yes, since more is better for these two indices.

Socrates: Do these two indices require the process to be stable?

Mathitis: I see. No, they don't.

Kokinos: In fact, they are based on the actual variation, combining special cause and common cause variation or assignable and unassignable causes of variation.

Socrates: In that discussion, didn't we group C_p and C_{pk} together as capability indices and P_p and P_{pk} as performance indices?

Mathitis: Then, you don't need the process to be stable to improve it, using these measures of process performance.

Peismon: But the other process capability indices do need the process to be in control to determine them. Moreover, improving the process performance means improving the process capability by getting at the cause system.

Mathitis: But we saw that we can determine those reliably and validly using response surfaces.

SHEWHART'S COUNTEREXAMPLE

Peismon: You're talking theory. Where are your examples?

Mathitis: Was Deming wrong about theory being necessary for deep or profound understanding?

Peismon: No, but doesn't he also say that you need to stabilize before improving or that stabilizing is not an improvement? Even, Juran, who thought control charts were statistical tests of hypotheses. His trilogy consists of three phases: quality planning, quality control, and quality improvement (Juran and Gryna 1988). This suggests that the aspect of control is different from that of improvement.

Socrates: But not necessarily that it is required. Let's go to Shewhart as you accept his evidence as real examples. He used the megohm data to show that points beyond the control limits were strong indicators of assignable causes existing. Was the process in control when he did that?

Peismon: No. By definition, it had to be unstable.

Socrates: What did he do when he saw the process was out of control?

Kokinos: He searched for assignable causes and found some. Removed them—that's what he said he did—and then collected more data to test the opposite: when there are no alarms, are there no assignable causes?

Socrates: Was the process in control then?

Peismon: Yes. He showed that, after removing the assignable causes, the process stabilized.

Socrates: Did the process improve as he went from out-of-control state to in-control state by removing assignable causes?

Peismon: We don't know, do we?

Socrates: We do know, by looking at the graphs, that the average of subgroups of size $n = 4$ went from 4498 megohms to approximately 4410. We don't know which direction is better.

Peismon: It's likely that this is a case of a nominal and two specifications.

Socrates: If the nominal is 4400, then we improved.

Peismon: We don't know if 4400 was the nominal.

Socrates: If the new mean is closer to the nominal, we know the process improved.

Kokinos: The control chart limits also narrowed. When it was out of control, the limits ranged from approximately 4005 to 4990, or 885, versus a range of approximately 4150 to 4700, or 550. The standard deviation shrunk by approximately 38%, given that $550/885 = 0.62$.

Peismon: What if the nominal is closer to the old mean? What if the nominal was 4500?

Socrates: Then, not all causes were bad. He should not have removed the good assignable causes. That would defeat the purpose of removing assignable causes if it makes the process worse.

Mathitis: Of course. Some points were below the LCL and some were above. Removing the assignable causes of the ones above lowered the average, taking it farther away from 4500—if that was the nominal.

Socrates: Didn't Shewhart have another example before the megohm one?

Kokinos: Yes. In that one, he specifically cites that the defective rate decreased from 1.4% to 0.8% as more and more points fell within the control limits.

Socrates: Was that process stabilized before the reduction of the defective rate?

Kokinos: No. In fact, it stayed out of control. He said that the number of points in control increased from 68% to 84%.

Socrates: Then, he improved without even stabilizing after the changes.

Kokinos: In these examples. Then, it's true that you can improve a process that is not in control and that it is possible to get a process to be in control by improving it.

Socrates: Of course, the ideal is to improve a process while stabilizing it. Certainly, after improving it, as there are advantages in having a stable process.

Peismon: That's the economic value Shewhart wrote about.

Mathitis: It makes sense to me.

27

Myth 27: Stability (Homogeneity) Is Required to Establish a Baseline

Scene: *Break room (next day after process stability needed for improvement discussion)*

BACKGROUND

Peismon: Socrates, I was rethinking our discussion yesterday and I forgot to mention the reason why stability is required for establishing a baseline.

Socrates: What is that?

Peismon: The data need to be homogenous to establish a baseline.

Socrates: Given our discussion about homogeneity (Myth 20), which is needed, homogeneity or stability, since they are not the same?

Peismon: If the process is not stable or the data are not homogeneous, then there is no single population or universe so the baseline is meaningless.

Kokinos: That's what I was taught. Stabilize the process first, then determine the baseline, then find how to improve, and finally implement the changes.

Socrates: Can we just say either stable or a single population, since homogeneity is checked by displaying stability?

Mathitis: But given the counterexamples to the claim that stability shows one universe, is the issue really just whether there is one universe and not even whether the process is stable?

Peismon: The issue is that you need one universe; otherwise, baselines are meaningless.

PURPOSE OF BASELINE

Socrates: What is the purpose of a baseline?

Agrafos: To compare. You have a baseline of process performance before the changes and you compare that to the performance after the changes to see if it is better. That's what we do with process improvement projects.

Mathitis: I agree. A baseline is a point of reference, by definition. It's a level or quantity or position for the purpose of comparing.

Socrates: If stability is required to have a meaningful baseline for the purpose of comparing, then, without the baseline, we cannot compare or that the comparisons would be wrong.

Agrafos: That's logical. How can you compare without a reference or with a wrong reference?

Peismon: Yes. A baseline is necessary to know you improved after making a process change.

Mathitis: The before-the-change results are used to establish the point of reference, which is the baseline. The after-the-change results are used to make the comparison.

Socrates: Peismon, your claim is that the performance of processes cannot be compared unless the process is stable or the resulting data are from a single universe.

Peismon: That's right.

Socrates: And that requirement applies to all processes under all conditions? No exceptions.

Peismon: Yes—no exceptions.

JUST-DO-IT PROJECTS

Socrates: What about comparing the time to deliver regular mail versus electronic mail?

Mathitis: No data are needed. We already know one is faster.

Kokinos: I don't need to have a stable process to know whether one pay-check is more or less than another paycheck.

Peismon: Those aren't processes.

Socrates: Your biweekly paychecks are the result of a process. Walking is a process. Communicating is a process. We used to communicate with regular mail and millions have switched to electronic communication for at least one reason: it is faster. Did you need a control chart or process behavior chart to determine that?

Mathitis: I agree with Socrates. We never determined whether the process was stable. We can improve the regular mailing process performance time by replacing it with electronic mail.

Kokinos: There are hundreds of such cases where we know that the process is better in some way during one period than another period. We replace walking with riding a bike, which is replaced by a bus, which is replaced by car, and the car by a plane. We use an elevator rather than the stairs to go up several flights.

Peismon: But those are just-do-it improvements.

Socrates: What's just-do-it?

Peismon: It means there is no need to discuss what to do and that we know doing "it" will result in an improvement.

Socrates: Then, you agree then that there are exceptions to your claim that baselines are required to know you have improved a process.

Peismon: Yes, in those instances.

Mathitis: Then, stability is not always required to establish a baseline.

Kokinos: I can see that as being an exception.

NATURAL PROCESSES

Socrates: Is growth a process?

Kokinos: Yes, and so is decay. All organisms go through both processes.

Agrafos: Don't all existing objects, even nonorganisms, decay?

Socrates: Let's consider an organism's length, weight, height, or age. If we plot the measures of those characteristics on a control chart, will they show a stable process?

Kokinos: By definition, no. Those processes change the organism with respect to those characteristics.

Mathitis: Growth trends will go in one direction and decay trends in the opposite direction. With enough points, each will display instability whether the trend is slow, fast, or in spurts.

Socrates: How many universes, lists of possible results?

Mathitis: The list could be the same with changing probabilities, or it could be changing as well as the probabilities. For example, humans will not be 5 feet tall at 2 months or 1 year of age and the range of possible heights for humans is limited, although infinite.

Socrates: But in either case, can we make comparisons? Can we correctly and validly compare these changing measures from one period to another?

Mathitis: Obviously. We can compare my average ages in primary school, high school, and college. Of course, we will always be older as time passes.

Kokinos: My daughter is taller today than a year ago and her average height in her teens is greater than her average height preteen.

Peismon: All that is true but they are silly examples, obvious examples as Mathitis said.

Socrates: Your claim about needing stability doesn't apply to obvious process cases? These are examples of meaningfully and validly comparing process performances with baselines from unstable processes. And there are many more.

Mathitis: Sure, all growth processes. All organisms grow once they are born. That would also show instability and yet we make comparisons at different times. Kokinos' daughter is taller this year than last year. Her average height the last 12 months is greater than her average height the first 12 months of her life.

Socrates: There are billions of real examples from nature of growth and decay that contradict the claim that stability is always required to establish baselines and make comparisons.

PROCESSES WHOSE OUTPUT WE WANT TO BE "OUT OF CONTROL"

Kokinos: As you were talking about natural processes of growth and decay, it occurred to me that there are processes that people create where we want the performance to change. If changing performance, such as trends, indicate instability, then we want them to be unstable.

Peismon: What processes are those?

Mathitis: My retirement fund and pension account, for two. I want the value of the fund and account to increase and even preferably increase in jumps, not incrementally.

Agrafos: So you can retire sooner! I'm for that.

Kokinos: For that to occur, you want the value of the assets, like stock or mutual funds or real estate or social security, to grow. The more dramatic in the right direction, the better!

Socrates: We discussed natural processes and now human-made processes where we want each to not be stable. And yet we make valid comparisons despite their instability. Kokinos, you mentioned social security and Mathitis mentioned his pension and retirement fund. The amount you will get from social security depends on the amount you contribute, which depends on your salary. Do you not want your wages to increase, Peismon?

Peismon: Of course. You are not trying to improve those processes. You need data that are homogeneous from a stable process to establish a baseline for improvement purposes. For the purposes of improving process, the average of data when they are not stable is meaningless.

Socrates: Are you saying that if we don't want to improve, we can compare process performance between two periods even if the process is not stable during either period but if we want to improve we can't compare?

Agrafos: That doesn't make sense. Why not just compare and after the comparison decide you want to improve?

Mathitis: I also disagree that we don't want to improve these processes. I want to improve the performance of my retirement and pension funds.

Peismon: You want the value of the fund to increase but that can occur without any improvement in the process itself. You simply add to it.

Mathitis: That does change the process—it improves it. Also, when I invest in stocks and the price decreases more than what I add to it. If I contribute a fixed amount to the funds so it will grow, I deliberately change it. And that change is viewed—at least by me— as an improvement. Also, I would like less variation, because it makes me very anxious. We said reducing variation, in the right conditions, is an improvement.

Peismon: Without stability, the baseline is meaningless.

Mathitis: The amount in my retirement fund is not meaningless to me.

Kokinos: There are business processes we want to improve. For example, we want sales to increase. They could have been erratic, out of control, for the past 24 months. We can establish a baseline for those 24 months. Even if sales steadily increased and so a control chart would show instability, we can compare any two periods.

MEANING OF "MEANINGLESS"

Socrates: I'm curious. Why do you say statistics on data from an out-of-control process are meaningless? A control chart of a person's monthly height in their first 36 months, 5 years, or 10 years displays instability. Yet, are people's actual and average heights meaningless?

Peismon: When the process is unstable and you want to improve the average performance, the average cannot be determined as it doesn't exist.

Mathitis: But that doesn't answer the question. We aren't talking averages. We compare the single measured height on the last day of each month. We can compare any 2 months to see whether they are different. We can compare the maximum height for the first 6 months to the maximum height for the next 6 months. Those are all meaningful comparisons.

Socrates: We *can* calculate the average of any 6-month period and compare—they do exist.

Peismon: I mean that it is meaningless. Yes, it is the average of that "pile" of data, but it has no meaning.

Mathitis: What if I find it meaningful? Da Vinci had a secret code that was meaningful to him but not anyone who didn't know it. Is it meaningless—it wasn't to him.

Socrates: What does *meaningless* mean?

Kokinos: This dictionary defines *meaningless* as having no significance or purpose or value.

Peismon: The average of data that do not come from a stable process has no significance. It can't be used as a baseline, so it has no purpose. It has a specific numerical value but that is not what value means in this context.

Mathitis: Growth amounts from one period to another do not exhibit stability when displayed on a control chart. Yet, they have value. For example, pediatricians use ranges of expected height by gender and age to determine whether a baby is growing normally.

Socrates: Peismon, would you agree that a calculation based on a meaningless quantity would still be meaningless?

Peismon: Of course, meaningless plus anything else is still meaningless.

Socrates: The result would also have no meaning.

Peismon: That's correct.

Socrates: Let's consider this control chart (Figure 27.1). Does it display stability?

Peismon: No.

Socrates: On the basis of what?

Peismon: There are points beyond the control limits; some above the upper limit and some below the lower limit.

Agrafos: There is also an upward trend. But it may not violate the six or more consecutive points increasing or decreasing.

Socrates: What if these were the daily closing prices of a stock I owned, are you saying that the average for these 24 days is meaningless in the sense that I cannot compare this average to the average of the next 24 days?

Peismon: Yes, because the price is not in control.

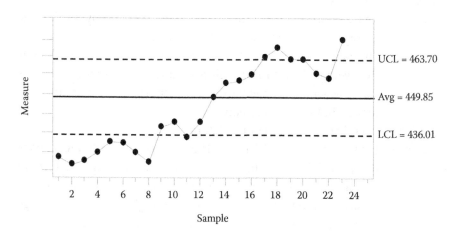

FIGURE 27.1
If the average (solid line) for unstable/nonhomogeneous data is meaningless, then so are the control limits (dashed lines).

Agrafos: Then if during the next 24 days the price rose in similar out-of-control manner to the 500–600 range with an average of 550, the comparison of averages would be meaningless?

Peismon: Yes, that's what I am saying.

Agrafos: Remind me never to use you as a financial planner.

Socrates: I agree, Agrafos. I bought the stock during the first 24 days shown here and sold them during another period where the price had reached 562, more than a 24% gain.

Agrafos: That's certainly meaningful!

Socrates: Peismon, remind us how we determine the control limits.

Peismon: We use a formula to set them at 3 standard errors from the center line on each side.

Socrates: And the center line?

Peismon: That is the average of all the data on the graph.

Socrates: Without the limits, would you be able to assess stability?

Peismon: No.

Mathitis: You can assess stability with a run chart and it has no limits.

Socrates: True. Without the center line, would you be able to determine the limits?

Peismon: No.

Socrates: So, the center line and the limits serve a purpose.

Peismon: Yes, to assess stability.

Socrates: They are valid and meaningful for that purpose.

Peismon: Of course.

Mathitis: That's contradictory. On the one hand, you are saying that the average is meaningless when the chart does not display stability. On the other hand, you are saying that the center line, which is the average, is *not* meaningless because it is essential to assess stability—and homogeneity. For run charts, we also use the center line but it is the median.

Agrafos: Even I agree that is contradictory.

Socrates: Let's summarize, Peismon. The average of nonhomogeneous data, which is displaying lack of control, is meaningless. It would stay meaningless if combined with other quantities, the results of such combinations would be meaningless, and the average cannot have any meaning to anyone. Yet, it has meaning to you and everyone as it serves the purpose of setting control chart limits to determine whether the data come from a stable process. And determining stability/homogeneity is meaningful.

Mathitis: In fact, he and Wheeler and others say that the center line or average and the control chart limits work regardless of whether the process is stable. That gives it significance, purpose, and value. It has meaning.

Peismon: But what I meant was that it is meaningless for establishing a baseline.

Socrates: Are points above the upper control limit different from the points between the control limits?

Peismon: Yes, they indicate lack of homogeneity and stability.

Socrates: And the same with points below the lower control limit.

Peismon: Yes, they are also different.

Socrates: Then, are you not comparing two sets of values when they lack homogeneity as stability? To make that comparison, are you not using those within the limits as a baseline to claim that these points beyond the control limits are different?

Kokinos: If we use other signals of instability, like six or more consecutive points on one side of the center line, then we are comparing points to the center line or average.

Mathitis: I agree, Socrates. Even on a control chart, we compare individual and groups of data even when the data display instability.

Kokinos: Then, for all control charts, whether displaying stability or not, the average of the points is not meaningless. It is used to calculate the control limits, which are used to make comparisons for determining stability, and therefore, homogeneity, and therefore, how many universes.

DAILY COMPARISONS

Socrates: Do you need more than one datum to establish stability?

Peismon: Of course. Typically 20–25 points—fewer if it shows lack of stability.

Socrates: Then, one datum per condition would not be enough. One result before versus one result after would not be enough because you need more than that to establish stability.

Peismon: That's right.

Socrates: Which would take longer, crawling across the room to get from one side to the other or walking fast?

Mathitis: Walking fast.

Socrates: How many times do you need to do each to know that? In other words, do you need to (a) collect more than one datum per condition and (b) establish a baseline of more than one point to make a comparison?

Peismon: But you are not comparing process performance.

Mathitis: I disagree. Those are processes and we make such comparisons.

Socrates: What other comparisons do we make without requiring stability?

Mathitis: We judge meals whether good or bad or just okay.

Kokinos: We judge restaurants on the basis of whether the meal was good or bad.

Mathitis: And recommend or not recommend the place often on the basis of a single meal.

Socrates: And you think that judgment is meaningful—has a purpose and value?

Mathitis: Yes.

Socrates: Peismon, when you judge a restaurant, do you check for stability first?

Peismon: Well, no.

Socrates: And if you ate there a second time, did you compare? Have you ever determined that one time wasn't as good as the previous, or vice versa, it was better?

Peismon: Yes, we all do that.

Socrates: Exactly.

Kokinos: We make comparisons numerous times every day using only two instances and we never check for process stability. We do not wait until we've eaten at a restaurant 20–25 times before determining whether it is bad, okay, good, or great.

Socrates: Then, doesn't that contradict the claims that (a) more than one datum is needed to compare, (b) stability is required to establish a baseline, and (c) a baseline is needed to make a comparison? How many control charts do you have for your processes outside of work?

Peismon: Actually, I have a couple for my running. I race and so I have one for my races and another for my practices.

Socrates: Why do you use those charts?

Peismon: To compare my races and see if I have improved.

Socrates: Have you improved your times?

Peismon: Yes.

Socrates: How do you know?

Peismon: I compare my time from one race to a previous race.

Socrates: Do you make other comparisons each day?

Peismon: Like what?

Socrates: Prices, times, tastes, amounts, distances, and so on. Do you compare without ever checking or even thinking about whether the results came from a process that was stable?

Mathitis: I see what you are getting at, Socrates. Peismon's claim would have more credibility if he used process behavior charts for all his comparisons and not just a couple. We can and do compare single or groups of results even when the data do not show homogeneity or stability.

Kokinos: Then, these types of comparisons do not require data to be stable or homogeneous.

Peismon: That's why those cases are different from using an average from a control chart. You need more than one data point.

Kokinos: But we showed that the average from an unstable process has meaning—even to you!

"TRUE" PROCESS AVERAGE: PROCESS, OUTPUTS, CHARACTERISTICS, AND MEASURES

Peismon: But, as I've said various times, we are looking at the process average, which means looking at the "true" average, not from a finite period.

Socrates: Consider an oak tree to see if we can understand the issue. An oak tree produces thousands of acorns yearly, varying from 1 to 6 centimeters long. We measure several acorns from a 45-foot-tall oak tree. We plot them in the order in which they were produced on a control chart. The data display stability, with an average of 4.3 centimeters. Which has that average: the tree or the acorns?

Mathitis: Clearly, it is the acorns not the tree.

Socrates: What is a process?

Peismon: It is the tasks and actions and decisions used to produce something. They transform inputs into outputs. The process also includes the resources, materials, and methods used.

Kokinos: We use the 6Ms to identify process components: manpower, mother nature, methods, materials, machines, and measurements.

Peismon: I use that often. We also use SIPOC, which stands for Supplier–Input–Process–Output–Customer, to define the components of a process.

Socrates: When you say process average, what does that mean? You said a process consists of actions. What is an average of several actions? You said a process consists of resources. What is an average of those resources? Using Kokinos' 6Ms, what is the average of methods or materials?

Peismon: The process average is the average from the data when the data are homogeneous.

Socrates: What do we plot on a control chart: suppliers, inputs, the 6Ms, outputs, customers?

Kokinos: None of those. Outputs have characteristics, called critical to quality or CTQs. We measure the CTQs and plot the results.

Socrates: We plot either those values, for example, on an individual control or *I* chart, or statistics of those values, for example, on a mean chart or standard deviation chart or proportion chart or count chart.

Peismon: That's right.

Socrates: Processes are not outputs or characteristics or measures. Each is not the other.

Peismon: That's correct.

Socrates: Let's summarize: Processes produce Outputs with Characteristics, which, when Measured, produce data that are plotted directly or as statistics on the control chart.

Agrafos: That sounds like the statistics are of the measures of the characteristics, not of the process.

Peismon: We say it is the process that has an average.

Socrates: Is it the process that has an average or does the process produce outputs whose average characteristic is such-and-such? For example, is it the process of producing acorns or acorns that have an average length of 4.3 centimeters?

Mathitis: It is the latter.

Peismon: I don't know if I agree with that. Wheeler and Stauffer refer to the process average when discussing an unstable process: "The average is not useful as a baseline because there is not one distribution, but many. The data is not homogeneous so the average does not characterize a process mean. The average does not represent where the process is today. There is no single value for the 'true' process mean, so an observed average statistic does not generalize to represent anything useful about the underlying process" (Wheeler and Stauffer 2014). One way to improve a process is to improve the process average. That requires a stable process to have a meaningful average.

Socrates: An output could have several characteristics on each output. Suppose we measure two different characteristics on the same outputs. For example, length and weight of acorns. A control chart on one characteristic shows that the process is stable and a control chart on the other shows that it is not. For example, length and weight of acorns. Is the process stable or unstable?

Peismon: It is stable with respect to one measure but not the other. There is no contradiction.

Socrates: How did you remove the contradiction?

Peismon: We are talking about different things.

Mathitis: The different things are the characteristics, not the process.

Kokinos: Then why do we say process average? I understand what you are saying but I was taught what Peismon said.

Socrates: As with other phrases, we have shortened a very long, cumbersome phrase. We might say "percent good" for "percent of the outputs that are good" or use acronyms, like "nimby" instead of saying "not in my back yard."

Agrafos: What is the long phrase for process average?

Socrates: Process average is a shortcut for the average measurement of characteristic C of outputs O produced by process P during time period T. Even that might not be all.

Peismon: But we are talking about baselines for process improvement.

Socrates: And that is another distinction that needs to be clarified. Recall the distinction Mathitis made between the process as the cause system and the data as the measures of the effect. It is the effect that has that average not the process, which is the cause. When we are dissatisfied with the effects—the measures of a characteristic of the output, then we change the process to produce the same output with different measures of those characteristics. If the average belonged to the process, then we wouldn't have to evaluate the output's characteristics.

Kokinos: Isn't there a third reason to say it is not the process that has the "true" average? We determined that we can only know about the process during the period covered by the data, say t_1 to t_k. What is the "true" process average if the same process at different

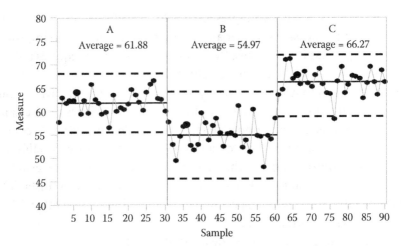

FIGURE 27.2
What is the "true" process average when the process is stable during different periods with different averages?

periods displayed stability each with a different average like in Figure 27.2? Which is the "true" process average? On the other hand, by realizing there is a completely different set of outputs produced each period, we can understand that each set can have its own average characteristic.

Agrafos: If processes waver, then how can there be a "true" process average?

Mathitis: That raises a fourth reason with which Kokinos' point can be explained in more detail. We saw this when we discussed using response surface studies. The same process can operate under different conditions during each period. So, a "true" process average—or more accurately, a "true" average measure of a characteristic—applies to the conditions under which the process is operating. We can determine what that average would be using response surface studies. For example, if I am baking bread, I know that the average time to bake will change by changing the temperature of the oven. It is still the same process but the temperature—a causal factor—is adjusted.

Kokinos: That was the benefit of response surface studies. They not only tell us what those measures can be and what that "true" average of the measures will be, but they tell us the conditions—factors and levels—that will produce them.

Mathitis: When we talk about the measures as coming from a probability distribution, we are talking about the list and their probabilities.

The process is neither. It is what generates that list with those probabilities—possibly changing either or both. If the process was the population, it would only be the list and there would be no distribution because there are no assigned probabilities.

WAYS TO COMPARE

Socrates: The average is just one way of comparing. There are others, each with their own baseline or reference.

Kokinos: Like what?

Socrates: For example, we can use definitions, compare to a reference, use paired matching, and, similar to averages, we can use medians or other percentiles.

Agrafos: I'd like to know more about these other ways.

Kokinos: How do you use definitions?

Socrates: We did that before when we compared ages. Mathitis said that, by definition, every year we are older. For another example, we know that the height of skyscrapers is greater than the height of ranches. By definition, skyscrapers are multistory buildings while ranches are single-story buildings. Skyscrapers are therefore taller by definition.

Peismon: How is that related to the question of baseline for process improvement?

Socrates: Suppose a builder built 20 consecutive multistory buildings with different number of stories in such a way that a control chart of the heights was not stable, for example, each one was smaller than the previously one built or he built one skyscraper that was the highest in the world. Then, the builder built 20 consecutive ranches. In this case, we do not even need the data to know that the comparison will show the heights of the multistory buildings to be greater.

Agrafos: That's similar to replacing regular mail with e-mail.

Mathitis: How do you use a reference?

Socrates: Consider Peismon's trend example. While he says that we cannot use the data as a baseline because it lacks homogeneity, we can compare—which is the point of a baseline.

Peismon: How?

Socrates: Use the upper control limit. Is there a point in time when all the subsequent points are above the upper control limit?

Peismon: Yes.

Socrates: Mark that time on the UCL as U. That is our reference. Before U, were all points below the upper control limit? After U, were all the points above the upper control limit?

Peismon: Yes and yes.

Mathitis: I see. Then, we know that the process performance after that time is greater than the result before that time—independent of homogeneity as stability. If we can validly claim that, after U, we are above the upper control limits, then we can compare the process behavior before and after U.

Kokinos: Let me see if I understand this way to compare. We use all the data in Peismon's example to determine the maximum during that period. Then, we make a change to the process and use all the data during the next period to show that all the results are greater than the maximum result before the change. Then, performance is greater even if the data are not stable before or after the change. Like the stock you bought—all the prices later were higher than before.

Agrafos: Or, like comparing Kokinos' daughter's heights. We know that her heights before age 13 are less than her heights after age 13, so she is taller in her postadolescent years.

Socrates: Yes to both of you. A reference can also be specifications. In improving process performance, we are trying to improve the percent of product or service that meets requirements. In Peismon's example, we can calculate the percent of results that meet specifications before and after a change and see if it is greater.

Peismon: Here's the problem. If the process is unstable such that the data are not homogeneous, then you don't know what happened at times when no data were collected. That's why it doesn't work.

Socrates: If the control chart *displays* stability, do you know what happened at those times?

Peismon: We are more confident.

Socrates: Even though we know that while false alarms are unlikely, false nonalarms are much more probable?

Peismon: We have more confidence when there are no alarms.

Mathitis: No, it's the reverse. You have more confidence that an alarm is a true alarm than a nonalarm is a true nonalarm. We saw that there are numerous changes that a control chart with 3 standard error limits will not detect. Also, Wheeler's simulations show that when a signal of a point beyond the 3 standard error limits occurs, it is very likely that the distribution changed.

Kokinos: Then, the difference is merely one of confidence and not of impossible or meaningless in one case and possible and meaningful in the other.

Socrates: Even if the process is unstable, we can make comparisons using a reference. For example, if we have all the data from one period, we can determine the maximum or minimum value. If all the results from a subsequent period are all below the minimum or all above the maximum, then we can correctly conclude that the process performance for one period was more or less than the other period.

Kokinos: What about paired matching comparisons?

Socrates: Paired matching is a variation of the reference approach. It can sometimes be used when the difference is not so clear.

Mathitis: For example?

Socrates: We saw that when we discussed the power of parametric and nonparametric tests (Myth 12). We looked at the differences of paired comparisons to see if the proportion that one group was greater than the other was significant.

Peismon: That was a statistical hypothesis test and I still think control charts are not.

Mathitis: How would it apply to baselines and comparisons?

Socrates: Suppose we had two periods v_1 to v_k and w_1 to w_k. We form the k differences $v_i - w_i$ and check whether the proportion of negative or positive differences is significantly greater than 50%. That would provide evidence of a change in performance—even average performance.

Kokinos: If the performance did not change on average, then we would expect those differences to be negative half the time and positive half the time. I get it.

Socrates: That approach can be used even if the data are not homogeneous in a stability sense.

Mathitis: That makes sense. I understand using medians as they are measures of central tendency like averages. But how do you use percentiles?

Socrates: The median is what percentile?

Mathitis: The 50th. But we use that because the average is close to the 50th percentile. It will be the same when the distribution is symmetrical.

Socrates: Percentiles can be used if we are interested in the tails, for example. In international educational competitions, we compare the top students in math from different countries. These top students may represent the top 1% or less from each country. We conclude that one country's educational system produces more "brighter" students than another country when its top students outperform the other country.

UNIVERSE OR POPULATION AND DESCRIPTIVE STATISTICS

Peismon: I disagree. When the data are not homogeneous, there is not a single universe or any universe per Deming. And, as Wheeler and Stauffer said, data from an unstable process does not tell us how the process operates today.

Mathitis: But we discussed that before (Myth 20) and showed that homogeneous cannot mean a single universe, especially if the only way to show homogeneity is by displaying stability. The Empirical Rule establishes that real and displayed stability can occur with data from multiple universes since they can come from multiple distributions.

Socrates: I'm not sure what "by today" means. We know there is a single universe—a list of actual results—when we consider all the output during the period t_1 to t_k covered by the data.

Mathitis: That is why determining the quality level of a lot of product or material constitutes a single universe. That's why we can do incoming inspection or release testing of lots.

Peismon: But we are interested in the process because our purpose for baselining is to improve.

Socrates: Improvement is a subsequent action. We can always compare without ever intending to improve. I can calculate meaningfully any statistics, for example, the average, proportions, standard deviation, percentiles, for that "lot" to establish a valid and meaningful baseline for the period during which the lot and only the lot was produced. Do the same for all the output for a different period, whether there was a change or not and whether I wanted to improve or not.

Mathitis: I accept that. The total output from each period constitutes a single universe.

Peismon: But the statistics can only be meaningful if the original data are meaningful. The Social Security Administration office produces social security numbers. We can calculate all sorts of statistics on them, but those statistics do not make sense since the numbers are merely identifiers.

Socrates: I completely agree—and to your astonishment, Peismon. Similarly, plotting social security numbers on a control chart won't tell us whether the process is stable or homogeneous or anything else about the process for the same reason.

Peismon: I agree with that. That's twice we have agreed.

Socrates: We can compare two different periods even when unstable—assuming that the data we are using are meaningful. We only need to have the entire output for each period.

Peismon: But that is not how control charts are typically used. We take periodic samples.

Mathitis: Not always. I've used all the data for low-volume production.

Agrafos: But just because they may not be used that way does not mean they can't be used that way. When discussing whether control charts are tests of hypotheses, we said we could just look at the output for the period covered by the data as the universe. Why not for this?

RANDOM SAMPLING

Socrates: Let's consider the bowl of beads that you and Wheeler and Stauffer say constitute a single universe. When you have only a sample, we need to make an inference from a sample to the total output when we do not know the content of it.

Peismon: That's right. If the sample is not homogeneous, then you don't know how many "bowls" or universes you have. You just know you have more than one. So which one are you describing with the descriptive statistics?

Socrates: How many universes are there in the original bowl of 5000 beads?

Peismon: One. It's a population.

Socrates: If you don't know how many white and black beads there are in that bowl, how many universes?

Peismon: Still one. That is where inferential statistics applies, to enumerate that single universe.

Socrates: This would be the Inference question (Myth 19). Recall that we said a universe is just the list of possible choices. We have that list—the bowl of beads. If the process that produced those beads was unstable, or, if we select nonrandomly from that bowl or order them in a manner so the results display alarms, how many universes are there?

Mathitis: Still only one because there is only one bowl. Since it doesn't matter whether they came from two or more sources, once they are in the bowl, there is only one universe.

Socrates: Then, we can randomly sample from that bowl. We do that with lots or batches to inspect their quality because those lots or batches are single universes. We make valid inferences about those universes of lots and batches. We can do the same for baseline comparisons, just as we can do for statistical hypothesis testing (Myth 22). We can randomly sample from all the output from each period and make valid and meaningful inferences as to whether the results are different or the same. We don't need to actually or physically put all the output in a bowl. But what makes it a universe is exactly what you Peismon said. We can enumerate the entire output.

Mathitis: Then you can use the entire output as your universe and establish a baseline on it. Or, you can randomly sample from that entire output, treating it as a single universe, and establish an inferred baseline with its probability of error such as with confidence intervals.

Socrates: Don't forget that we don't have to use only the average to compare. Consider a customer that buys a lot of size 100 units from you. You plot all 100 values or a sample of 25 but the control chart shows instability. The customer inspects all 100 units, finding 8 defective units. She complains and asks for a new lot. She doesn't care about the stability. You produce another lot of 100; again a control chart shows instability. This time, she finds no defective units. She accepts the replacement. Is she right that it is better? Is it an appropriate comparison?

Mathitis: Yes, it's appropriate because it *is* a valid comparison. We do it all the time: one unit versus another unit or a batch versus another batch. We make those comparisons daily without ever asking or checking about stability or homogeneity.

Kokinos: I agree. She was interested in the percent good. It changed for the better.

Agrafos: I see. Given that, we do not have to sample as we might for control charts, using rational subgroups or periodically, we can randomly sample from that finite sequence to make inferences to the entire finite sequence.

Mathitis: But we already discussed that issue. There is no "true" process average as there is no process average (Myth 20). There is an average of the measurements of characteristics of outputs produced by the process. And that average depends on the conditions in which the process was operating and the period covered by the data.

WHEN IS HOMOGENEITY/STABILITY NOT REQUIRED OR UNIMPORTANT?

Socrates: So we have discovered that, for natural processes whose performance changes or processes that humans want the performance to change, we do not need stability to make comparisons. In fact, we do not expect stability in, for example, natural growth or decay processes and man-made processes like retirement funds.

Mathitis: To require stability would contradict the purpose of those processes.

Socrates: We also make daily comparisons between single values without requiring stability because our interest is in only those outputs. If we have causal control over the results, we don't need homogeneity as stability to make comparisons. We don't use control charts or require homogeneity as stability to drive our car consistently between the lines that mark the two sides of the lane.

Mathitis: For process improvement, even for those processes that we want not to change, we do not need homogeneity because we can always use the entire output for the baseline period to establish the baseline.

Kokinos: Or, we can take a random sample from that "bowl."

Socrates: Because the question we want to answer with a baseline is "How much/many?" during that period of interest.

Mathitis: It is not the process but the results of the process that is of interest when establishing a baseline.

Socrates: Is there a case when homogeneity as stability is required?

Peismon: Balestracci (1998) has an example of three data sets with about the same average, standard deviation, sample size, percent meeting specifications, and histograms. All passed a test for normality. Yet, when the data are plotted on a control chart, they look vastly different. One has an upward trend, another has three stable periods, and the third displays stability.

Socrates: That only says that the answer to "Are the averages the same?" is yes and the answer to "Are the processes stable?" is no. The issue we are discussing is when homogeneity is important or necessary for comparisons.

Peismon: The statistics don't represent the process. Those who create histograms from data lacking homogeneity are doing meaningless analysis.

Socrates: How does that show that the averages or any other statistics are meaningless?

Peismon: It does not tell you what happened to the process over time.

Socrates: Then, it is meaningless with respect to that issue: what happens over time? Your car and my car are the same with respect to number of wheels but different with respect to model. It doesn't mean we can't compare number of wheels or model or color or any other characteristics. For example, does a control chart tell you how satisfied customers are?

Peismon: No.

Socrates: Should we therefore say control charts are meaningless? Or, should we say control charts are meaningless with respect to customer satisfaction determination and comparison?

Mathitis: The latter.

Socrates: Do control charts tell us the frequency of each value?

Mathitis: No. But we shouldn't say they are meaningless; only meaningless with respect to that question.

Socrates: But histograms do tell us frequencies. If we put specifications on the histograms, we can compare defect rates *for those data* whether the processes from which they came were stable or not.

Mathitis: So they are meaningful with respect to that issue.

Peismon: But those defect rates are meaningless when the processes are unstable. We don't know what happened at times we lack data.

Socrates: Why don't we know that?

Peismon: Because the process was unstable.

Socrates: How could we know?

Mathitis: By looking at or randomly sampling from all the data during that period.

Socrates: Then is the issue lack of stability or that the sample when unstable is not representative and therefore, the inference is invalid?

Mathitis: The latter. But even when stable it may not be representative when the subgroups are not random.

Socrates: Then the issue of histogram versus control chart is not relevant to whether a baseline can be established with nonhomogeneous data. To show the contrary, you must show the following:

- What meaningless means with respect to making comparisons
- A single example showing that a comparison of two periods was wrong when the data lacked homogeneity as stability

Kokinos: Balestracci did not show either. But we did show that the average, even from unstable processes, is meaningful since Peismon, Wheeler, and everyone use them to establish control limits and to make decisions about stability or homogeneity. They can't have it both ways: say it is meaningless and then proceed to use it meaningfully.

28

Myth 28: A Process Must Be Stable to Be Predictable*

Scene: *Break room (continuation of homogeneous discussion)*

BACKGROUND

Mathitis: The primary reason to use a control chart is for the economic value of having a controlled process.

Kokinos: That's my understanding of what Shewhart said.

Socrates: What did he mean by a controlled process?

Kokinos: He meant that it was stable, no special cause or assignable cause variation. That only the inherent or common cause variation of the process existed.

Mathitis: The purpose was to determine when to make changes to the process and when to leave it alone. Deming also viewed another critical benefit of control charts. He thought that when special or assignable cause variation existed, then process operators were responsible for taking action. But if only common cause variation exists—the process is under control—then management is responsible for taking action on the process.

Peismon: But a key benefit is that a stable process is predictable.

---

* Parts of this dialogue are based on: Castañeda-Méndez, K., Beyond control charts for predicting and improving processes, *isixsigma*. Available at http://www.isixsigma.com/tools-templates/control-charts/go-beyond-control-chart-limitations-to-predict-and-improve-processes/.

Socrates: You mean, for example, if you have a six-sided die with the numbers 1 through 6, if the results of rolling of the die is in control, you can predict the number each time you roll the dice?

Mathitis: Not the exact value. You can predict within limits what the result will be, that's what Shewhart meant. For example, subgroup means are plotted on an $\bar{X}$ chart. When it's stable—no special cause variation occurs—we predict that the process will produce results within defined limits. Those limits are the 3 sigma or standard error control limits.

Socrates: By prediction, do you mean a guess?

Mathitis: Not a guess—that isn't predicting.

Peismon: Especially if you are wrong.

Socrates: Predicting doesn't always mean predicting correctly. Don't we often check what futurologists have predicted to see if they were right? Haven't people predicted the end of the world several times? We say they are predicting. One sense of prediction doesn't mean being right—it just means making a claim about what will happen. In that sense, I can make predictions without a stable process.

Peismon: You have to have some basis for it—so it isn't just a guess.

Socrates: Isn't my belief a basis for it, especially if I have that belief because of my experiences?

Kokinos: That seems reasonable to me. We often do that, make predictions based on our beliefs. For example, who will win a game. All betting involves predicting.

Peismon: With control charts, we have a sounder basis than just belief. It has the backing of the data—there is reason for the prediction. The stability of the process indicates continued similar results. We make an inference about future results on the basis of that continuation.

Socrates: Are you saying predictions from stable processes are valid but not when it is unstable?

Peismon: That's right.

Socrates: Is that logical validity? That what will happen in the future follows logically from the data showing a stable process?

Mathitis: It can't be a logical inference because we do get out-of-control points at some point after a process is stable. Stability doesn't last forever. We said all processes waver (Myth 22).

Peismon: I'm saying that it is a reasoned prediction supported by evidence.

Socrates: Is the prediction for just the next point or for a certain number of points? How long can we wait before we collect that one point or more points? How far into the future is the prediction still appropriate or reasonable?

Mathitis: It's not indefinitely—things change.

Peismon: Wheeler and many others say the same thing. Wheeler describes process stability as homogeneity of the data. When you have homogeneous data, then you can predict. He gives several examples of that.

Socrates: Is this the same meaning of homogeneity as we discussed before, that the order in which data were produced is random or comes from only one universe?

Peismon: Yes, it's the same meaning. Only a process under control is predictable.

Socrates: That applies to all processes?

Peismon: Yes, all. Shewhart, Deming, Wheeler, and many experts in the field say the same thing: uncontrolled or unstable processes are unpredictable.

TYPES OF PREDICTIONS: INTERPOLATION AND EXTRAPOLATION

Socrates: Does this apply to both types of predictions?

Agrafos: What are the two types of predictions?

Socrates: Interpolation and extrapolation. In both cases, we use data to predict or estimate results for conditions under which we do not have data.

Agrafos: For example?

Socrates: When we do linear regression analysis, we infer that the linear relationship applies to the entire range of the data not just at the points for which we have data.

Kokinos: But don't we advise not to extrapolate beyond the range of the data?

Socrates: Yes, we do. In this example, we only predict by interpolating. But, it appears that you are saying control charts are used to make both types of predictions.

Agrafos: When is interpolation done?

Socrates: For example, we take samples of three every 15 minutes, plot the means on an $\bar{X}$ chart, and if there are no special cause indicators, then we interpolate to the times we didn't sample. We assume the process is stable during the entire time, not just every 15 minutes.

Mathitis: We said stability applies to a period t_1 to t_k even when no data are collected at times between these k moments in time, for example, between t_1 and t_2.

Peismon: We are also saying that we can extrapolate to the next point or points, but only if the process is stable. We then predict that the process should produce results that occur within the control limits of the stable process.

Kokinos: So, we do both types of predictions, interpolation and extrapolation, when using control charts. That seems like a great advantage of control charts.

Socrates: And, my question is "Why is stability required for either type of prediction?"

INTERPOLATION: STABILITY VERSUS INSTABILITY

Peismon: Wheeler analyzed data on major hurricanes to show that you can't interpolate accurately with an unstable process. Statistics are misleading when data are not homogeneous. He says "All data are historical. All analyses of data are historical. Yet all of the interesting questions about our data have to do with using the past to predict the future" (Wheeler 2013).

Mathitis: But we agreed that it was a fallacy that statistics are meaningless when the data lack homogeneity as stability (Myth 26). Since the statistics can be used to correctly distinguish between stability and instability, then they are not meaningless.

Socrates: Can you tell us about the hurricane example?

Peismon: He plotted the number of major hurricanes for each year from 1935 to 2013 on a control chart. He shows that without homogeneity of data or stability, predictions can't be made.

Socrates: How does he show this?

Peismon: He shows that an unstable process does not have a single population (Figure 28.1) and therefore fitting distributions to such data is misleading. He fits a Poisson distribution to the histogram of the data (Figure 28.2) and shows that it doesn't accurately predict the results during those years. The model predicts that we should

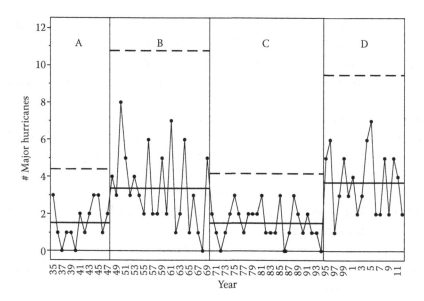

FIGURE 28.1
Number of major hurricanes in the years 1935–2012 showing four "universes."

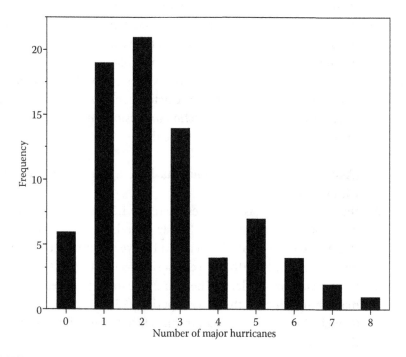

FIGURE 28.2
Histogram of the number of major hurricanes in the years 1935–2012.

have had only 1.24 years with seven or more major hurricanes during the period 1935–2012, but there were in fact 3 years.

Socrates: Does that show that predictions cannot be made or that predictions from unstable processes are inaccurate or less accurate than predictions from stable processes?

Peismon: He says the data are misbehaving—they are not homogeneous.

Socrates: But we now understand that if homogeneous means a single universe or distribution, then lack of stability doesn't mean lack of homogeneity and being stable doesn't guarantee homogeneity per Wheeler's own simulations.

Mathitis: But he says the predictions are inaccurate. What is the prediction from a stable process?

Peismon: The process wasn't stable, so we don't know.

Socrates: That's an odd way of showing that homogeneity or stability is required.

Peismon: Why?

Socrates: First, it contradicts the quote you gave. If all interesting questions are about predicting the future, why did he predict the past?

Kokinos: You mean he interpolated rather than extrapolated.

Socrates: He did neither. Clearly he didn't extrapolate but he didn't interpolate either. He knows the results for each year already, so why would he model it to predict what he already knows?

Mathitis: Doesn't that show that the model is wrong, not that the data are wrong and not that the data are heterogeneous or misbehaving? That's the argument of *reduction ad absurdum* we talked about before. His argument seems to be this:

1. (Assumption): H_o: hurricane data have a Poisson distribution with mean 2.56.
2. (Prediction): If H_o is true, then there are 1.24 years with seven or more major hurricanes in a 78-year period.
3. (Inference from 1 and 2): There are 1.24 years with seven or more major hurricanes in the 78-year period 1935–2012.
4. (Fact based on all data): There are 3 years with seven or more major hurricanes in the 78-year period 1935–2012.
5. (Conclusion: inference 3 contradicts fact 4): Therefore, (assumption) 1 is false.

Socrates: That would be my conclusion.

Kokinos: That's what we do with scientific studies. We assume that a new treatment is no different from a control or placebo. Then, we show that the difference is statistically significant. Therefore, the assumption of no difference is rejected.

Agrafos: That matches what we discussed about statistical hypothesis testing. We want to reject the null hypothesis.

Socrates: That's correct. However, what did Wheeler conclude?

Mathitis: We know the actual number of hurricanes for each year. If, by assuming a Poisson distribution, we get predictions that don't match the facts, the conclusion should be that the data do not follow a Poisson distribution, i.e., the assumption is false. Wheeler concluded that the facts were wrong.

Kokinos: But we knew the assumption was false anyway. The Poisson distribution is a model.

Mathitis: All models are wrong—some are useful. This one is not useful for predicting.

Agrafos: Why is it a model and not reality?

Socrates: Consider what a probability distribution is. It has a list of possible outcomes and a probability associated to each. What is the list for the Poisson distribution?

Mathitis: All positive integers, from zero to infinity. Each integer has a nonzero probability of occurring. The 78 years of data show only a finite number of choices. So, it will never match it exactly as there were no years with more than eight major hurricanes. But the Poisson distribution model says there is a nonzero probability of more than eight major hurricanes.

Peismon: But he also shows that the problem is with the data not being homogeneous by showing that a 95% confidence interval of the mean is from 2.2 to 3.0, which doesn't describe the mean of any of the four stable periods.

Socrates: Is that how you use a control chart to interpolate when it is stable?

Peismon: No, but it shows a problem when the process is unstable.

Agrafos: Why is that not a valid issue?

Socrates: For several reasons. First, there is no need to interpolate or model data to predict what you have complete knowledge of. Second, if a control chart is about predicting future results or extrapolating, then showing that interpolating from an unstable process is inaccurate, is irrelevant. Third, when you have complete knowledge, it doesn't matter whether the results were from a stable process or not for describing those results. Fourth, when we predict with a stable process, we predict that the next point will be within the control chart limits, not what the exact value will be.

Peismon: That's just your opinion, Socrates.

Mathitis: I don't think these are opinions—they're facts.

Socrates: Consider the first point. Why use a theoretical probability distribution when we have an actual probability distribution to model the data to make unerring interpolations (or, in this case, to predict what is already known)?

Agrafos: What distribution is that?

Socrates: The histogram of the 78 years.

Agrafos: But a histogram is not a probability distribution.

Socrates: It's true we usually use histograms of samples to visually determine if the sample fits a theoretical probability distribution, for example, a normal distribution. But remember that a probability distribution consists of a list and the probabilities assigned to each member of the list.

Mathitis: The horizontal axis is the list of the possible number of major hurricanes in a year. The vertical axis is the frequency of each possible number, which is a probability.

Peismon: That's just a made-up distribution.

Socrates: You said that histograms show actual distributions. This is the actual distribution of the data. We can use it, just as we use every theoretical probability distribution that is "made up," to model any data we wish. In this case, we use it to model itself. It will do it without error.

Mathitis: Perfect interpolation.

Socrates: The point of predicting is when we don't know. We develop a method for predicting and check it against what actually occurred. When the prediction is an interpolation, we need to have gaps. Rather than use all the data from 1935 to 2012 to predict what we already know, we should use some of the data and develop a method for predicting what we didn't use. That's one way models are developed and tested.

Mathitis: Then, to show that stability is required for better prediction or for any prediction, we compare when the data display stability versus when they are unstable.

Socrates: That's right. Suppose we arrange the data so that they display stability (Figure 28.3). We can often do this by randomizing the

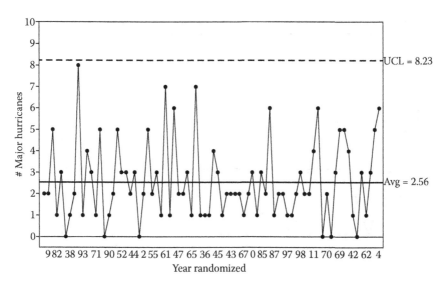

FIGURE 28.3

Number of major hurricanes from the years 1935–2012 randomly ordered to display stability.

order of the data even from an unstable process. Now, for this order and the original, we take a subset and see if we can interpolate the missing values.

Agrafos: How do we select the subset and how do we predict?

Socrates: For the second question, let's ask Peismon. If you collected subgroups every 15 minutes for an 8-hour period and those data displayed stability on a control chart, what would you predict for the data you did not collect during those 8 hours?

Peismon: We would predict that whatever subgroup statistic you plotted on the control chart would be within the control limits for the other times. You can't predict what each individual value is exactly—just within a range determined by the control chart limits. Remember, that's how Shewhart defined statistical control.

Socrates: Wonderful, Peismon. Let's do that analysis for both conditions. Rather than using all the data, let's take a sample, such as every fifth value. Then compare. Here is the result using the original order (Figure 28.4). What do you notice?

Mathitis: The first thing is that the chart displays stability. So, using a sample tells us the opposite of what actually occurred.

Kokinos: But wasn't that point made earlier? Just because a control chart displays stability from a sample, it doesn't mean that the process was actually stable during the entire period.

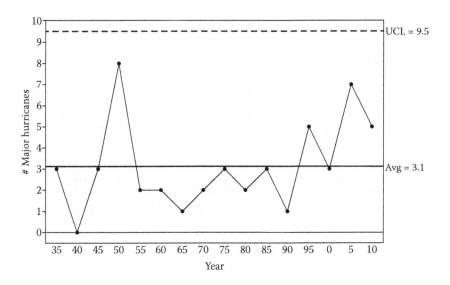

FIGURE 28.4

Number of major hurricanes during 1935–2012 for every fifth year from the original order showing stability.

Socrates: Alarms and control charts are not infallible. Also, as with the control chart of reordered data showing stability, just because a control chart displays stability does not mean there is only one cause system. The original data were interpreted as having four stable periods occurring from two (or more) cause systems. One cause system had a mean of approximately 1.5 major hurricanes per year and the other had a mean of 3.5. Both charts (Figures 28.3 and 28.4) showing stability hide the fact of two cause systems when they are mixed.

Kokinos: That's like what happened when we considered using two dice with numbers {1, 2, 3 4, 5, 6} and {2, 3, 4, 5, 6, 7}. We couldn't tell them apart using an I chart or an $\bar{X}$-chart with $n = 2$.

Mathitis: Also, while the false alarm rate is significantly reduced with 3 standard error limits, it increases the false nonalarm rate.

Socrates: In either case, using every fifth year, how would you interpolate given that the chart displays stability?

Peismon: We would predict that the number of major hurricanes in each of the other years would be between 0 and 9.51. That's 100% correct.

Socrates: Let's take a different subset so it shows instability (Figure 28.5), for example, every third year.

Peismon: Good. Now we see that you can't predict with unstable processes.

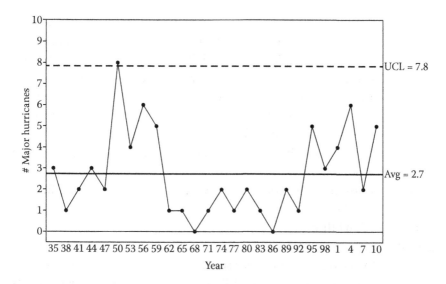

FIGURE 28.5
Number of major hurricanes during 1935–2012 for every third year from the original order showing instability.

Kokinos: I don't see that. The control chart limits are 0 to 7.84, which means we would predict all the other years to have seven or fewer major hurricanes. That's 100% correct.

Peismon: But you are wrong for 1950.

Mathitis: We didn't predict that one. But using Wheeler's Empirical Rule, we would expect approximately 1% beyond the control limits, and for some distributions, even as much as 2%–4%. One out of 78 is only 1.3%. His simulations produced more than that several times.

CONDITIONAL PREDICTIONS

Peismon: But we're talking about extrapolation or predicting the future with control charts, not interpolation.

Socrates: How are those predictions made?

Peismon: As we said before, if the control chart shows stability, then we use the limits to predict results within those limits. That's what Shewhart has said and still applies today.

Kokinos: Shewhart specifically stated that the "fact that an observed set of values of fraction defective indicates the product to have been controlled up to the present does not prove that we can predict the future course of this phenomenon. We *always* have to say that this can be done provided the same essential conditions are maintained, and, of course, we never know whether or not they are maintained unless we continue to experiment" (Shewhart, 148). He also believed from his experience that once you have a stable process, normally it can be maintained.

Peismon: Without sustainability, we wouldn't get much benefit of a stable process.

Socrates: With that understanding, we can always make predictions whether stable or not.

Peismon: I don't think so.

Socrates: First, even with a stable process, we don't know it will continue to behave the same since someone or something can change the process. You might get different raw materials, an operator sets the machine to wrong settings, environmental conditions may change, and so on. Are you saying that, if the process was stable, the prediction would be accurate if these changes occurred?

Peismon: No, of course not.

Socrates: The prediction is a conditional prediction, even according to Shewhart. If the conditions C that produced result X continue, then result X will continue to occur.

Agrafos: In fact, the opposite is how insanity is defined: doing the same thing but expecting a different result.

Socrates: The same can be said when the process is unstable or "insane." The original hurricane data displayed instability with one point beyond the upper control limit. We reanalyzed it to show four stable periods. We also showed that we can reorder the data to display stability. What is the range of results of all the charts?

Peismon: The numbers range from 0 to 8.

Socrates: But I just want to predict. I can do that exactly how Shewhart said it could be done: If the conditions that produce annual major hurricanes from 0 to 8 continue, then I predict that the number of major hurricanes annually will be from 0 to 8.

Peismon: But what confidence do you have of that?

Socrates: The same confidence you have when the process is stable. The reason I can say that is because your confidence is not a statistical confidence like a confidence interval unless it is based on probabilities. What would you need to have that kind of confidence?

Mathitis: Probability limits, which we originally said they aren't. So, it would be contradictory to say you have a certain statistical confidence in your prediction but the limits are not probability limits.

Peismon: But we said they could be if we assume a probability distribution. Also, we know approximately from Wheeler's analyses and Empirical Rule what those probabilities are.

Socrates: *Now* you want to use them as probability limits. We also have various theorems like Tchebycheff's that allow us to make predictions whether the data display stability or not. The fact is that my prediction is what occurred for the last 78 consecutive years.

Mathitis: Are you saying that we can use the maximum and minimum values to predict what will happen merely by assuming that the conditions that produced those values are present?

Socrates: Yes.

Kokinos: That's a lot easier than calculating the charts and the limits.

EXTRAPOLATION: STABILITY VERSUS INSTABILITY

Peismon: But how do you know that works?

Socrates: Let's take the hurricane data and compare the two approaches. Let's call the approach I described as the min–max approach. Tell us again what the requirement is for your approach.

Peismon: There's only one. The control chart must show that the process is stable. There can't be any special cause variation or out-of-control signals.

Socrates: Is it okay to call your approach the control chart stability approach or CCS then?

Peismon: That's fine.

Socrates: Let's use the four stable periods. The first stable period of 14 points for the years 1935–1948 has a mean of 1.54, the lower control limit (LCL) is −1.34, and the upper control limit (UCL) is 4.42. What do you predict will occur next?

Agrafos: But on the chart, the lower limit is 0, not −1.34.

Socrates: We know that the measure is a count, so we can set LCL to zero if the calculation is negative. But that is another interesting advantage of the min–max approach versus the CCS.

Kokinos: What is that?

Socrates: That, using the equations, we do get negative results for the LCL, which we know is not possible for certain measures, for example, counts. That will never occur using the min–max approach. A key question is whether we use an I-MR chart of individual values and moving ranges or a C chart of counts. Both can produce negative LCL values but are replaced with zero.

Mathitis: That means that the min–max approach is more accurate on the lower end for some situations.

Socrates: That's right. Peismon, what do you predict assuming we use zero as the LCL?

Peismon: That there will be no more than 4.42 major hurricanes annually, from 0 to 4.42.

Mathitis: Doesn't the same thing occur with the UCL? Using the CCS approach, we will get a fraction of a major hurricane, which is not possible but never with the min–max approach.

Socrates: True. We know we can't have 4.42 major hurricanes in a year.

Peismon: That just sets the boundaries. As with LCL, we can adjust that if we wanted.

Socrates: And how many subsequent years is that prediction accurate?

Peismon: The next 2 years.

Socrates: How often is it inaccurate?

Peismon: I'm not sure, but once the process is unstable—starting with year 1950—the prediction stops anyway.

Kokinos: I counted 14 years of the next 64 or approximately 22% were wrong predictions.

Socrates: Now, use the min–max approach.

Peismon: That approach is wrong the very next year! The maximum was three major hurricanes from 1935 to 1948, but there were four major hurricanes in 1949. How can it be better?

Socrates: True. But now the maximum becomes four and we can continue predicting, unlike the CCS approach. Now, how many years is the prediction accurate?

Peismon: Again, it does worse than CCS. The next year's prediction is accurate but not the year after that, 1950. There were eight major hurricanes that year. The min–max approach missed twice in three tries.

Socrates: Again, replace the maximum of four with eight and continue predicting.

Mathitis: I see what you are doing. Now the min–max approach correctly predicts all subsequent years while the CCS approach stops predicting or is wrong 22% of the time.

Peismon: That's just arbitrarily changing the rules.

Socrates: The purpose is to accurately predict, regardless of what the rules are.

Peismon: The range will just widen and widen, making it useless, meaningless. Your limits are zero and eight while the control chart limits are −1.42 to 4.42. Much narrower.

Socrates: True. But as Mathitis noticed, the CCS limits have fractions and negative limits that cannot occur in reality, although these are changed to zero when that occurs. Let's consider the next stable period. Since the process has to be stable, then using CCS, we abandon the data from 1935–1947. We wait until the data show a stable process again, which are the years 1948–1969. What are the limits now?

Kokinos: They are much wider than before, LCL = −4.06 or 0, and UCL = 10.88.

Mathitis: And wider than the min–max limits of zero and eight. So, Peismon's complaint about wide limits for the min–max approach is now applicable to the CCS approach.

Peismon: But the control chart limits accurately predict the next years.

Socrates: Except that according to the stability requirement, after 8 or 9 years, the process is no longer stable and therefore the requirement is not met. While there are no points outside the control limits, there is the rule of certain number of consecutive points on the same side of the center line.

Mathitis: That number varies from seven to nine points.

Socrates: Let's be conservative. Using nine, we see that in the years 1970–1978, the number of major hurricanes was each less than the average of 3.41 for the period 1948–1969. In addition, the analysis and the chart state that those years are different from the previous years with respect to stability.

Mathitis: That means that the CCS approach would stop predicting after 1978 but the min–max approach would continue.

Kokinos: Then, wouldn't the CCS wait for another stable period to predict again?

Socrates: It appears so.

Kokinos: But the next stable period has narrower limits, similar to the first stable period. The years 1970–1994 have UCL = 4.18 and wrongly predicts 7 of the 18 subsequent years.

Mathitis: But the min–max approach correctly predicts all the years after 1950.

Kokinos: It appears that, in this case, for extrapolation or predicting the future, requiring homogeneous-as-stability data isn't better. It either doesn't predict more accurately or doesn't predict at all. It has limitations.

FALLIBILITY OF CONTROL CHART STABILITY

Socrates: There is another issue with the CCS approach.

Mathitis: What's that?

Socrates: As we showed before, lack of special cause indicators or alarms does not guarantee that the process is stable. When we sampled every fifth year, we would have erroneously concluded that process was stable despite the four distinct stable periods.

Peismon: We are quite sure that a special cause signal means the process is not stable using Wheeler's Empirical Rule. No more than 1% for 3 standard error control limits.

Mathitis: Yet, some of his examples showed as much as 3%–4% probability of being beyond the 3 standard error limits. But the question is about the assurance of stability, not instability.

Socrates: When there are no special cause signals, you believe the process is stable. You don't check in other ways to confirm that?

Peismon: No, we don't. We check for signals. If they don't occur, the process is said to be stable.

Socrates: And you make your predictions that the future results will be within the control chart limits when it lacks alarms.

Peismon: That's right. And, it works.

Socrates: Then, in those cases when there are no special cause signals and the conclusion that the process is stable is wrong, you do make predictions that "work."

Kokinos: Doesn't that contradict the claim that only stable processes are predictable?

CONTROL CHARTS IN DAILY LIFE

Peismon: Those cases are when minor fluctuations occur. When major changes occur, the kind we are seeking to identify with a control chart, predictions cannot be made or are wildly inaccurate because the process is unstable. A control chart is the only way to determine stability and homogeneity of data.

Kokinos: How do you know the fluctuations are minor if you don't check?

Socrates: Do you buy more food than you will need for your next meal?

Peismon: Doesn't everyone? I plan ahead and buy food.

Socrates: Do you often buy the wrong amount?

Peismon: No.

Socrates: So you can predict accurately how much you need.

Peismon: It isn't all that hard to do.

Socrates: Do you use a control chart to establish that your buying or consumption process is stable so predicting is possible?

Peismon: No.

Mathitis: I see what Socrates means. You don't know whether your process is stable but you can predict accurately.

Socrates: Do you drive erratically, or can you stay in your lane when you want or need to?

Mathitis: We all drive in our lane except when switching lanes, turning, or are distracted. If not, there would be many more accidents.

Socrates: Do you use control charts to manage that?

Mathitis: Of course not. Besides, we aren't supposed to text and drive.

Agrafos: It could be fatal if we had to plot points on a control chart to check whether we were driving under control.

Socrates: Why can you control your driving, maneuvering the car as the road curves or as you need to switch lanes or avoid obstacles? How do you control that process without control charts?

Kokinos: From experience. We know that first-time drivers have difficulty maneuvering a car. That's why state and federal governments require driver's licenses and they require passing a driving test and a written test.

Mathitis: And a visual test.

Socrates: What about other activities? Do you use control charts for predicting the time you will arrive for appointments? Or, do you use control charts to predict the results of repetitive behaviors, for example, walking, reaching, grabbing?

Mathitis: Of course not. That would require hundreds of charts.

Peismon: We don't predict those results.

Socrates: We don't say what those results are, but we act expecting certain results and are surprised when they don't occur. For example, we expect to successfully walk up a flight of stairs and are completely surprised when we misstep and stumble. We act confidently because we are predicting that our foot will rise the right amount—not too little and fail to clear the rise or not too much and lurch forward.

Agrafos: I agree. For example, a prank consists of extending your hand as if to shake hands with another person and then pull your hand away when the other person extends their hand. They are surprised because what they predicted would happen did not.

Mathitis: We don't need control charts because we control those processes in other ways.

STATISTICAL VERSUS CAUSAL CONTROL

Socrates: Then are you suggesting that because processes can be controlled in other ways than statistically, we can make predictions when processes are statistically unstable?

Mathitis: I hadn't thought about it that way, but that seems reasonable.

Socrates: Let's consider these other ways of controlling processes. When my daughter wanted to learn how to ride a bicycle without the training wheels, she managed after several tries. The first time she fell, fortunately on the grass. By the fifth time, she was wobbly but staying on the sidewalk. After that, she was able to ride controllably.

Peismon: How does that relate to predictions?

Socrates: At first, she wasn't sure whether she would fall or not. She could not predict the result.

Peismon: Then that confirms that stability is required for predictions.

Socrates: Not quite. It also supports the hypothesis that without control—and not in a statistical or control chart sense—prediction is possible. Once she was able to ride comfortably, she could act as if she was out of control but predict where she would go. She could make the ride wobbly, ride toward a tree, and swerve right before hitting it. The path was erratic but under control.

Mathitis: I see what you are saying. It's the same thing with my son and throwing a ball. Without control charts, kids learn how hard or soft and how much arc to throw a ball to travel a specific distance. The distances they throw may look like out of control on a control chart but they can predict where the ball will land.

Agrafos: What's the difference between statistical control using a control chart and this kind of control?

Socrates: First, the control chart is based simply on patterns. Special cause variation or out-of-control signals are nonrandom patterns. When we don't see any patterns or we think the results are random, we conclude that the process is in control—statistical control. This is not based on understanding the causes of the results but simply whether or not there are patterns on the chart.

Mathitis: That reminds me of other types of predictions. Investors use patterns of the stock market or of the price of an individual stock to predict which direction prices may go. They use moving averages, such as ARIMA models or auto-regressive integrated moving averages.

Socrates: Why do they use those types of models?

Mathitis: I imagine because we can't or it's illegal to control the price of shares. So, we try to find factors that correlate with price.

Socrates: That's one type of prediction—which control charts do but in a more simplistic way than those models. More complicated models are weather forecasting models of precipitation and temperature. Do they use control charts?

Mathitis: No. My understanding is that they use models developed over years of the different factors, for example, wind currents, seasonality, and geography. They do simulations to see what results are likely.

Kokinos: But they also use science, like gas laws and force field analysis, to determine barometric pressure and direction of winds.

Socrates: Do we control those factors?

Kokinos: No. As Mark Twain said, everyone complains about the weather but no one does anything about it.

Peismon: That's why weather forecasts are not always accurate.

Mathitis: But when we toss a ball, ride a bike, or go to work, we do control many of the factors.

Socrates: We can make this clearer. If we are plotting the results of applying scientific laws, then it may not matter whether the process is stable or not. Suppose I work as a recycler. I take my truck to consecutive houses in a neighborhood and pick up the recycling bins and dump the content in the truck. I plot on a control chart the force required to lift the recycling bins. We can imagine that it looks exactly like the results of control charts of major hurricanes—except the vertical axis is the force.

Kokinos: We know that force = mass times acceleration. Therefore, as long as I know the mass of recycling bins and the acceleration desired, we can calculate the force. Knowing those two quantities, anyone can predict the force required.

Mathitis: I see. If we collected on four streets, it could be that on the first street all the recycling bins are about the same mass and they are similar to the third street, while the second and fourth street are similar but the bins have greater mass. It could look like the patterns of the 78 years of major hurricanes. But we could predict the next point.

Kokinos: The point is that when the results follow a scientific law, we can predict the results not only when the conditions stay pretty much static but also when they change and we know what those changes are and the relationship between them and the results.

Socrates: That is a critical point, Kokinos. We know the causal relationship. The predictions made with a statistically controlled process are simply that if the unknown conditions that produced stability persist, we predict that the results will be the same as before.

Mathitis: Then, aren't there two problems with that kind of extrapolation prediction? One, we can only predict that the results will be the same but we cannot predict when the conditions change. Or, using the cause–effect framework, predictions based on stability (a) restrict the predicted effect to being the same as before assuming the unknown causes stay the same and (b) prevent predicting the effect when the causes change.

Kokinos: With scientific laws, we can predict different results with different causes.

Socrates: That is the difference among these various ways of predicting. They can be ordered as follows depending on the knowledge and control of the causes: simple guesses, patterns of historical results, understanding some or most of the causes but not being able to control them, understanding and controlling the causes. Which type or level are control charts?

Mathitis: I think they are the second level: patterns of historical results.

Socrates: Then, do we need stability of data or homogeneity of data to predict?

Kokinos: Not if we use the other, higher levels: correlation models and causal controls.

Socrates: Even more than. What is the purpose of an analytical study?

Kokinos: To determine why we got the enumeration we did.

Mathitis: Specifically to determine the critical causal factors and levels that produced the past results and then determine what levels those factors should be at to produce a more desirable result or enumeration.

Socrates: If we accomplish that, have we not moved from statistical control to causal control?

Mathitis: Yes, I see. Then we don't need the control charts or monitor for stability. We only need to monitor that the causal factors are at the optimal levels.

29

Myth 29: Adjusting a Process Based on a Single Defect Is Tampering, Causing Increased Process Variation

Scene: *Break room (continuation of stability-predictability discussion)*

BACKGROUND

Agrafos: I happened to be rereading Deming and I came across a warning that was related to our discussion about improving processes without the need to stabilize them first.

Socrates: What was that warning?

Agrafos: That you must be sure that you are not tampering with the process.

Mathitis: I agree. You can't just make any adjustment to the process.

Socrates: Why not?

Peismon: Deming defines tampering as management by results.

Socrates: How else would you manage if not by the results?

Peismon: He said in *Out of the Crisis*, "What is required for improvement is a fundamental change in the system, not tampering" (Deming 1986, 327).

Socrates: Does he define fundamental change and tampering?

Peismon: He gives an example: "A common example is to take action on the basis of a defective item, or on complaint of a customer" (Deming 1986, 327).

Socrates: If you were a coffee shop owner and had a customer complaint, do you not want to take action to address the complaint? If she

Socrates: ordered a latte and got a cappuccino, would she not complain and want it corrected immediately?

Agrafos: I would, certainly.

Socrates: If you had a defective item, would you not want to take action on it? Remember that Deming defined an enumerative study as a study whose purpose was to take action on the frame, for example, a lot with defects. That action could be scrapping the lot, or inspecting the lot for the purpose of sorting the defects and either scrapping or reworking them.

Mathitis: No. He means action taken on the process, the cause system.

Socrates: I'm still confused. If your process is producing defects, do you not want to improve the process?

Peismon: The issue is tampering with a process when there is only common cause variation, when the process is stable. Deming says that tampering would increase the variation, doubling it at least or even having it blow up.

Mathitis: Nelson's funnel activity (Deming, Chapter 9) shows this effect and the problem with tampering.

Peismon: The simulation requires establishing a stable process and then making adjustments to the process based on an undesirable result. The variation gets worse and worse.

Socrates: Then, tampering can only occur when the process is stable.

Kokinos: That's what I learned. It's treating common cause variation as special cause variation. When a result occurs within the control limits of a stable process, but it is undesirable or even desirable, responding to that result by adjusting the process is tampering.

Socrates: Then are you saying that for all measures and all processes, all adjustments of a stable process result in the process worsening with respect to performance?

Kokinos: I hadn't thought of it so encompassing but I suppose so.

DEFINITION OF TAMPERING

Peismon: Why doesn't Deming use the dictionary definition of tampering?

Peismon: He is using that definition. That's why tampering is bad, makes the process worse.

Agrafos: Peismon might be right. The dictionary definition of tampering is specifically negative: harmfully interfering, adjusting for the purpose of subverting, or acting improperly for the purpose of influencing.

Socrates: That definition does not mention anything about responding to an undesirable or desirable result or treating common cause variation as special cause variation.

Peismon: But reacting to a single undesirable result is tampering because you are purposefully influencing the process in a way that makes it worse.

Mathitis: But I can see Socrates' point. We don't do it with that intention—just the opposite. We react with the intention of making it better.

Socrates: Besides, simply adjusting based on a single defect does not imply it is negative.

Peismon: The funnel activity shows tampering based on a single undesirable result leads to increased variation.

Socrates: Tell me how it is done so I can understand.

Peismon: You have a funnel, a marble that will fit the funnel with some room to spare, a large piece of paper, and a marker. Lay the paper on a flat surface and identify the middle of the paper by crossing centered vertical and horizontal lines. There are four rules for dropping the marble through the funnel. You mark the spot where the marble stops. The objective is to have the marked spots as close as possible to where the lines cross.

Using rule 1, the process consists of holding the funnel about 4–6 inches above the "X," dropping the marble through the funnel, and marking the spot where the marble stops rolling. Repeat for a total of 50 drops. This establishes the performance of the process as shown here (Figure 29.1).

Mathitis: This shows that the process is stable.

Peismon: Yes. Now comes the tampering, using the other three rules. Rule 2 is move the funnel in the opposite direction from its last position relative to the center to compensate for the deviation. Continue adjusting the process in this manner 49 times. The result will look something like in Figure 29.2.

Mathitis: I've done this activity. It always leads to increased variation.

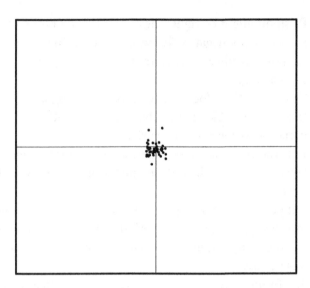

FIGURE 29.1
Funnel simulation establishing stability.

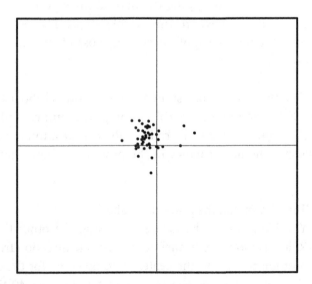

FIGURE 29.2
Funnel simulation applying rule 2: move the funnel in the opposite direction with respect to the center as far as the last observation was from the center.

Socrates: Is each movement of the funnel tampering?

Peismon: Yes, because you are adjusting the process after each undesirable result.

ZERO VERSUS ONE VERSUS MULTIPLE DEFECTS TO DEFINE TAMPERING

Socrates: This is Deming's example of tampering, taking action when a single undesirable or desirable result occurs?

Peismon: Yes. By definition, that's tampering when the adjustment is based on a single result. Deming says "If anyone adjusts a stable process to try to compensate for a result that is undesirable, or for a result that is extra good, the output that follows will be worse than if he had left the process alone (attributable to William J. Latzko)" (Deming 1986, 327).

Mathitis: In fact, Deming believes that tampering will at least double the variation and he uses the funnel activity to show that.

Socrates: Certainly, if you define tampering that way, then it is tampering. Why not define improving or succeeding or dancing or flying "adjusting based on a single defect" rather than tampering?

Peismon: That is not the definition of those words.

Socrates: Why is it okay to define tampering that way but not other words?

Peismon: Because tampering is negative, something to be avoided.

Kokinos: That's what I was taught. Are you suggesting Deming is wrong?

Socrates: Note that the dictionary definition says nothing about how many desirable or undesirable events must occur to be tampering. But if that's the definition, we can easily avoid tampering. Wait until you have two undesirable or desirable results.

Mathitis: I can see that. If the definition explicitly requires only one instance to be tampering, then having two instances is not tampering.

Socrates: Is adjusting the process based on two defects tampering?

Kokinos: Deming doesn't say.

Socrates: Consider an FMEA. Why do we use the tool Failure Mode and Effects Analysis?

Mathitis: To prevent defects from occurring before they occur.

Socrates: How many defects must we have before we make process adjustments using FMEA?

Mathitis: None. That's why it's preventive—it's before the defects occur.

Socrates: Is it tampering?

Mathitis: No. I don't know if it's contradictory, but it sure is odd to say that adjusting the process based on exactly one defective unit is tampering but not if the adjustment is based on zero or more than one defective unit.

Agrafos: Maybe Deming was only using "responding to a single instance" as an example.

Socrates: How many defective units does it take to no longer be tampering? Whatever that number is consider this. You wait until you have that number of defective units and make a change. I make the very same change after only one defective unit. If the same action is taken after multiple defects as after one defect, then the adjustment "works" in both cases or in neither. Why is the very same action not tampering when based on multiple or zero defects and tampering when based on a single defect?

Agrafos: It doesn't make sense to me.

Mathitis: Nor to me.

Peismon: I'm not completely convinced. I still accept what Deming said.

Mathitis: Because he said it?

Socrates: Consider this. I adjust a process upon getting a single defective unit and the process improves as shown by a reduced defect rate. Later, with respect to a different kind of defect, I wait until several defective units occur and adjust the process but it gets worse—the defect rate increases. Which is tampering and why?

Mathitis: I can see that the number of undesirable or desirable units or cases is not the issue. It's the effectiveness of the adjustment to the process.

Socrates: We agree, then, that tampering has nothing to do with the number of desirable or undesirable instances that trigger the action to adjust the process.

ROLE OF THEORY AND UNDERSTANDING WHEN ADJUSTING

Peismon: The issue is the adjustment. Tampering is overadjusting.

Socrates: Nothing in this definition implies that there is any overadjusting, regardless of the number of undesirable or desirable

Socrates: units on which it is based. If I underadjust, would it not be tampering?

Peismon: Over- and underadjusting are both tampering.

Socrates: If I adjust not too much and not too little but just right, would that be tampering? Call it the Goldilocks tampering approach.

Peismon: Ha-ha. No, but adjusting based on a single defect or complaint would not be adjusting just the right amount.

Socrates: Why not?

Mathitis: I think the idea is that you need more information. Deming said that without theory or knowledge or understanding, the adjustment will only lead to over- or underadjustments.

Socrates: Which do you mean: will always lead to over- or underadjustments or is likely to or may?

Mathitis: Not always. I could be lucky.

Socrates: Then, not all adjustments are bad. Are all adjustments based on a single defect without theory, knowledge, and understanding?

Peismon: How much knowledge can you have if there is only one defect?

Socrates: Doesn't that depend on how much knowledge of the process you already have? The purpose of R&D, for example, is to gain that knowledge. If you have characterized a product and the process that produces it well, then you may have sufficient understanding to make knowledgeable adjustments based on a single defect. We do this often in our daily lives. How much knowledge do we have when we adjust without any defects using an FMEA?

Agrafos: You mean, for example, a child only needs to burn their hand once to know not to touch the stove. Animals learn that way—so do people but not always. An animal may try to eat a plant or other animal that tastes bad and avoid it after that, especially if it got sick.

Socrates: Yes. Remember the advice "Learn from your mistakes." The question is "How many times do you need to make the same mistake before you learn?"

Agrafos: I think that any case where the consequence of the defect to the customer or business is deleterious, catastrophic, or fatal, we want to learn from only one occurrence. Preferably, even before a single bad occurrence. Like using an FMEA.

Peismon: But FMEA is a thorough analysis of the process.

Socrates: Then, an adjustment triggered by a single occurrence but based on a thorough analysis of the process would not be tampering.

Mathitis: I think I agree with that.

Socrates: Doesn't that mean that not all adjustments are bad and not all adjustments triggered by a single defect are based on lack of information? Making an adjustment because one defect occurred does not mean that you can't do a thorough investigation to gain the theory, knowledge, and understanding or to investigate based on existing theory, knowledge, and understanding (Castañeda-Méndez 2013).

Kokinos: Then, it isn't necessarily tampering if the cause of the defect is known or readily knowable or the action taken is known or readily knowable to have no unwanted side effects.

Mathitis: More importantly, is to know it will have the desired effect.

Kokinos: Yes. For example, if I am baking, I know that temperature and cooking time are critical factors. I can adjust for a single defect based on that knowledge. If I smell something burning or if I look at my watch and see how much time has passed, I can make adjustments.

Agrafos: I can think of many instances where we make adjustments based on single instances. While driving, the car crosses the center line. We hope that one scary near-miss when texting while driving is enough to learn and change my driving process by not texting anymore.

Mathitis: My son carries a glass of water filled to the top and spills some, so I tell him not to fill it so high. When he spilled it and had to mop the entire floor, he learned.

DEFECTS ARISE FROM SPECIAL CAUSES: ANOMALIES

Peismon: Those are special cause variations and therefore you would not be tampering.

Socrates: Why do you say it's attributed to special cause variation?

Peismon: If you have only one defect, then it has to be special cause variation. It's an anomaly.

Socrates: Are all anomalies the result of special causes or all defects are the result of anomalies?

Peismon: I'm saying that if the process is stable and you get one defect, then it has to be because of a special cause since there were no other defects. Without other defects, it is an anomaly.

Kokinos: I agree. An anomaly is something rare.

Socrates: I think you are confusing "getting one defect" with "only getting one defect."

Kokinos: But you are responding when there is only one defect.

Socrates: What if I just started the process and the first unit is defective. I have only one defect, but is it rare? My best guess so far is that it occurs 100% of the time. Is that rare?

Peismon: No, but if the process is stable, then you will have collected enough units to establish stability. Now, a single defect is attributed to special cause variation.

Mathitis: Are you saying that a process can't be stable unless we confirm it be a control chart? That would contradict your previous claim that before control charts we invented people coincidentally improved when the process was stable.

Socrates: I agree with Mathitis. Suppose I collect 24 units to establish stability using an *I* chart. The next unit is defective. My defect rate so far is 5%. Is that rare?

Kokinos: Not necessarily.

Socrates: Can a stable process produce defects?

Mathitis: I think we are discussing the difference between stability and capability. A process can be stable but not capable. It can be stable, having only common cause variation, and produce 0% defective units or 100% defective units. Deming himself says that.

Agrafos: Then, can you have a defect that is not caused by special cause variation and have special cause variation that does not cause a defect?

Mathitis: That's right. Remember the four states of a process (Myth 23).

Peismon: But a stable but not capable process will produce many defects on a regular basis.

Mathitis: But we could be talking about being on the threshold: stable but not capable. Then the defect rate could be any amount greater than 0% or 1%.

Socrates: If my defect rate is 1%, how regularly will I see a defect? If my *I* chart displays stability based on 50 units, how many defective units should I see?

Mathitis: Less than one. You won't necessarily see one with only 50 units.

Socrates: Now suppose I start with 25 units to establish the control chart limits. The process displays no alarms. I continue my sampling and plotting the points and the process continues to display no alarms. On the 35th unit, I get the first defect even though the point is within the control chart limits, that is, a false special cause variation? Is it an anomaly?

Kokinos: I guess not. Or, we don't know.

Mathitis: I would also have to say it is neither.

CONTROL LIMITS VERSUS SPECIFICATION LIMITS

Socrates: In the context of an anomaly, we have confused control limits and specification limits. An anomaly can be with respect to specifications, that is, a rare defect. An anomaly can also be with respect to the control chart limits, that is, a false special cause variation signal. A commonly occurring defect may be a rare special cause and vice versa.

Kokinos: Even Deming says that there is no "logical connection" between the two.

Socrates: When we consider the case of responding to a single defect, which are we using: control chart limits or specification limits?

Mathitis: Specification limits.

Peismon: But tampering is still treating common cause variation as special cause variation.

Socrates: Then, tampering can only occur when the process is stable.

Peismon: Yes. Although it is also a problem when you treat special cause variation as common cause variation. That's a second mistake that Deming says can be made.

Socrates: How are common cause and special cause variation determined?

Peismon: I know you know how, Socrates. But I will humor you. A control chart tells you which you have. When the pattern of points on the control chart is random, we have common cause variation. When it displays a nonrandom pattern, especially a point beyond the control limits, then we have special cause variation.

Socrates: Then the control limits are used not to categorize the variation as either common cause or special cause but categorize whether the point is an alarm or not.

Peismon: That's right.

Socrates: As we have noted several times before, is this method of categorizing infallible?

Mathitis: No, it's not.

Socrates: What action do we take when we have a point beyond the control limits?

Mathitis: We investigate to find whether it is a true alarm or a false alarm. It can be either although the limits are designed to have very low false alarms. But, you are right, it is still not infallible.

Socrates: Is that why you call these nonrandom patterns alarms?

Mathitis: Yes. They categorize the pattern as alarm or no alarm. The patterns do not categorize the variation as common or special cause variation.

Socrates: Recall our previous conversations. The purpose of an alarm is to signal something. It is not that thing. A fire alarm signals the possibility of a fire; it doesn't tell us that, in fact, there is a fire nor is the alarm the fire. The nonrandom pattern of a point beyond the control limits is a signal that special cause variation may have occurred. The point is not the special cause. Similarly, doesn't a lack of nonrandom patterns only signal that common cause variation may be occurring? In other words, the correct reading of a control chart is that it displays *evidence* of common cause variation or process stability rather than it displays process *stability*.

Kokinos: We've talked about this before. We saw that there can be false alarms and false nonalarms. In fact, we agreed that the false nonalarms would be more prevalent than false alarms. While Deming says that we can't attach a specific probability to it, we saw that all it took was an assumption of a probability distribution to do that.

Socrates: Do you, Peismon or Mathitis, know what the false nonalarm rate is?

Mathitis: I don't. And I agree with you. These signals or lack of them could be wrong.

Peismon: Still, we do not want to treat common cause variation as special cause variation.

Socrates: Although it appears that since false nonalarms do occur more frequently, that you do treat special cause variation as common cause variation. Before, you said that whenever there were no special cause signals, you always assumed that the process was stable. But what I am interested in is clarifying the role of the alarms. They serve to signal that special cause variation may have occurred not that it has occurred.

Peismon: I accept that.

Socrates: Even if they differentiated 100% between special and common cause variation, the control chart limits do not, by themselves, differentiate between defective units and good units.

Peismon: We agree on that. We cannot stress enough that control chart limits are not specification limits. We teach people not to put specification limits on control charts because they serve different purposes. And, while I didn't agree with the case of the individual control chart, I do see that that might be the only chart where both limits could be put on the chart.

Socrates: So, control chart and specification limits are limits with independent purposes.

Mathitis: Yes, of course. That's why we said earlier that stability using control chart limits is independent of capability using specification limits.

Socrates: This is my confusion. Control chart limits only tells us whether we (might) have common or special cause variation independent of whether there are defects. Specification limits tell us whether we have a defect independent of whether the process is stable. Then why would responding to a defect determined by specification limits be treating common cause as special cause variation, which is determined by control chart limits—especially when we don't even know which variation it is?

ACTIONS FOR COMMON CAUSE SIGNALS VERSUS SPECIAL CAUSE SIGNALS

Peismon: The actions taken when you have common cause variation are different from when you have special cause variation. You would want to know under which condition the defects occurred.

Socrates: But the issue here is not what action to take when you have each type of variation, but what action to take when you have an alarm or no alarm. I can understand that using the fire extinguisher when there is no fire is bad. And perhaps that's what Deming really meant: responding to a false alarm as if it were real is bad rather than treating common cause variation as special cause variation is bad. The distinction is critical. Recall our modified flowchart (Figure 19.2).

Peismon: But he clearly says tampering is treating common cause variation as special cause variation.

Mathitis: Common cause variation is what occurs naturally in a process, the inherent variation of the process. Special cause variation is extra variation, not typically associated with the process. Like a new source of variation. Therefore, the search for the cause would be different.

Socrates: When do you know which it is?

Peismon: That's what the nonrandom patterns are for.

Socrates: But recall that they are not infallible. The nonrandom patterns are only signals that there may be special cause variation and the lack of the patterns is only a signal that there may be only common cause variation. If you do not know which it is, or believe wrongly that it is one rather than the other, will you not do exactly what you say you want to avoid—treat one type of variation as the other?

Agrafos: It seems to me you would.

Peismon: Assuming you knew which it was, you would take different actions. Besides, we are pretty sure when the alarm goes off by having a nonrandom pattern we have special cause variation.

Kokinos: But aren't we discussing the case when there are no special cause variation signals? We said it was treating common cause variation as special cause.

Socrates: Exactly. Mathitis and Peismon said that tampering was when the process displayed—but perhaps falsely—evidence of stability, adjusting it would be tampering. We do not know what the false nonalarm rate is and therefore we cannot say that we are pretty sure there is only common cause variation. But even assuming that's the case, the question is what action is taken? Let me change that. What is the purpose of the action taken when there is only common cause variation and what is the purpose of the action when there is special cause variation?

Mathitis: When it is truly common cause variation, we do not want to adjust the process but improve it. That will require a fundamental change in the process. The purpose is to reduce the inherent variation of the process.

Socrates: We know that many processes can be made better, such as reducing the defect rate, by adjusting the average. If the process is stable and only off target or nominal, why would we want to reduce the variation?

Mathitis: We can do either. But in general, we want to reduce variation.

Peismon: When there is special cause variation, the purpose of action is to remove that noninherent variation to stabilize the process, to get it back to where it was before the special cause variation occurred.

Socrates: Do you see that neither of these actions is to reduce the defect rate?

Kokinos: But why would we want to reduce variation if it isn't to improve process performance? By process performance, we mean the defect rate. I think addressing the defect rate is assumed when we say reduce variation.

Socrates: Perhaps. But having a stable process seemed to be a goal on its own. We talked about that with respect to improvement, calculating capability, predicting, and so on.

Mathitis: I agree. We sometimes just address special cause variation signals to merely stabilize, without regard to the defect rate.

Socrates: What about actions taken on special cause variation? You said stabilize, which isn't necessarily reducing the defect rate. Recall that we said not all special cause variation signals indicate a worsening of the process (Myth 22). Some are signals of improved performance. In these cases, would not stabilizing to the previous levels actually increase the defect rate?

Kokinos: Yes, not all special cause variation is bad. But when it is, we want to eliminate them. So, we investigate them to learn how to reduce its probability of occurring again.

Socrates: Then, regardless of whether it is common cause or special cause variation that produced the defects, the action is to investigate to find the cause and reduce the probability of those defects recurring.

Agrafos: It doesn't seem like there is any difference.

Peismon: The cause is different. One is inherent to the process, the other is not.

Socrates: That doesn't change the actions: investigate, find the cause, and address the cause. When we say tampering is action taken when there is only a single defective unit, are we not assuming that

(a) no investigation to find the root cause precedes that action, (b) no investigation can find root causes, or (c) all such actions to address root causes are maladjustments or result in making things worse? Logic and real examples tell us these assumptions are often false.

Agrafos: I agree and the examples I gave earlier show that.

Socrates: It is possible that knowing whether a defect coincided with common or special cause variation may help in the investigation to find the cause or help in determining how to address it, but those are details, not changes in the actions or barriers, to effective actions. The cause could be a machine, raw materials, or a procedure. It doesn't change the general actions but the details.

Peismon: But it is the details that are critical.

Socrates: You said that when we have special cause variation the purpose of the action is to stabilize the process. But when we have a single defective unit and we adjust the process, are we adjusting the process to stabilize it?

Kokinos: No, to reduce the probability of that defect recurring.

Socrates: Is not that the action taken when we have a stable process whose performance we want to improve?

Kokinos: Yes.

Mathitis: Then, adjusting a process because of a defect where the action is to reduce the probability of its reoccurrence, we are taking the very action we take when we have only common cause variation.

Socrates: Then how can we be treating common cause variation as special cause variation?

Mathitis: We aren't. I think I see the mistake we made. When we think of adjustment when the process is stable, we think of actions to restabilize the process, but some actions are to reduce defect rates. We said these are independent.

Peismon: I'm not sure I agree.

Socrates: Perhaps we can see it more clearly with an example. I start a process and I have no intention of using a control chart. We do this for almost all of our processes outside of work. I mentally track my performance, even crudely. The first unit is defective. I know that if I continue, all I will get are defects. I also know what to do to reduce the probability of it recurring. What should I do? Continuing processing units to determine whether the process is stable?

Agrafos: The process can't be stable if it is producing defects 100% of the time.

Mathitis: It really is hard to separate stability from capability. Remember, Agrafos, that a stable process can have any defect rate from 0% to 100%.

Socrates: Here is a concrete example. You want toast for breakfast. You put a slice of bread in the toaster. It's burnt.

Peismon: That's a special cause.

Socrates: No. I know that if I toast 24 more slices, they will all be burnt. We can measure how burnt by the darkness or the amount of the bread that is black or simply burnt/not burnt.

Kokinos: I've been there. The point is that, without changing the settings, you will get exactly the same result. The process is stable but performance is horrible: 100% defectives. We adjust the process by adjusting the light–dark knob.

Socrates: Is it necessary to know whether special cause or common cause variation caused this defect of burnt toast, to adjust the process appropriately? Is it necessary to continue collecting units to get more than one defective unit so it is not tampering? Third, is adjusting the light–dark knob tampering if it improves the process by reducing the defect rate to zero?

Mathitis: No to all your questions. I agree that there are many cases where, with enough knowledge and understanding of the cause–effect relationships of the process, responding to a single defect would not make the process worse.

Socrates: Then, do we not want to make the definition of tampering: adjustment of the process that makes the performance or behavior worse? Is this a better definition than the ones we considered early that were based on any adjustment whether it improved performance or not, any number of defective units, and treated common cause as special cause or vice versa?

Peismon: We don't want to treat common cause variation as special cause variation or vice versa.

Socrates: I agree in general. But if you want to call treating common cause variation as special cause variation tampering, then don't assume that actions taken to reduce the defect rate are always tampering regardless of the number of defective units that triggered those actions.

Agrafos: It makes more sense to me now. Of course, if we had adjustments that made the process worse when we based it on a stable process or on a single defective unit, I can see why people would think those examples of tampering would be the definition of tampering.

Kokinos: The examples should have been used to warn about how to adjust processes.

IS REDUCING COMMON CAUSE VARIATION ALWAYS GOOD?

Socrates: Does reducing variation always improve process performance?

Peismon: That is a fundamental tenet of process improvement methodologies like Six Sigma.

Mathitis: But we said that was not true. Reducing variation doesn't always improve processes and increasing variation can improve processes.

Socrates: Can you illustrate the two actions Deming says are required to improve processes?

Peismon: We can have these three possibilities when there are two specifications (Figure 29.3). In case (a), the spread of the distribution is too wide to fit inside the tolerance or specification window. So, we need to reduce variation. In case (b), the spread does fit inside the tolerance but the distribution is outside the specifications.

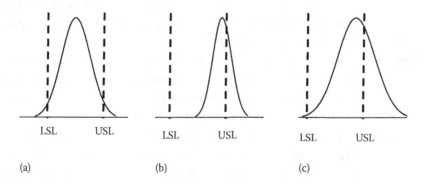

FIGURE 29.3

Three conditions for improving process performance: (a) reducing variation, (b) shifting the mean, and (c) both reducing variation and shifting the mean.

Shifting the mean will improve performance. In case (c), we have both situations.

Socrates: Excellent. Why did you use two specifications?

Peismon: That's just an example. We could consider the other two cases where we have only a lower specification limit (LSL) or only an upper specification limit (USL).

Socrates: If we have just USL in case (a), will reducing variation alone improve performance?

Peismon: No. That's similar to the second case. Shifting the mean will accomplish that.

Socrates: Will increasing variation toward the lower values improve performance?

Peismon: Yes, but not as much as shifting the mean.

Socrates: Doesn't that mean that increasing variation can improve performance? Or, said differently, increasing variation is not always bad or worsens performance and decreasing variation is not always good or improves performance?

Agrafos: I see that now and agree.

Socrates: If the variation is in the right direction, then the increased variation can be good in that it reduces the defect rate—or increases the good rate.

Peismon: I don't agree. Increased variation can never be good.

Mathitis: I don't understand why these examples don't convince you.

Socrates: There's a simple test of your convictions and of your belief.

Peismon: What's that?

Socrates: You now get periodic paychecks that are approximately the same. They are based on your annual salary. Raises or bonuses would be special cause variation that would increase the variation. Do you not want them?

Agrafos: Of course he wants raises and bonuses! We were just talking about that the other day. I want my company to go wild and crazy out of control and pay in any nonrandom amount it wants that is greater than what I get now. I'm all for that.

Mathitis: That would apply to other financial processes: our pensions, 401ks, social security fund, investments, all of our assets. As long as they keep increasing in value, we don't care if it is stable or not. In fact, we want them unstable in a particular way.

Kokinos: We talked about this before (Myth 27). There are processes whose results we do not want to be stable. Increasing variation or having special causes in the right direction are not inherently bad—certainly not bad for these types of processes.

FUNDAMENTAL CHANGE VERSUS TAMPERING

Socrates: Kokinos, when I asked about the actions taken when the process is stable, having only common cause variation, you said that, to improve, a fundamental change was necessary.

Mathitis: That's what Deming says. "Improvement of a stable system requires fundamental change in the process" (Deming 1986, 327).

Peismon: One enhancement to statistical process control made by Deming is the distinction between who is responsible for taking action. He said that the additional value of separating common and special cause variation is that it determines who is responsible.

Socrates: How does it do that?

Peismon: Since special cause variation is not an inherent or natural part of the process, then we do not expect it to occur often. But more importantly, Shewhart called it assignable cause variation. Deming viewed special cause variation signals as alarms for the process operators. They would be the ones investigating and addressing these special causes because they are assignable or discoverable.

Socrates: Then who would be responsible for common cause variation?

Peismon: Management. Once a process is stable, then it would be up to management to decide whether further improvements were needed or wanted. That typically involves a major change to the process. A fundamental change.

Socrates: What is the difference between adjustment and fundamental change?

Kokinos: I understand fundamental change to mean an extensive change in the process. Although we have acknowledged that you can improve an unstable process and that special cause variation can be good so you don't always want to eliminate it but make it common cause variation.

Mathitis: Deming says that improvement "almost always" involves reducing the common cause variation and sometimes it involves shifting the average.

Socrates: Presumably, he means shifting the average so that the process improves.

Peismon: Yes, shifting in the right direction, increasing or decreasing the average as appropriate.

Socrates: Then any adjustment of these two types, whether based on a single defective unit or multiple defective units, whether the process is stable or unstable, is a fundamental change.

Mathitis: In fact, Deming says that fundamental change can be as simple as brightening a room or it can be a more complex action.

Agrafos: What's the difference between change and fundamental change?

Socrates: Perhaps we don't need the value-laden adjective *fundamental* since it implies more than just a simple change. We can call them process adjustments. But in any case, we see that any adjustment that improves the process is good and one that worsens the process is bad. I think Kokinos' comment about taking Deming's examples as cautions can be stated more explicitly with respect to these two actions.

Kokinos: Yes. We run the risk of making things worse when we make changes without really knowing or having evidence that it will have one of the two results: reducing variation or shifting the mean.

Agrafos: But we should always include the phrase *in the right direction* when we speak of these types of changes.

Mathitis: We have assumed with only one defective unit or undesirable result, we don't have the knowledge necessary to do that. But we've seen that that assumption is not always true. Especially since he can have that knowledge without defects when using an FMEA.

FUNNEL EXERCISE: COUNTEREXAMPLE

Peismon: I still think of the funnel activity that shows that tampering based on a single undesirable result—adjustments without basis—leads to increased variation.

Socrates: Is each movement of the funnel an adjustment?

Peismon: Yes. It's tampering.

Socrates: It increases the variation. Does it change the average?

Peismon: Yes it does that too.

Socrates: In this case, since neither change of the variation nor the change in average improves the performance, they are not fundamental changes, according to Deming. Correct?

Peismon: That's correct.

Socrates: The example you quoted from Deming involved adjusting the process based on a single defective unit using the funnel simulation (Figure 29.2). Where are the specifications so we can determine which results are defective and which ones are not?

Peismon: We don't use specifications for this activity.

Socrates: Then it doesn't really simulate the case we are discussing: adjusting the process based on a single undesirable result, for example, a defect. But we can rectify that. Let's draw a dotted circle with its center on the center of the paper aimed at in the first round and a diameter wide enough to include all the first 50 results. This will be the specification limit.

Mathitis: But then you have a stable process and no defects. So, that doesn't describe the example either.

Socrates: I haven't started the process, Mathitis. Now suppose that the requirements change. Let's illustrate that by moving the circle in any direction so the center of the paper crosses the circumference (Figure 29.4). Continue dropping the marble using rule 1. But as soon as you get one result outside where the circle is now, we will adjust the process. Any ideas on where to move the funnel so no more undesirable results occur?

Mathitis: Obviously, over the center of the circle.

Socrates: Do it and drop the marble. According to our original definition of tampering, this would be tampering. You adjusted the process based on a single defective unit. If you did not adjust the stable process, would you continue to have defects?

Kokinos: Yes, but you're saying don't continue with a stable process when we have knowledge and understanding of the cause–effect relationship. Instead, make an adjustment that reduces the predicted or projected defect rate based on your knowledge.

Agrafos: Is it then a fundamental change?

Socrates: I don't know. It depends on the definition. I'm not fond of the term since it is simply an adjustment. The same as when we were tampering. The only difference is that one adjustment made performance worse and the other made it better. That is why I am

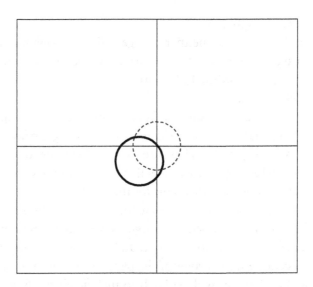

FIGURE 29.4

Funnel simulation: move the specification (dotted circle) to the position (solid circle) where the circumference crosses the center of the paper. Continue using rule 1 until the first observation is outside the new circle (solid); then, move the funnel to the center of that circle and continue dropping the marble for a total of 50 drops.

inclined to call them all adjustments and say that some are good and some are bad, or worse and better.

Mathitis: It's true. Deming mentions two other improvements: (1) lower the height of the funnel from the paper and (2) replace the paper with a cloth so the marble doesn't roll as much. He doesn't say whether these are fundamental changes or just adjustments.

Socrates: There is another adjustment he doesn't mention that reduces the variation.

Kokinos: What's that?

Socrates: Lower the funnel so it is on the paper. The marble will not roll and the variation will be no wider than the diameter of the funnel.

Peismon: But you can't do that.

Agrafos: Why not? If Deming allows lowering the funnel, why not all the way to the paper?

Socrates: This is just a simulation. The purpose was to show that tampering (using his definition) *could* increase variation. Unfortunately, perhaps some have interpreted the funnel simulation as showing that all "tampering" always leads to increased variation. That would be a mistake. It does not show that all processes and all adjustments will always increase variation. Deming says that lowering the funnel is acceptable. If you don't like the funnel on the paper, then place it above the paper a distance equal to the radius of the marble. It's not touching the paper but prevents the marble from rolling. If you can replace the paper with cloth, replace it with clay or a sticky substance like a coat of honey so the marble doesn't bounce or roll.

Kokinos: I would call changing the funnel height an adjustment but changing the paper to cloth or clay or coating it with honey a fundamental change. But I agree that Deming is not clear.

Socrates: In any case, do all adjustments—even those based on single undesirable or desirable results, for example, a defect—increase variation? And more important, does making adjustments based on a single defect always lead to worse performance, more defects?

Kokinos: No.

Socrates: Let's do one more simulation. Replace the circle specification with the vertical line down the middle of the paper. This is our specification now. Defects occur when the marble stops to the left of the line. The "tampering" rule is as follows: (a) if a defect occurs, move the funnel to the right of the last spot an amount equal to the distance the marble stopped to the left of the line; (b) if there is no defect, keep the funnel where it is. What do you expect to happen?

Mathitis: I suspect the variation will initially increase and then decrease once you get far enough away to the right from the line. It will then stabilize.

Peismon: That's just an artificial example.

Socrates: Peismon, you can't have it both ways. You can't use an artificial example with specialized rules to show what you want claiming it is valid and then reject artificial examples with specialized rules contradicting what you want to show as invalid. But we can create actual examples. Do you drink coffee?

Peismon: Yes, we all do.

Socrates: Put the coffee cup on your kitchen table. Then stand over the sink and start pouring the coffee. Will it go into the cup?

Peismon: Of course not. They are not anywhere near each other.

Mathitis: I see what you are doing. As soon as you see that a defect occurs—the coffee did not go into the cup—we know what to do. Move the coffee pot over to the cup and pour there.

Agrafos: Or move the cup to the sink—it's either going to the mountain or bringing the mountain to you. I understand now. If you continue pouring in the sink, you have both a stable process and 100% defects. Moving the pot over the cup changes the defect rate but destabilizes the process.

Socrates: Exactly. Don't we have dozens or more examples in real life where we are pouring a liquid or adding a solid to a container? If we have a defect—the liquid or solid does not fall into the container—do we adjust to end the spilling defects? We never call that tampering. And we know that, if we don't adjust, we will continue to have defects.

Mathitis: I'm convinced that we need clearer definitions of tampering and fundamental change, and better yet, different terms that do not imply more than the words suggest.

Agrafos: I would stick with the dictionary definitions rather than inventing technical but misleading definitions.

Mathitis: Or using value-laden terms.

Kokinos: Certainly, with the appropriate knowledge and understanding of the causal relationships or with the appropriate investigation, process adjustments made on zero or one undesirable or desirable result can improve process performance as measured by the defect rate.

30

Myth 30: No Assumptions Required When the Data Speak for Themselves

Scene: *Break room (after class lesson on graphical display of data)*

BACKGROUND

Socrates: What did you think of today's lessons on graphical display of data?

Agrafos: I liked the lesson. It makes so much more sense to just look at the data.

Kokinos: I did too. Given that we are learning about statistics, I was surprised to hear the instructor say that statistical analysis should be the last thing to do to understand the data.

Agrafos: I also liked the idea of starting with the practical question you want to answer, then presenting the data graphically, and if the answer isn't clear from the graph, then resort to statistical analysis. I like this example (Figure 30.1) with clear statistical differences: the means, the percent meeting specifications, and the variation are all different. We were testing whether changes in one factor, a potential root cause, produced changes in the response measure.

Socrates: What did you understand was the point of this example?

Agrafos: To show that sometimes no statistical analyses are needed.

Peismon: I think the graph tells the story. The only question I have is whether the sample size is large enough to reach the conclusions you did.

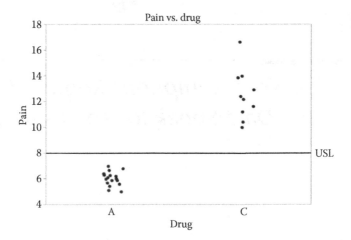

FIGURE 30.1
Letting the data speak for themselves—two groups with clear differences in averages, meetings specifications, and variation.

Kokinos: I think what made me believe this approach was when he showed that the same statistical analyses can result from very different data, which is evident when graphically presented (Anscombe 1973; Chatterjee and Firat 2007). Like these graphs (Figure 30.2). See how the statistics are the same but the graphs differ?

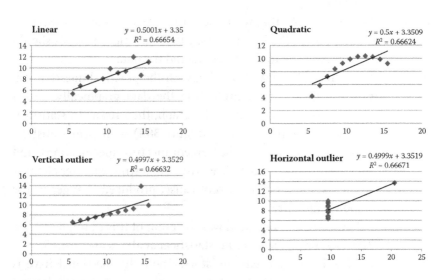

FIGURE 30.2
Letting the data speak for themselves. Anscombe's regressions showing that having the same statistics does not mean the data are the same.

Kokinos: That clearly showed that looking at just the statistics can be misleading. You need to understand the data. As they say, "a picture is worth a thousand words."

Socrates: It is important to see the data before understanding or interpreting the statistics. The more perspectives you have of a situation, the deeper your understanding.

Agrafos: I'm all for letting the data speak for themselves—graphically or otherwise.

Socrates: That's fine, but to make any interpretation or inference, you must make some assumptions. The data and the graph don't actually talk!

Kokinos: That's true, the data don't talk.

SIMPSON'S PARADOX

Mathitis: But we have to be careful when interpreting graphs and tables. Have you heard of Simpson's paradox?

Agrafos: No. What's that?

Mathitis: It's when there is a reversal of a conclusion when the same data are looked at from different perspectives, thus creating contrary conclusions or a paradox.

Agrafos: How does that occur?

Mathitis: It occurs when, for aggregated data, a measure Y is greater under condition A than condition B, but for a particular disaggregation of the data, Y is less under condition A than condition B for each disaggregated group. Here is one example (Figure 30.3). It shows two sets of data, one by the squares and the other by the circles, on the same graph.

Agrafos: There appears to be a linear relationship for each symbol type. The squares have one linear relationship and the circles another.

Peismon: I agree. The slopes are positive: Y is proportional to X and increases as X increases.

Mathitis: If we assume there is this relationship, we could do regression analysis and determine the lines.

Kokinos: Where is the paradox?

Mathitis: Now analyze all the data combined (Figure 30.4). The fitted line has a negative slope. This reversal is one way Simpson's paradox can occur. Separately, the lines for the squares, say L1, and for

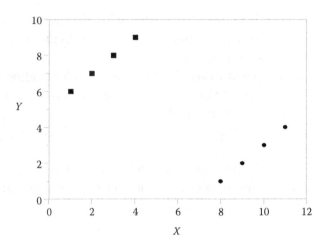

FIGURE 30.3

The squares as one group and the dots as another each appear to show a positive slope between *X* and *Y*.

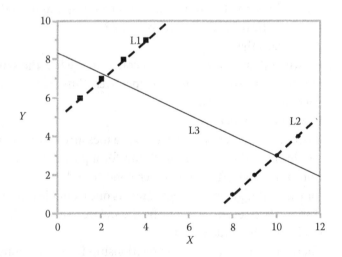

FIGURE 30.4

Reversal of slope when data are aggregated (solid line L3 with negative slope) versus disaggregated (dashed lines L1 and L2 with positive slopes).

the circles, say L2, have positive slopes while the combined line L3 is negative.

Kokinos: How did that happen?

Socrates: What do the data *appear* to tell us? Which is the appropriate conclusion *if we let the data speak for themselves?*

Kokinos: I don't know. As Peismon noted, each individual group says that as *X* increases so does *Y*, but combined, the data tell us that as *X* increases *Y* decreases.

Agrafos: I see why it is called a paradox. We have two contradictory interpretations of the data. Which is correct?

Mathitis: There are other examples in the literature. One consists of different patient groups and different treatments (Table 30.1). Patients with kidney stones were classified according to the stone size, either less than 2 centimeters or at least 2 centimeters long. Within each class, there were two types of treatments: three kinds of open procedures or combinations of them versus percutaneous nephrolithotomy (Charig et al. 1986).

Peismon: Look at the data by kidney stone size. Treatment A works better for each size separately, although for sizes of at least 2 centimeters, the treatment is not as effective as it is for smaller stones. But for the aggregate, Treatment B is better.

Mathitis: Another actual example occurred with a 1973 lawsuit. The aggregated data were used by a group of women to claim bias against women in admission to the University of California Berkeley (Table 30.2). But the disaggregated data showed that females were admitted at the same percentages or higher, suggesting bias for women, by several departments (Table 30.3).

Agrafos: How can that be? It doesn't seem that the math is correct.

TABLE 30.1

Kidney Stones Treatment Successes by Stone Size and Treatment: Example of Simpson's Paradox

Stone Size	Treatment A (Open Procedures)	Treatment B (Percutaneous Nephrolithotomy)
<2 cm	Subgroup 193% (81/87)	Subgroup 287% (234/270)
≥2 cm	Subgroup 373% (192/263)	Subgroup 469% (55/80)
Aggregate	78% (273/350)	83% (289/350)

TABLE 30.2

Aggregate Numbers and Percentages of Admissions per Application by Gender

Gender	Applicants	Admitted (*n*)	Admitted (%)	Rejected (*n*)	Rejected (%)
Male	8442	3738	44	4704	56
Female	4321	1494	35	2827	65

TABLE 30.3

Disaggregated Numbers and Percentages of Admissions per Application by Gender and Department

	Male		Female	
Department	Applicants (*n*)	Admitted (%)	Applicants (*n*)	Admitted (%)
A	825	62	108	82
B	560	63	25	68
C	325	37	593	34
D	417	33	375	35
E	191	28	393	24
F	272	6	341	7

MATH AND DESCRIPTIVE STATISTICS: ADDING VERSUS AGGREGATING

Socrates: One reason it seems like a paradox is that we think we are adding fractions or percentages when we are not. We are aggregating quantities. How do we add fractions, for example, 3/7 + 2/5?

Agrafos: You have to find a common denominator, convert the fractions to that common denominator, and add the numerators. In this case, the common denominator is 7 × 5 = 35. So, 3/7 + 2/5 = 15/35 + 14/35 = 29/35.

Socrates: You add the numerators but divide by the common denominator. But when we aggregate data, both numerators and denominators are added. Consider this scenario. You and I buy a pizza with eight slices. If I eat one slice and you eat two, what fraction did we each eat and how much was eaten?

Agrafos: You ate 1/8 and I ate 2/8 so together we ate 3/8 of the pizza. We added the fractions.

Socrates: But now consider this scenario. We each buy a personal pizza with eight slices each. We eat the same number of slices as before. What fraction did we each eat *of our personal pizza* and what is the fraction of *all pizzas* eaten?

Agrafos: The fraction eaten for each of us is still the same, 1/8 and 2/8. But the fraction of all slices is now 3 divided by 16 slices total, not 8 for each pizza.

Kokinos: So, when we aggregate data, we also add the total amount in each set, or the denominators. It's combining quantities. Like 1 ounce of olive oil and 1 ounce of vinegar produces 2 ounces but now each is 50% of the total not 100% as in the original sources.

Peismon: Then, it also depends on the amounts combined. If we had 3 ounces of olive oil and 1 ounce of vinegar, the combination is 75% olive oil and 25% vinegar out of four ounces total.

Socrates: Exactly. When we determine fractions or percentages *after* aggregating data, we are actually calculating a weighted average. What is the percent of olive oil in a mixture from two containers where one container has 4 ounces of oil and 1 ounce of vinegar and the other has 6 ounces of oil and 9 ounces of vinegar?

Peismon: We have a total of $4 + 1 + 6 + 9 = 20$ ounces and $4 + 6 = 10$ ounces of oil. The percent is 50–50 oil and vinegar.

Socrates: But what about the percentages for the total and for the containers?

Peismon: The first container had 80% oil and the second had 40%, but the total has 50%.

Socrates: Is there a paradox? Is it contradictory that one container had four times the percent of oil but the combined had only 50% oil?

Agrafos: No. It's clear to me.

Socrates: What were the total ounces for each contained?

Peismon: The first had 5 ounces total and the second had 15 ounces.

Mathitis: So, the second with 80% oil is weighted three times the first with 40% oil, for a total of four units. We could have done the math this way: $(1 \times 0.8 + 3 \times 0.4)/(1 + 3) = 2.0/4 = 0.5$.

Socrates: When we start with the disaggregated data and aggregate it, we tend to think in terms of adding fractions because we are comparing percentages. This gives us the impression of a paradox. When you take the aggregate data and partition it, you don't see paradoxes because you realize that you are creating different groups that will have different percentages. Consider the oil and vinegar example. If you know the combined container has 50% of each, do you know the percentage of oil and vinegar in the source containers?

Kokinos: No. They can be almost anything and they can be different volumes.

Mathitis: If we already know they can be almost anything from 0% to 100% of one ingredient, then why should we be surprised when we get some particular combination?

TABLE 30.4

Kidney Stones Treatment Successes by Stone Size and Treatment: Resolving Simpson's Paradox with Equal Subgroup Sizes

Stone Size	Treatment A (Open Procedures)	Treatment B (Percutaneous Nephrolithotomy)
<2 cm	Subgroup 193% (93/100)	Subgroup 287% (87/100)
≥2 cm	Subgroup 373% (73/100)	Subgroup 469% (69/100)
Aggregate	78% (166/200)	73% (156/200)

Peismon: I guess we shouldn't. To me, it seems more like an optical illusion. I look at it one way, aggregating, and I see a paradox but I look at it another, disaggregating, and I can explain the mathematical results.

Socrates: One way to avoid the appearance of a paradox is to have equal subgroup sizes or volume for each subgroup. Then, the weighting is one. Suppose there were 100 cases for each subgroup for the kidney stone example and the percentages were the same (Table 30.4).

Peismon: Then, the quantities would be 93, 87, 73, and 69 for subgroups 1–4, respectively. The aggregates would be 166 and 156, respectively, for Treatments A and B. There is no paradox.

Kokinos: But we can't always have equal subgroup sizes.

Agrafos: I still don't understand how the weighted average resolves the paradox examples Mathitis showed us. We concluded that the slopes were both positive and negative for the fitted lines, we said that one treatment is better for each size of kidney stone but worse for any type, and the lawsuit said there was bias for the university as a whole but not by any department. Each example contains a pair of contradictory claims—both can't be true.

INFERENCES VERSUS FACTS: CONDITIONS FOR PARADOXES

Socrates: Good point, Agrafos. We need to make a distinction between descriptive statements about subgroups and aggregates that are not contradictory and inferences that are.

Agrafos: What is an example of a descriptive statement since that is what I thought we were doing by letting the data speak for themselves?

Socrates: Descriptive statements just repeat the facts about the aggregates and subgroups. In other words, they are statements only about the data at hand. It's the difference between descriptive statistics versus inferential statistics and predicting as discussed before (Myth 19).

Mathitis: Then, inferences could be statements beyond the data.

Agrafos: When we make inferences, we have to check that the argument is good. But if it's a logical argument, doesn't that make the inferences correct, true?

Socrates: But not necessarily. Remember the deductive argument about my pet being a swan assuming all swans are white and my pet being white (Myth 19).

Mathitis: That argument was invalid and one assumption was false since you don't have any pets.

Socrates: Arguments consist of premises and conclusions. The premises and conclusions are statements that can be either true or false. Premises can be statements of facts or assumptions. We would like to have sound arguments, where the premises and conclusion are true and the conclusion is a valid inference from the premises.

Mathitis: Often we use assumptions as tests, like we do with hypothesis testing.

Kokinos: I don't see that we are making any invalid inferences or have false premises in these examples of Simpson's paradoxes. Aren't we just summarizing the data—letting them "speak?"

Kokinos: If that's right, how do the paradoxes occur?

Socrates: We may not know we are making false assumptions or making invalid inferences. But what can happen when we do?

Kokinos: We could have paradoxes.

Socrates: Apparent paradoxes rather than actual ones.

Agrafos: What's an actual paradox? Or is that a paradox in itself?

Socrates: You may have answered your own question! Consider this common one. Here's a piece of paper. On the front write "The statement on the other side is false." Turn the paper over and right: "The statement on the other side is true."

Agrafos: Hmm. I see the paradox. Assuming one statement is true makes the other one contradict that assumption. The same thing occurs if we assume one is false. This is an example of *reductio ad absurdum* but with no resolution.

Kokinos: If this is an actual paradox, why aren't the other examples actual ones also?

TABLE 30.5

Conditions for Actual and Apparent Paradoxes

Paradox	Statement 1 (Example: Percentage for Aggregated Data)	Statement 2, Contradicting Statement 1 (Example: Percentage for Disaggregated Data)
Real	Fact	Fact
Apparent	Fact	Valid inference from false assumption or invalid inference
Apparent	Valid inference from true assumption	Valid inference from false assumption or invalid inference
Apparent	Valid inference from false assumption or invalid inference	Valid inference from false assumption or invalid inference

Note: Arguments with false assumptions are unsound, whether valid or invalid.

Socrates: Are the calculations wrong? Do we not understand that combining percentages requires calculating weighted averages?

Agrafos: No, the math is right and I understand the aggregate percentages are weighted averages of the disaggregated percentages.

Kokinos: But aren't you saying that when we make inferences, the reason for paradoxes or apparent paradoxes is more than just not understanding that aggregating data requires calculating a weighted average?

Socrates: That's right. Let's identify and describe the various situations (Table 30.5) comparing facts and inferences. We have two contradictory statements, such as a reversal. If both are facts, it's a real paradox. If at least one is an invalid inference or inference from a false assumption, we could have an apparent paradox.

Kokinos: Then the trick is determining which situation we have.

Socrates: Exactly. When we make these inferences, it is no longer the data speaking to us.

Agrafos: So you think the errors of Simpson's paradoxes are apparent? I don't see them so I agree with Kokinos. Where are the invalid inferences or false premises?

ASSUMPTIONS FOR MODELING

Socrates: Not apparent but yes, they are there. Let's analyze the fitted lines example first. A few years ago, there was an article authored by two scientists claiming to show that female marathon runners would eventually run faster than male runners (Whipp and Ward 1992). Since this conclusion was thought to be false by many, there was much questioning. One scientist explained to a *New York Times* reporter "This is not me talking, it's the data" (Angier 1992).

Kokinos: Was that true?

Socrates: Using recent data, they showed that at the rate female runners were improving, they would eventually pass male runners on the basis of their rate of improvement over the same period. In this case, the authors assumed a linear relationship, just like we did in the square and circle example.

Peismon: Clearly, they were making an assumption. They assumed that the rates of increase for both men and women would continue. That's an assumption we caution students about when doing regression analysis. Do not extrapolate beyond the data because we don't know if the relationship holds.

Kokinos: Then, in the example of the squares and circles, we are making an assumption by fitting a line to the data whether each set of data is separate or combined.

Socrates: Why do we fit a line or any other equation to data?

Kokinos: Typically to make predictions.

Mathitis: We already know that both types of predictions, extrapolation and interpolation, require assumptions.

Socrates: Are we not simply modeling the data?

Mathitis: Yes. And all models are wrong; some are useful. So, we already know that the inference is based on a false assumption about whatever relationship is fit to the data.

Socrates: Can marathon times be negative?

Mathitis: No. So if the rates were extended, the times would eventually be negative. We also make assumptions about the possible values that the variables can have. In this case of the circles and squares, we are comparing two inferences on the basis of assumptions that are or may be false.

Kokinos: But there is a linear relationship between X and Y.

Mathitis: But we cannot conclude that X causes Y. This just shows a correlation.

Kokinos: I'm not saying that X causes Y, just that there is a linear relationship.

Socrates: Why do you say the relationship is linear?

Mathitis: I agree with Kokinos. Since for $X = 1, 2, 3$, and 4, $Y = 6, 7, 8$, and 9. Thus, Y is just five more than X: $Y = X + 5$, which is a linear equation.

Kokinos: This is what I mean about letting the data speak for themselves. We can show that relationship by drawing a line through the points.

Socrates: I agreed with both of you until you said to draw the line.

Kokinos: That's a paradox of facts. Drawing the line is the same thing as what Peismon said.

Socrates: When you draw the line *through* the points, does that mean the same relationship holds for values between the points?

Kokinos: Yes.

Mathitis: Then we are interpolating. That's an inference requiring an assumption.

Socrates: The line through the points no longer describes just the data on this graph.

Agrafos: Why not?

Socrates: Do we have values of X between 1 and 2?

Agrafos: No. So, a line assumes we can have $X = 1.3$, for example.

Mathitis: And, it also assumes Y can have values other than the eight shown.

Socrates: What if at $X = 1.5$ and 3.5 the equation is $Y = X + 5.5$ and at $X = 2.5$ the equation is $Y = X + 4.5$? Then, is the relationship linear?

Agrafos: No, but we don't know that's what it does.

Mathitis: That's Socrates' point. We don't know, and therefore, we make an inference when we draw a line through the points based on an assumption.

Socrates: Consider Mensa puzzles. What number is next in the sequence 1, 2, 3, 4?

Kokinos: 5, of course.

Socrates: Why not 3?

Kokinos: Why?

Socrates: Maybe the rule is increase by 1 starting at 1 until you get to 4 then decrease by 1 until you return to 1 and repeat indefinitely.

Peismon: How about alternating sets of three consecutive prime numbers followed by three consecutive composite numbers? The next number would be 6.

Kokinos: How would I know that?

Mathitis: That's the point. You can't know with a finite number of items in the sequence. You must make an assumption because you are extrapolating.

Agrafos: Then, to know the next number, you have to make an assumption.

Socrates: Yes, for any finite series without more explanation, e.g., you are given the rule or equation, the choices are infinite.

Kokinos: So, any number is correct for these types of Mensa or IQ puzzles?

Socrates: (winking) Yes, but they won't tell you that.

Agrafos: Then, the same is true for this paradox since there are only four points for each case.

Socrates: Because it is an inference based on an assumption not part of the data, the statement about a line "through" the points is not the data speaking but you speaking. When we fit a line to the data and claim that the relationship holds for the values of X between the values for which we have data, we are interpolating, which is also an inference.

Kokinos: So, it could be a sinusoidal wave connecting the points; that would not be linear.

Peismon: If you only collect data near the peaks or troughs, you might conclude that the relationship is a straight line—when it isn't.

Agrafos: What are we limited to? Can we say there are four points and that for just these four points, the relationship is $Y = X + 5$?

Socrates: Yes, and you can say what the mean is for the Xs and the Ys and calculate all sorts of descriptive statistics for the Xs and the Ys and for any relationship between the four X values and the four Y values but nothing more than that for each set or the combination. Beyond that requires an assumption.

Mathitis: Unless we have more information.

Socrates: Yes, and if that additional information is not on the graph or part of the data, then it is not these data and graph that are "speaking to us" but inferences from assumptions of additional information. So far, we have identified three assumptions for modeling:

M1: The applicable values of the variables in the model.

M2: The relationship fit to the data also applies to values between the data.

M3: The relationship fit to the data also applies to values beyond the data.

Agrafos: How did we use these assumptions to make inferences?

Socrates: Since we only have eight values of X and Y, then any conclusion about other values than these eight would require the first assumption, M1.

Mathitis: Assumptions M2 and M3 are used for interpolation and extrapolation, respectively. For example, if Peismon's equation, L1, is the line that describes the relationship between X and Y, then what do we predict for other values of X than these eight? If the predictions are wrong, then we reject the assumptions.

Peismon: The scientists that predicted female marathon runners would surpass male runners fit a line to the data and made the assumption that the improvement rate would continue. It hasn't. Clearly, it would have meant that eventually women would do it in no time or negative times. That's impossible, so their assumption was false. That's a *reduction ad absurdum* argument showing why the assumption is false.

Socrates: When we make inferences, the arguments could be valid but not sound if a premise is false. Peismon showed how we check whether an assumption is true. For example, one logical argument for checking the assumption of a linear relationship might start like this:

Premise: If the line L1 is the relationship between X and Y, then for $X = 4.5$, $Y = 9.5$.
Assumption: The line L1 is the relationship between X and Y.
Fact: $X = 4.5$.
Inference 1: $Y = 9.5$

What's next?

Peismon: We observe what Y is. If $Y \neq 9.5$, then we reject the assumption of linearity.

Kokinos: We have to be careful not to conclude that if Y does equal 9.5, it proves L1 is the relationship.

Agrafos: Then, the same is true when we say the relationship is linear with a negative slope for the combined data. We are making some of the assumptions M1–M3.

Kokinos: Even more so since that line just does not seem to fit the data at all. At least six of the eight points don't fit the line. Each set of values supports the relationship, L1 or L2, for just those values of X for which we have data. But they don't support the linear relationship L3 for the combined data. We could show that by this argument:

Premise: If the line L3 is the relationship between *X* and *Y*, then for *X* = 4, *Y* = 6.1.
Fact: For *X* = 4, *Y* = 9.
Conclusion: Therefore, the line L3 is not the relationship between *X* and *Y*.

Socrates: Excellent, Kokinos. This indicates that for inferences from modeling to be valid, we may need two more assumptions:

M4: A model must "fit" the data; if comparing models, it must fit the data "better" than other models.
M5: The definition of "fit" is ...; the definition of "better" is ....

Mathitis: I see. We can't just use any model and make an inference from that regardless of how poorly it matches the data. We could consider other models to see if they fit better.
Socrates: Since the values of *Y* increase, then decrease, and then increase, any odd-degree polynomial might fit better.
Peismon: With only eight pairs of values, this can be done quickly with software I have. Here are the results for a cubic and fifth-degree equation (Figure 30.5).
Mathitis: We can approximate the values of *Y* for the eight values of *X*.
Peismon: The fifth-degree polynomial seems to fit better and I have calculated equation, L5. The logical argument would be this:

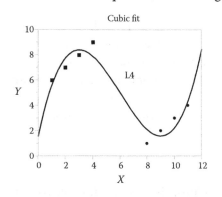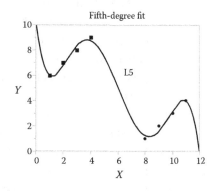

FIGURE 30.5
No reversal of slope when data are aggregated but fit to equations of higher order, for example, third (left curve) or fifth (right curve).

Premise: If L5 is the relationship between X and Y, then for $X = 4$, $Y = 8.75$.

Fact: For $X = 4$, $Y = 9$.

Conclusion: Therefore, the equation L5 as the relationship between X and Y is supported by the data.

Mathitis: It doesn't fit it exactly, but is much better than L3. We already know that all models are wrong, some are useful. For the combined data, L5 is more useful for predicting the actual data.

Socrates: But why fit any single equation to the combined data? A table showing the value of Y for each X would tell us all about the data. When we use those data to make statements about other values of X and Y, those statements are inferences and we need assumptions on which to base the inferences.

Mathitis: Any graph that shows a line or curve between points is an assumption or inference. So, it's subject to being false, and that would resolve the apparent paradox. If we could check the values of Y for various other Xs, we could test which equations better describe the relationship between X and Y.

Kokinos: But in this case, we checked the fit of L3 only for the eight pairs of X and Y values and saw that it was not a fit.

Socrates: If an assumption leads to a false statement or contradicts a fact, what do we conclude?

Kokinos: The assumption is false. This is what we did before. We saw that assuming a sample size of $n = 30$ contradicted the statements that larger sample sizes were needed for nonparametric and discrete data analyses.

Mathitis: We used this approach several times and especially for the modeling of the hurricane data with a probability model. The correct conclusion when the probability distribution did not correctly predict the facts is to reject the assumption of the model fitting the data.

Agrafos: Then, this example of Simpson's paradox is a case of inferences based on false assumptions.

Socrates: What are the inferences?

Agrafos: Linear relationships?

Mathitis: I think those are the assumptions: the relationship between X and Y is linear for the aggregated data and for the partitioned data. Then, the inferences are about the slopes. Those statements are contradictory.

Kokinos: We are also assuming that X and Y can have values other than these eight. We have used all three assumptions M1–M3 to make inferences about the slopes.

Mathitis: Then, the apparent paradox of the inferences is resolved by showing that the assumptions on which the inferences are based are false, so the argument is unsound.

Socrates: Is it valid to assume linearity if we don't know if X and Y can have other values?

Mathitis: No. The inferences also come from invalid arguments.

ASSUMPTIONS FOR CAUSAL INFERENCES

Kokinos: There is no modeling in the example with kidney stones.

Agrafos: I agree. Also, aren't these three statements true?

> T1. Treatment A is better than Treatment B for patients with kidney stones <2 centimeters.
>
> T2. Treatment A is better than Treatment B for patients with kidney stones ≥2 centimeters.
>
> T3. Treatment B is better than Treatment A for patients with any size kidney stones.

The first two together contradict the third one. If these are inferred, they are based on facts not false assumptions. So we have an actual paradox.

Socrates: Let's make sure we understand these three statements. Does "for patients with kidney stones…" mean just these 700 patients or other patients not in the study?

Kokinos: We do such studies to determine what action to take on other patients who present with these conditions. So, I would say it does mean more than these 700 patients studied.

Socrates: Do you see that you are making an inference from what is true about the 700 to what will be true about others?

Agrafos: Okay, but aren't these statements true about these 700 patients?

Socrates: Rather than say "Treatment X is better," do you agree with the following statements?

1. Of the patients with small kidney stones given Treatment A, 93% were successfully treated while those with small kidney stones given Treatment B, 87% were successfully treated.
2. Of the patients with small kidney stones given Treatment A, 73% were successfully treated while those with small kidney stones given Treatment B, 69% were successfully treated.
3. In aggregate, 78% of the patients given Treatment A were successfully treated and 83% of the patients given Treatment B were successfully treated.

There is no actual paradox because we know that the percentages for the third statement are a weighted average of the percentages in the first and second sentences.

Agrafos: Those statements seem the same to me.

Mathitis: I think the difference is in the tense of the verb. You can say Treatment A *was* more successful than Treatment B for these patients, but without making an inference you can't say that Treatment A *is or will be* more successful than Treatment B *for other patients.*

Socrates: Note that if by "better" you mean something else than 93% is greater than 87%, then you need to define "better," as we mentioned when modeling, assumption M5. In addition, if the assumptions for the first two statements are true, what happens to the paradox when the subgroup sizes are equal?

Peismon: It disappears.

Kokinos: But we want to make the inference to other patients. That's a statistically valid inference.

Socrates: To be statistically valid, what assumptions must be true?

Mathitis: That the sample came from the population to which the inference is made and that the sample was randomly selected. Both assumptions are false when applied to a patient other than these 700. This study did not randomly select patients or randomly assign them to treatments.

Socrates: In addition, these 700 patients are treated patients. A new patient is untreated. Can a new patient come from a population of treated patients?

Peismon: No.

Kokinos: While clinical trials do not randomly select participants, they are randomly assigned to treatments.

Mathitis: The inference doesn't appear to be statistical because we want to say that the treatment caused the percentages or success rates in each subgroup. Pearl (2009) and others* have said that statistics only deals with correlation. Therefore, the issue can only be resolved by understanding the causal relationship. They distinguish between the probability of a result given an action was taken and the probability of the result without that action. That's why we have the paradox or appearance of a paradox because we are dealing with causation, which statistics does not address.

Kokinos: But all patients were initially untreated. So, it could include a new, untreated patient.

Socrates: Does the table represent untreated or treated patients? If treated, what happened to make them so?

Kokinos: Treated because all 700 were give Treatment A or B.

Socrates: Does that explain why there is an assumption when applying the results to a new patient? We had an intervening event that we are assuming caused the success rates. As Mathitis noted, there is a causal relationship between treatment and success. Why not say that size is the cause of success? Or, if we switch the role of Treatment and Success in the table, does Success then cause Treatment? The data do not tell us which is the cause, which is the condition, and which is the effect.

Agrafos: Okay, I see that we are making assumptions because the data do not tell us which factor is the cause. We insert or assume that information. Then, this is a case of two inferences based on causal assumptions.

Mathitis: That would require us to check those assumptions.

Kokinos: But that's how studies on new drugs or vaccines or treatments are interpreted. I always thought that was a statistical inference because of the probabilities.

Socrates: That is a misunderstanding of statistical inference. The inference that this treatment will work because it worked before cannot be a valid statistical inference. Future predictions are not

---

* See references in: Pearl, J., An introduction to causal inference, *The International Journal of Biostatistics*, vol. 6, no. 2 (2010): Article 7.

valid statistical inferences because the sample can't come from the future. Therefore, it violates the necessary assumption of randomly selecting the sample from the population to which the inference is made.

Kokinos: Then this is a causal inference. Science—our understanding and belief, so assumption, that all humans are the same in certain relevant ways—is the basis for the inference.

Agrafos: But isn't math and statistics the language of science?

Socrates: Yes, but we are now making inferences. There is a difference between a mathematical inference and a statistical inference even if both use calculations. Recall our discussion on estimation (Myth 3).

Agrafos: There are no correct formulas or equations for estimating. That is what we are doing here with the statistics.

Socrates: Mathitis is right that statistics only tells us about correlations. Statistics can answer the question "What happened to Y when X changed?" If there is a change in both, then there is a correlation; if one changes and the other doesn't, there is no correlation. Science answers "Why did Y change (or did not change) when X changed?" Science explains the correlation. That makes it a casual inference.

Mathitis: This sounds like the difference between enumerative and analytic analyses. The percentages and quantities are merely enumerative analyses. Stating why they occurred, such as why Treatment A works better or would work better for a new patient, is an analytic analysis.

Kokinos: Explain the causal inference since I want to know what treatment to apply. I want to know that a treatment causes a cure.

Socrates: There are tools that help confirm the causal relationship, because it is an analytic study, as Mathitis noted. We spoke about those before: controlled studies or designed experiments. We don't need statistical testing to show causation. The point of enumerative versus analytic studies is that the former tell us "How much?" while the latter tell us "Why?" When we talked about descriptions of the data, like the percentages, which were we doing?

Kokinos: Enumerative statistics or math.

Agrafos: And if we want to know why one subgroup had a higher success rate than another subgroup, that's the analytic study. The data suggest it is because of the treatment.

Peismon: We know it is not because of the size of the kidney stones because the patients came with that condition. The only intervening factor was the treatment.

Socrates: However, not just any intervening factor is a cause.

Peismon: Why not?

Socrates: If we bought two bags of candy, mixed them, and then wrapped half the candies in one bag with a red wrapper and the other half in a second bag with a blue wrapper, we have an intervening factor—an action taken on the candy. If all the candies in the bag with the blue wrapper are chocolates, did the wrapping cause them to be chocolate?

Peismon: No. That makes it clearer that we are adding assumptions when making even a causal inference.

Mathitis: It also makes clear the distinction between correlation and causation and why statistics is limited to correlation. We can show a correlation between wrapper color and chocolate but neither caused the other.

Socrates: Exactly. The discipline of science is to explain by determining causes. For the inference from a sample to a population, specifically where the sample does not come from the population to which the inference is made, we need something other than statistics.

Mathitis: Both Shewhart and Deming say the same thing, as it relates to control and prediction.

Socrates: In the kidney stone case, we need an understanding of the disease, the subject organism, and other characteristics. With that understanding or knowledge or just an assumption, we infer that this treatment will have a good chance of succeeding when applied on this patient with this condition. Note, for example, that we don't infer that Treatment A will work for all organisms capable of having kidney stones. We know that cures for one species can be toxic for another. Science, not statistics, allows the inference to other individuals or species.

Peismon: Does that resolve the apparent paradox of statements about other populations beyond what is in the table or data? We still have the question of what treatment do we provide given what we know or don't know about the patient.

Mathitis: Is the inference that because Treatment A is better for these 700 patients, it is better for any other human with kidney stones?

Socrates: If that is the inference, it requires an assumption. We then need to identify it and confirm that it is true.

Mathitis: What assumption is that?

Socrates: I can only guess. Perhaps something like this: "What likely cures one human with disease D, will likely cure any other human with disease D?" But even that assumption would need additional data to support it. Whatever that assumption is, if it is false, it can lead to apparent paradoxes between two causal inferences or a causal inference versus a fact. All causal inferences require an assumption that explains the relationship between a cause and an effect and under what conditions it applies:

C1: A causes B under condition C.

Mathitis: I see. In the kidney stone case, we are assuming that A is a treatment, B is a cure, and C is kidney stones. We need some assumption like that to make an inference from the 700 to any other untreated patient with kidney stones.

Kokinos: But neither treatment was 100% successful. How does that work?

Socrates: Why do you think it is not 100%?

Kokinos: In part because we don't understand completely how the cause works. Or, maybe I should say we don't understand why it doesn't work.

Agrafos: How can we show that the assumption of a causal relationship is false?

Mathitis: I think that if the cause is supposed to work 100% of the time, then a single exception or counterexample is enough to show that the assumption is false.

Socrates: Consider this. I hold a balloon filled with air, extend my arm, and let go. I repeat this with a tennis ball and a 1-lb. weight. What happens in each case?

Kokinos: Each drops to the floor.

Socrates: One hundred percent success. You want to show someone else this causal relationship. You use a balloon filled with helium instead of air, do it outside with a strong wind, or do it where there is no gravity. What happens?

Kokinos: It doesn't work. An object "drops" when in the presence of a gravitational force that is stronger than any other countering force. That's why no object "drops" in a spacecraft, the helium-filled balloon rises, and the air-filled balloon might rise and fall and go various directions before landing depending on the wind.

Mathitis: Then, for Treatments A and B, although we don't have 100% success, there might be conditions in which each is 100% successful.

Socrates: We should seek to understand when the cause does and does not work. Let's return to your question about what to do when a new patient presents with unknown size kidney stones. We can't make decisions based on information we don't have. But that does not remove the obligation to get that information. If only the aggregated results are known, such as what physicians knew before the study, then Treatment B is preferred.

Agrafos: Even though neither is 100% successful?

Socrates: Yes, but what happens when further studies are done? What do we learn?

Agrafos: That Treatment A is better.

Mathitis: Additional information counters the original assumption of Treatment B being better regardless of the size of the kidney stones.

Socrates: Suppose we do further studies on the cases that weren't successful to understand why. We look at differences between ages and gender and suppose we get these results for the aggregate (Table 30.6). What is your conclusion about Treatment A?

Kokinos: You would not want to be a female younger than 50 years and get Treatment A.

Agrafos: But we don't know that.

Socrates: Before the kidney stones study by size, did we know that Treatment A appeared to be better?

Mathitis: No. That's the purpose of further studies.

Socrates: Note that if we know that the causal relationship is valid, then we know there is no paradox. We can show this by simply keeping

TABLE 30.6

Hypothetical Distribution of Successes for Treatment A Applied to Stones ≥2 centimeters by Age and Gender

Age	Male	Female
<50	Subgroup 179% (50/63)	Subgroup 20% (0/25)
≥50	Subgroup 3100% (100/100)	Subgroup 456% (42/75)
Aggregate	92% (150/163)	42% (42/100)

the percentages as they are but making the subgroup sizes equal. The paradox disappears.

Agrafos: But without data, you can't change the subgroup sizes. How do you know the percentages will stay the same?

Socrates: It's contradictory to claim that the causal relationship holds according to these percentages when applying to a new patient but when adjusting the subgroup sizes they don't. Nevertheless, your question confirms that it is an assumption whether applying to new patients or adjusting the subgroup sizes.

Agrafos: Okay, that makes sense.

Socrates: Also, an inference based on a causal relationship that is not 100% effective could be based on a false causal relationship. What could be false?

Mathitis: It could be A′, rather than A, that causes B under condition C. Or, it could be that A causes B but under condition C′, which has some common characteristics with C.

Agrafos: Then, in the kidney stone study, other conditions, for example, race, gender, age, number of stones, other medical conditions, and so on, could be factors explaining the less than 100% effectiveness for either treatment.

Kokinos: We know this is true. It used to be that medical studies would only have male participants and sometimes only white males. Now clinical trials are required or are encouraged to include females and other groups that have been shown to respond differently to treatments and have different rates of disease, especially in the area of cancer (Fejerman et al. 2014; Woodward et al. 2006). This further explains why there is an inference that may be based on a false assumption when we state that a treatment is better than another for patients not included in the study.

Mathitis: One reason given by some for the apparent paradox is that doctors tended to use the traditional treatment B, thereby increasing the number of patients in subgroup 2.

Socrates: That explains why the weighted average percent appears contrary to the disaggregated percentages. But it does not explain whether the inference that Treatment A is better for new patients is valid or is based on true premises.

Kokinos: And that's because further details on the patients may reveal that we are mistaken about either the treatment being the cause or the conditions under which it is a cause. The inference of which treatment to use depending on stone size is an apparent paradox because it is based on false assumptions or is invalid.

Socrates: The article also included aggregate data on age and gender, but not race of the patients. We don't know what subgroups by each of these factors and in combination tell us about the effectiveness of each treatment. Also, why use 2 centimeters to differentiate between small and large kidney stones? What are the success percentages if 1.5 centimeters or 2.5 centimeters is used, or more than two sizes of stones? If we further partition the data by these factors and choices, does it change which treatment is better? If so, what inference would we make for a new patient given age, gender, and kidney stone size?

Kokinos: Then, when a new patient presents with kidney stones, this study only suggests that Treatment A is better. We don't know if it is in fact better for that individual as it is possibly based on a false assumption.

Socrates: We certainly don't know it based on statistics as it is not a statistical inference, as we defined it.

ASSUMPTIONS FOR INFERENCES FROM REASONS

Agrafos: What about the gender discrimination lawsuit? No model is fit, but it does seem to involve a causal relationship. Women sued because the aggregate data showed that proportionally fewer women were admitted than men. They claimed that the cause was gender discrimination against female applicants.

Socrates: Reasons and causes are sometimes the same but not always. We can, for example, say that one person hit another because of jealousy. Do we treat that reason as a cause in the same sense as a treatment or gravity?

Mathitis: No, because we expect the treatment and gravity to be 100% effective under the appropriate conditions while with jealousy we don't. We use reasons as ways of understanding or explaining without proscribing a cause.

Socrates: We expect treatments and gravity to be effective in the future. Do we expect the same with bias or jealousy, for example? Is there an inference to future applicants?

Mathitis: Not in this case. The claim by the women who sued was with respect to the applicants for the fall quarter of 1973. They were not making any claim about any past quarters or applicants or about any future quarters or applicants.

Agrafos: Is this case a statistical inference with false assumptions or invalid inferences?

Mathitis: The analyses were statistical. The authors said they would look at various analyses to identify strengths and weaknesses with each. They used a chi-square statistical test that compares actual to expected proportions. If the difference is significant, then the proportions are not equal.

Socrates: What assumptions are required for that statistical test?

Mathitis: First, each applicant contributes only to one of the proportions. The admission or rejection of one applicant is independent of the decision of other applicants. There is also the assumption about a minimum number of cases in each group.

Socrates: What is the null hypothesis?

Mathitis: The proportions are equal under the null hypothesis. If there is bias or discrimination, then the proportions admitted should not be equal.

Peismon: Don't we have the same issue we discussed before? If the results are not significant, it doesn't mean that the null hypothesis is false or should be rejected.

Kokinos: This is a lawsuit, so the rules for accepting and rejecting are legal.

Socrates: How was the analysis done?

Mathitis: The best approach the authors concluded was for the aggregate data to be tested statistically in a different manner from what was proposed by the women who sued. Rather than statistically testing whether the proportions admitted were equal or not (they were not) for the entire university or aggregated, they recommended the difference between the expected and the actual numbers of admissions be determined for each department or disaggregated.

Kokinos: That makes sense since admission decisions are at the department level.

Mathitis: Then these department differences are summed to get a total for the university. The aggregate difference was statistically significant favoring women.

Agrafos: You mean that, rather than a bias against women, it was a bias for women?

Mathitis: That's right.

Agrafos: Then, there is no paradox.

Peismon: If the statements are merely about the percentages using the aggregated data versus the disaggregated data, there is an apparent paradox that is resolved because the number of applicants of each gender varies from department to department, which are the deciding bodies with their own admissions procedure. We have a weighted versus an unweighted average.

Mathitis: In fact, that's what the authors noted. They said they made two nonstatistical assumptions. "Assumption 1 is that in any given discipline, male and female applicants do not differ in respect of their intelligence, skill, qualifications, promises, or other attribute deemed legitimately pertinent to their acceptance as students. Assumption 2 is that the sex ratios of applicants to the various fields of graduate study are not importantly associated with any other factors in admission" (Bickel et al. 1975). It was the second assumption being false that led the authors to conclude that bias occurred before applying.

Agrafos: How was that?

Mathitis: Women applied to certain departments more than men, and men applied to others more than women. The authors attributed this to social pressure on women against certain fields, like math and science and engineering.

Socrates: That's an assumption in itself. Regardless of the reasons, women and men applying in different quantities to different departments explains the apparent paradox between the aggregate and department percentages simply because the weighted averages will favor the rates with more applicants. But it does not explain nor is it relevant to the issue of bias and discrimination.

Mathitis: Why not? Assumption 2 is false and it was used in the analysis. And there is a reasonable and supportable explanation for it being false.

Socrates: How were the expected frequencies or proportions determined?

Mathitis: That was based on the proportion of male and female applications. If only 10% of the applications are female, then we should expect 10% admitted if there is no bias either way.

Socrates: If Assumption 1 is false, does that affect how the expected frequencies are determined?

Mathitis: Of course. That's in part what the authors are saying about Assumption 2. If women are less qualified for some departments, for example, missing a prerequisite, then those departments will have fewer female applicants. But, the authors admit that Assumption 1 is false. They described it via an analogy of two kinds of fish of the same size. However, one kind (female) tries to pass through a net with small mesh while the other kind (males) tries to pass through a net with large mesh. Clearly, the second kind will be more successful.

Socrates: The issue is one of bias and discrimination, not whether or why there is an apparent paradox.

Agrafos: I don't understand the difference.

Socrates: Once Assumption 1 is false, then the analyses the authors did and the conclusions they made are unsound—false premises. And, Assumption 1 is a critical assumption that makes Assumption 2 irrelevant for the issue of bias.

Kokinos: I see how the argument is unsound. But how does that make Assumption 2 irrelevant?

Socrates: Let's use their analogy first. If the two kinds of fish are not the same size (Assumption 1 false), it changes the conclusion. Let's keep the net mesh the same, but now the female kind is much smaller relative to the mesh size compared to the male fish. What will or could happen?

Kokinos: Just the opposite. More females will get through.

Socrates: If 80 men and 20 women apply to a department, but 40 men are less qualified than every woman, what should the expected number and percentages of admissions be for each?

Kokinos: The initial analysis based the expected frequencies on the proportion of male and female applicants. Now they should use the proportion of *qualified* male and female applicants.

Mathitis: That makes sense.

Socrates: Let's use department A as an example and suppose 40% (330 out of 825) of the male applicants were not qualified. Rather than having 825 male applicants and 108 female, we would have 495 qualified male applicants and 108 qualified female applicants, or a total of 603. What are the expected proportions or frequencies admitted by gender, assuming no bias?

Mathitis: For males it would be 495/603 = 82% versus 18% for females. But the actual percentages would still be 85.2% and 14.8%. Now the bias switches against women.

Peismon: It's worse than that. In this scenario, only 495 of the 512 male applicants accepted were qualified. So, 16 unqualified males were accepted but 16 qualified females were rejected.

Socrates: Do you recall Bakke's reverse discrimination lawsuit?

Peismon: Yes. Bakke sued because he was rejected while minority students who he considered less qualified based on grades and test scores were accepted. At that time, there were racial quotas, which accounted for these circumstances. He won his case in the Supreme Court.*

Kokinos: But the issue for the Supreme Court was about racial quotas. The fact that others with high grades and scores were also rejected was not considered. It could be argued that, yes, a less qualified applicant was accepted but also a more qualified applicant was rejected, so if one more person was to be accepted, it still did not have to be him.

Socrates: The relevant issue here is that the statistical tests of equal proportions must be based on the number of qualified applications, not just on the total number of applications.

Kokinos: In fact, we should note that the evaluation procedure categorizes applications into one of three categories: Admit, Maybe, Reject.

Mathitis: Then, the first analysis should use the proportion of males and females in the Admit category by department to determine the expected frequencies and percentages. If there are still openings, then the Maybes would be considered.

Socrates: What if the bias occurs with only the Maybes? For example, suppose that five criteria are used to categorize the Admit and Reject. If all five criteria are met, Admit. If not one of the five is met, Reject. Bias has an opportunity to occur when deciding the weighting of the five.

Kokinos: But we also know that as with quotas back in the 1970s, there are also exceptions. Children of professors or alumnae (depending on their donations), those with political influence, and athletes were put in the Admit category even if others were more qualified.

---

* Bakke v. Regents of University of California, 553 P. 2d 1152-Cal: Supreme Court, 1976.

Mathitis: I can now see that with Assumption 1 false, the analysis is useless in assessing whether bias and discrimination occurred. More recent studies have shown that two applications differing only in gender or name suggesting different races are not treated equally (Gaucher et al. 2012; Moss-Racusin et al. 2012).

Kokinos: Isn't this the same as the kidney stone case?

Socrates: It differs in two ways. Bias and discrimination can occur in many ways and need not occur even if a person is biased. There is no intent to make an inference to other applications. From a statistical analysis perspective, the issue is how to determine the expected frequencies to test the hypothesis of no bias. Wherever bias can occur, a test can be done in theory if we can determine the expected frequencies.

Mathitis: You mean, for example, if there are subgroups that have separate criteria for acceptance, does bias exist in the creation of those groups? Does the subgroup of athletes favor men over women, for example?

Socrates: Yes. Or, do the criteria for acceptance favor one gender or for rejection disfavor one gender? However, the authors missed the opportunity to make sound arguments when they knew their assumptions were false.

Mathitis: But they said it was a difficult task to check Assumption 1.

Socrates: There are two easy ways. Could they have asked the departments whether all applicants were equally qualified? Under the assumption of no bias, would the departments agree to change their admissions procedure to simply randomly select enough applications to meet their admission goal?

Mathitis: I see. If they use three categories, the answer is clearly no to the first question and falsifies Assumption 1. If the departments do not admit that not all are equally qualified or if they actually have biases, but they wouldn't agree to random selection, that would be evidence that Assumption 1 is false.

Agrafos: So this case has two apparent paradoxes. The one due simply to the admission percentages of the whole university are weighted averages of the departmental admission percentages. We know that women and men applied in unequal numbers to the departments, even if it is because of existing societal sexism.

Kokinos: But that only affects the weighting and does not address whether bias and discrimination occurred. The inference of gender discrimination against women using aggregated data and the inference that there is no gender discrimination are both based on false assumptions.

Mathitis: I see that now. It's not necessarily the causal relationship that determines how to resolve the apparent paradoxes. Whenever the apparent paradoxes are because of inferences, the tasks are to first understand that there are inferences and then determine whether the assumptions on which they are based are true.

Kokinos: Those do not seem to be easy tasks.

Socrates: Not at all in the case of bias. Why was the analysis by department? If aggregated percentages can be apparently contrary to disaggregated percentages, then what's to stop further partitioning? As we saw with the oil and vinegar, knowing the aggregate tells us nothing about the partitioning percentages and volumes, there could be another reversal.

Mathitis: But Pearl and others say we can't arbitrarily partition the data. It makes sense to look at the university data by department because each department has its own admission procedure.

Socrates: Even if true, it includes another assumption that the procedure is gender biased or that gender bias occurs by department.

Agrafos: How else could it occur?

Socrates: Suppose it is by race or age. Suppose the departments that have older members are gender biased against females. What should the partitioning be? What if it isn't the entire university but groups within the university but not by department?

Kokinos: In that case, it would be similar to the kidney stone case without being a causal relationship.

Epilogue

Deming defines profound knowledge as a system with four components, one being theory of knowledge. The foundation of knowledge is theory. The purpose of a theory is to provide understanding, thereby being able to explain. It is akin to being able to answer the analytic question "Why?" Profound knowledge enables analysis, synthesis, adaptation, and creation.

Yet, having profound knowledge does not mean that these capabilities are exercised. A rote learner with considerable knowledge may have these capabilities but does not apply them.

A profound learner is a philosopher in the original Greek sense of the word—a lover of wisdom. A profound learner develops the ability to exercise the capabilities of analysis, synthesis, adaptation, and creation not because of knowledge but because of the love of knowledge. A profound learner does not necessarily have substantial profound knowledge. Nevertheless, a profound learner is on a path to acquiring profound knowledge. For the profound learner, it's about the journey because the journey never ends for him or her. The profound learner has some level of metacognition, knowing that they do not know but wanting to know and understand more.

Rote and inflexible learners, however, are not disposed toward these activities. Instead, they learn facts, advice, rules, mantras—whether true or not—without the understanding and, critically, without an interest in understanding.

Paradoxically, a profound learner can have less knowledge of facts than both rote and inflexible learners but greater understanding than either. The dialogues illustrated differences between the profound learner and rote and inflexible learners. Peismon is an inflexible learner while Mathitis is more flexible. Kokinos was more academic, absorbing what was said, readily stating "I was taught…" as support for beliefs, as a rote learner would. Agrafos was much like Kokinos but listened with little participation, exhibiting other characteristics of the rote learner, for example, being swayed by what was most recently said.

No one is a profound learner on every subject and everyone is a rote learner on almost all subjects. There are exceptions: Aristotle, Archimedes, and da Vinci come to my mind. While the focus here was on some aspects of statistical analyses and concepts, we can apply the approach to any subject. I leave that to profound learners of those subjects, as I am not one.

This book described and illustrated a framework for being a profound learner as it applied to statistics and statistical myths. If you have a different view of one of these myths based on the dialogue, then I have achieved one purpose of the book. By having a different perspective—I hope because of a deeper understanding—then you have one of the building blocks for becoming a profound learner. Other building blocks include the love of learning, a desire to understand, and an ability to reflect on what one knows and does not know.

The dialogues illustrated these building blocks. For example, the profound learner would seek to understand the following:

- The inconsistency in the belief that $n = 30$ is adequate for all analyses (Myth 13) and the beliefs that certain conditions require n to be more or less than other conditions, for example, parametric versus nonparametric, discrete versus continuous (Myths 11 and 12)
- The benefits of a different taxonomy than merely discrete versus continuous upon recognizing that there are categorical scales that are infinite (Myths 1 and 2)
- How combining specifications and control chart limits can answer several questions simultaneously about stability and capability (Myths 23 and 24)
- Why capability indices C_p and C_{pk} are the reciprocal of the P/T index for assessing measurement systems and how to apply criteria for one to the others and understand that the weaknesses and strengths of one apply to the others (Myths 7, 9, and 24)
- The *reduction ad absurdum* argument that if a population or universe depends on stability, then to characterize even a bowl of beads or case of bottles requires one to know the order in which the beads or bottles were produced (Myths 20, 26 through 28)
- The modeling basis for control charts and their multiple uses (Myths 15 through 18, 21 through 26, 28 and 29)

Of course, the profound learner would not limit themselves to just these relationships.

As the Socrates of this book would say, "Consider these scenarios."

Scenario I: *Child returns from school.*

Parent: What did you learn today?
Child: Elephants are reptiles.
Parent: Where did you hear that?
Child: My teacher said it in science class today.
Parent: No wonder the education system needs to be overhauled.

Neither learned. The parent believes the teacher said "Elephants are reptiles" and believes the statement is false. The child accepted what the teacher said.

Scenario II: *Colleague returns from a conference where your favorite expert spoke.*

You: What did you learn at the conference?
Colleague: Elephants are reptiles.
You: Where did you hear that?
Colleague: [Expert] said it.
You: Of course [expert] didn't say that. You must have misunderstood or didn't hear correctly.

Neither learned. "You" believe "Elephants are reptiles" is false and do not believe the "expert" made the statement. The colleague accepted what was said.

Scenario III: *Colleague returns from a conference where your favorite expert spoke.*

You: What did you learn at the conference?
Colleague: Elephants are reptiles.
You: Where did you hear that?
Colleague: "Expert" said it at conference.
You: How can that be true? Elephants are mammals: live births, mammary glands, nurse young. Reptiles do not have any of these characteristics.
Colleague: That was the point. When questioned, "expert" said that, by elephant, "expert" meant reptile. Therefore, the statement is true.
You: In that case, I agree. Did anyone ask why that definition was used?

Colleague: "Expert" said that the word or label did not matter. It was the concept or definition that was important.

Both learned. You believe "Elephants are reptiles" is false using common definitions but understand that changing the definitions can make the statement true.

It depends on the meanings of the words. For the profound learner, a critical question is "Which meanings provide useful information or lead to deeper understanding?"

Each myth could be true, depending on the meanings of the words. Thus, the dialogues consisted of understanding the definitions of the believer and testing alternative ones to discover which definitions were richer, more fruitful with respect to understanding, applying, and developing theory.

A person I shared some myths with commented that several were not myths but merely lacked the assumptions for the statement to be true. He missed the point. When we make universal statements without stating the assumptions or conditions, we teach others, like the child or colleague, that there are no assumptions or conditions. Hence, we get the repetition of the statements that are wrong answers to certain problems, obstacles to finding more effective approaches, and barriers to understanding.

This book is for those who want to be more than rote or inflexible learners primarily with respect to statistics. If you have come to understand at least one myth more profoundly, then I have accomplished another objective. The book was designed to spark curiosity about learning. If you become more inquisitive about subjects that do interest you, I believe you will enjoy that subject even more.

GUIDANCE FOR PROFOUND LEARNING

Are you ready and willing to question—even the experts? While questioning is a type of interrogation, it is not of the person but of the words, meanings, and facts. Its purpose is to understand. Rather than debating to be right or to display one's knowledge, questioning should adhere to Covey's fifth habit: "Seek first to understand, then to be understood" (Covey 1989, 235–260).

The profound learner will seek understanding when there is disagreement. When there is agreement on the facts but disagreement on conclusions, the causes can usually be traced to differences in these areas: definitions, assumptions, questions to be answered, and purpose.

These areas of potential differences lead to the following guide for profound learning:

1. Seek definitions of terms
   a. Is this a dictionary definition or a technical definition? If technical, what is the reason for using it rather than the dictionary definition?
   b. Does the term have an unnecessary positive or negative connotation? Why?
   c. Are there other definitions that accomplish the purpose of the term while providing greater understanding and applicability and richer theory?
2. Question universal statements
   a. Is the statement always true, for all cases and all conditions?
   b. What would count as evidence against this statement?
3. Check logical consistency of universal statements on the same topic
   a. Do two or more statements appear contradictory?
   b. Are the statements facts or inferences?
   c. Do inferences contradict other accepted statements?
4. Check validity and soundness of arguments
   a. What are the assumptions?
   b. Are assumptions facts or supported by evidence?
   c. Are there conditions when the assumptions are false?
   d. Is the conclusion a valid inference or just based on a reasonable implication?
5. What question are we answering?
   a. How do the definitions help us answer the question?
   b. Can the question be reworded to increase understanding or applicability?
6. Why do we need that question answered?
   a. What is the purpose of answering the question?

One final, personal thought. Spanish has a verb I love and wish existed in English: *profundizar.* You can guess what it means.

References

Angier, N., 2 Experts say women who run may overtake men, *New York Times* (January 2, 1992).

Anscombe, F.J., Graphs in statistical analysis, *American Statistician*, vol. 27, no. 1 (1973): 17–21.

Balestracci, D., Data "sanity": Statistical thinking applied to everyday data, *ASQ Statistics Division Special Publication* (Summer 1998).

Bather, J.A., Control charts and the minimization of costs, *Journal of the Royal Statistical Society. Series B (Methodological)*, vol. 25, no. 1 (1963): 49–80.

Bickel, P.J. et al., Sex bias in graduate admissions: Data from Berkeley, *Science, New Series*, vol. 187, no. 4175 (1975): 398–404.

Bower, K.M. and J.A. Colton, *Why We Don't "Accept" the Null Hypothesis* (Milwaukee, WI: American Society for Quality, Six Sigma Forum, July 2003).

Box, G.E.P. and T. Kramer, Statistical process monitoring and feedback adjustment. A discussion, *Technometrics*, vol. 34 (1992): 251–285.

Box, G.E.P., W.G. Hunter, and J. Stuart Hunter, *Statistics for Experimenters* (New York: John Wiley & Sons, Inc., 1978).

Castañeda-Méndez, K., Single-unit process improvement, in *The Five Types of Process Problems* (Boca Raton, FL: CRC Press, Taylor & Francis Group, 2013).

Charig, C.R. et al., Comparison of treatment of renal calculi by open surgery, percutaneous nephrolithotomy, and extracorporeal shockwave lithotripsy, *British Medical Journal*, vol. 292 (1986): 879–882.

Chatterjee, S. and A. Firat, Generating data with identical statistics but dissimilar graphics: A follow up to the Anscombe dataset, *American Statistician*, vol. 61, no. 3 (2007): 248–254.

Chiu, W.K., Comments on the economic design of $\bar{X}$-charts, *Journal of the American Statistical Association*, vol. 68, no. 344 (1973): 919–921.

Covey, S.R., Habit 5 Seek first to understand, then to be understood, in *The 7 Habits of Highly Effective People* (New York: Simon & Schuster, 1989).

Deming, W.E., *The New Economics for Industry, Government, Education* 2nd ed. (Cambridge: MIT Press, 1994).

Deming, W.E., *Some Theory of Sampling* (New York: John Wiley & Sons, Inc., 1950).

Deming, W.E., On the distinction between enumerative and analytic surveys, *Journal of the American Statistical Association*, vol. 48 (1953): 244–255.

Deming, W.E., *Sample Design in Business Research* (New York: John Wiley & Sons, 1960).

Deming, W.E., On probability as a basis for action, *The American Statistician*, vol. 29, no. 4 (1975): 147.

Deming, W.E., On the use of judgment-samples, *Reports of Statistical Applications, Japanese Union of Scientists and Engineers*, vol. 23, no. 1 (1976): 29.

Deming, W.E., *Out of the Crisis* (Cambridge: Massachusetts Institute of Technology, 1986).

Duncan, A.J., The economic design of X bar charts used to maintain current control of a process, *Journal of the American Statistical Association*, vol. 51, no. 274 (1956): 228–242.

Duncan, A.J., *Quality Control and Industrial Statistics*, 4th ed. (Homewood: Irwin, 1974).

Fejerman, L. et al., Genome-wide association study of breast cancer in Latinas identifies novel protective variants on 6q25, *Nature Communications*, vol. 5 (2014): 5260.

Gaucher, D. et al., Evidence that gendered wording in job advertisements exists and sustains gender inequality, *Journal of Personality and Social Psychology*, vol. 101, no. 1, 109–128 (2011): 1–12.

Gill, J., The insignificance of null hypothesis significance testing, *Political Research Quarterly*, vol. 52, no. 3 (1999): 647–674.

Global Harmonization Task Force, *Quality Management Systems—Process Validation Guidance*. GHTF_sg3_fd_n99-10_edition2 (2004). Available at http://www.fda.gov /OHRMS/DOCKETS/98fr/04d-0001-bkg0001-10-sg3_n99-10_edition2.pdf.

Goel, A.L., S.C. Jain, and S.M. Wu, An algorithm for the determination of the economic design of $\bar{X}$-charts based on Duncan's model, *Journal of the American Statistical Association*, vol. 63, no. 321 (1968): 304–320.

Gosset, W.S., The probable error of a mean, *Biometrika*, vol. 6, no. 1 (1908): 1–25.

Grant E.L. and R.S. Leavenworth, *Statistical Quality Control* (New York: McGraw-Hill, 1980).

Hager, W., The statistical theories of Fisher and of Neyman and Pearson: A methodological perspective, *Theory Psychology*, vol. 23, no. 2 (2013): 251–270.

Hubbard, R. and R. Murray Lindsay, Why *p* values are not a useful measure of evidence in statistical significance testing, *Theory Psychology*, vol. 18, no. 1 (2008): 69–88.

Juran, J.M., *Quality Control Handbook*, 3rd ed. (New York: McGraw-Hill, 1951).

Juran, J.M. and F.M. Gryna, *Juran's Quality Control Handbook*, 4th ed. (New York: McGraw-Hill, 1988).

Kolata, G., At last, shout of "eureka!" In age-old math mystery, *The New York Times* (June 24, 1993).

Lehman, E.L., The Fisher, Neyman–Pearson theories of testing hypotheses: One theory or two? *Journal of American Statistical Association*, vol. 88, no. 424 (1993): 1242–1249.

Montgomery, D.C., *The Economic Design of Control Charts: A Review and Literature Survey* (Milwaukee, WI: ASQC, 1980).

Moss-Racusin, C.A. et al., Science faculty's subtle gender biases favor male students, *Proceedings of the National Academy of Sciences*, vol. 109, no. 41 (2012): 16474–16479.

Neave, H.R. and D.J. Wheeler, Shewhart's charts and the probability approach, *Ninth Annual Conference of the British Deming Association* (May 15, 1996): 3.

Neyman, J. and E.S. Pearson, On the use and interpretation of certain test criteria for purposes of statistical inference, *Biometricka*, vol. 28A (1928): 175–240.

Ott, E.R., Analysis of means—A graphical procedure, *Industrial Quality Control*, vol. 24 (1967): 101–109.

Overbye, D., Chasing the Higgs Boson, *New York Times* (March 4, 2013).

Pearl, J., *Causality: Models, Reasoning, and Inference*, 2nd ed. (New York: Cambridge University Press, 2009).

Perla, R.J., L.P. Provost, and S.K. Murray, The run chart: A simple analytical tool for learning from variation in healthcare processes, *BMJ Quality & Safety*, vol. 20 (2011): 46–51.

Richter, S.J. and C. Richter, A method for determining equivalence in industrial applications, *Quality Engineering*, vol. 14, no. 3 (2002): 375–380.

Schuirmann, D.J., A comparison of the two one-sided tests procedure and the power approach for assessing the equivalence of average bioavailability, *Journal of Pharmacokinetics and Biopharmaceutics*, vol. 15, no. 6 (1987): 657–680.

Shewhart, W.A., *Statistical Method from the Viewpoint of Quality Control* (Toronto, Ontario: General Publishing Company Ltd, 1986).

Shewhart, W.A., *Economic Control of Quality of Manufactured Product* (Milwaukee, WI: American Society for Quality, 1980).

Steven, S.S., On the theory of scales of measurement, *Science*, vol. 103, no. 2684 (1946): 677–680.

Tagaras, G. and H.L. Lee, Economic design of control charts with different control limits for different assignable causes, *Management Science*, vol. 34, no. 11 (1988): 1347–1366.

Tukey, J.W., Conclusions vs decisions, *Technometrics*, vol. 2, no. 4 (1960): 423–433.

Western Electric Company, *Statistical Quality Control Handbook* (Indianapolis: Western Electric Co., 1956).

Wheeler, D.J., An honest gauge R&R study, *2006 ASQ/ASA Fall Technical Conference*, 2006.

Wheeler, D.J., All outliers are evidence, *Quality Digest* (May 2009a). Available at http://www.qualitydigest.com/magazine/2009/may/department/all-outliers-are-evidence.html.

Wheeler, D.J., Two definitions of trouble, *Quality Digest Daily* (November 2, 2009b): 2.

Wheeler, D.J., Why we keep having 100-year floods, *Quality Digest Magazine* (June 6, 2013). Available at http://www.qualitydigest.com/inside/quality-insider-column/why-we-keep-having-100-year-floods.html.

Wheeler, D.J. and D.S. Chambers, *Understanding Statistical Process Control*, 2nd ed. (Knoxville: SPC Press, 1992).

Wheeler, D.J. and R. Stauffer, Of processes and project baselines: Why homogeneity matters, *isixsigma* (January 13, 2014). Available at http://www.isixsigma.com/tools-templates/statistical-analysis/of-processes-and-project-baselines-why-homogeneity-matters/.

Whipp, B.J. and S. Ward, Will women soon outrun men? *Nature*, vol. 355, no. 6355 (1992): 25. doi:10.1038/355025a0.

Wiles, A., Modular elliptic curves and Fermat's Last Theorem, *Annals of Mathematics*, vol. 141, no. 3 (1995): 443–551.

Woodall, W.H., Controversies and contradictions in statistical process control, *Journal of Quality Technology*, vol. 32, no. 4 (2000): 341–350.

Woodward, W.A. et al., African-American race is associated with a poorer overall survival rate for breast cancer patients treated with mastectomy and doxorubicin-based chemotherapy, *Cancer*, vol. 107, no. 11 (2006): 2662–2668.

Index

Page numbers followed by f and t indicate figures and tables, respectively.

C

Camp–Meidell theorem, 207, 222
Capability
  actual *versus* potential, 86–87
  checking for, 389
  defined by specifications, 370
  meaning, 92–96
  values for nonnormal distributions, 95
Capability index (Cp), *see* Process
    capability index
Capability indices, 85–86, 87–91, 93, 96
  assumptions, 91t
  descriptions, 91t
Causal inference, 500
Causal relationship, 499, 504
Causal *vs.* statistical control, 452–455;
    see also Stability and
    predictability
Cause-effect relationship, 316–317, *see also*
    Control charts
Central limit theorem, 142
Central tendency, measures of, 427
Change and fundamental change,
    difference, 476
Chaos, 360, 360t, 407
Characteristics, 421–425
Charts of individual values, *see* Individual
    value chart (*I* chart)
Chi-square distribution, 222, 282, 283, 332
Chi-square statistical test, 506
Classical yield, 87
Classification criteria, 297
Class tests, 321
Coding equation, 94
Colors
  as wavelengths, 5
Common cause signals *vs.* special cause
    signals, 468–473; *see also*
    Tampering with process,
    adjustment and
Common cause variation, 383, 458
  capability indices on, 373
  defined, 363
  determination of, 466
  evidence of, 467
  management and, 475
  and process stability, 365
  purpose of action, 469

reduction, effect of, 473–475, 473f
Conclusions, 255, 489
Conclusive sampling, 305
Conditional predictions, 445–447; *see also*
    Stability and predictability
Conditional probability, 105, 107, 113–114
Confidence interval, 115–116, 142–143,
    266, 268–270
  in enumerative study, 299, 301t
  estimate, 145, 264
  in interpolation, 441
  procedure for, 154
Consistent, defined, 284, 285
Continuous data, 5
  charts, 389
Continuous data statistical tests *vs.*
    discrete data, 111–121
  discrete examples when *n* = 1, 112–114
  factors that determine sample size,
    114–118
  relevancy of data, 118–121
Continuous probability distributions, 329,
    335
Continuous scale, 13–14
Continuous test, 112
% contribution *versus* % study, 65–66
Control and treatment data, 132t
  modified, 136
Control charts, 41–44, 181–192, 253, 273f;
    see also Statistical inferences
  analytic problem/study/solution,
    302–308, 308t
  axes of, 382
  background, 181–183, 291–292
  based on patterns, 453
  cause-effect relationship, 316–317
  in daily life, 451–452
  and defects, 468
  displaying stability/instability, 419
  economic design of, 232
  enumerative and analytic studies,
    procedures for, 308–309
  enumerative or analytic studies
    describing, 309–315, 312f, 316f
  enumerative problem/study/solution,
    296–301, 301t
  enumerative *vs.* analytic stage,
    distinguishing characteristics,
    293–295, 296t

Stability (homogeneity) for baseline
   background, 411–412
   baseline, purpose of, 412
   comparing, ways of, 425–428
   daily comparisons, 419–421
   just-do-it projects, 412–413
   meaningless, meaning of, 416–419, 417f
   natural processes, 413–414
   out of control processes, 414–416
   process average, 421–425, 424f
   random sampling, 429–431
   universe or population/descriptive
     statistics, 428–429
Stability *vs.* instability
   extrapolation, 447–450
   interpolation, 438–445, 439f, 443f,
     444f, 445f
Standard deviation control chart
   factors for S control chart limits, 42f
Standard deviation control chart (S chart),
     41–44, 332
   for capability calculation, 365
   defined, 41
Standard deviations, 26–30, 35–44,
     185–192, 235, 243t, 327
   background, 35
   and capability calculation, 365
   control chart, 41–44
   definition of bias, 39–41
   degrees of freedom, 36–37
   removing bias and control charts,
     41–44
   t distribution, 37–39
   and tolerance, 373
Standard error, 185–192, 244, 245f, 282
   limits, 200, 201, 208, 209, 220, 221,
     224, 227
   rule, 221
2 standard error limits, 246
3 standard error limits, 230–234, 310, 323,
     329, 340, 396
Standard errors, 328
Statistical hypotheses and statistical tests
     of hypotheses, difference, 326
Statistical hypothesis testing, 165
   on alpha, 328
   control charts and, 327–329, *see also*
     Control charts and tests of
     hypotheses

Statistical inferences
   about population parameters,
     264–267
   background, 253–254
   deductive
     estimation, 268–270
     hypothesis testing, 267–268
   and enumerative study, 299, 301t
   with false assumptions, 506
   induction *vs.* deduction, 256–258
   inductive inferences, cases of, 258–261,
     259f, 260f
   and probability, 499
   and probability distribution, 345
   probability distributions, 261–264
   to process, 273f
   real-world cases of, 271–273, 273f
   reasoning, 255–256
   and statistical hypothesis testing, 327
   to universe, 273f
   to unknown universes, 280
   vs. facts, conditions for paradoxes,
     488–490, 490t
Statistical process control, 292
Statistical risk analysis, 301t
Statistical solution, 271
Statistical test, 506
Statistical theory
   application of, 278
   defined, 261
Statistical *vs.* causal control,
     452–455, *see also* Stability
     and predictability
Statistics as a religion, xxix–xxxiii
   Six Sigma, xxxii–xxxiii
STDEV(), 49
Stratified random sampling, 375
% study ratio, 63–64
Subgroup Averages, 206
Subgroup Ranges, 206
Subgroups, 365
Subsets, 263n
Sum of squares (SS), 38
   and square roots: Pythagorean
     theorem, 46–48
Supplier–Input–Process–Output–
     Customer (SIPOC), 422
Sustainability, performance *vs.*, 372–373
Systematic sampling, 375

About the Author

Kicab Castañeda-Méndez is a business and process improvement professional with more than 30 years of experience as an internal and external consultant to manufacturing, service, health care, government, and nonprofit organizations from 21 industries on five continents. He is the author of three other books, including *What's Your Problem: Identifying and Solving the Five Types of Process Problems*, and more than 35 articles, spoken at more than 25 conferences, and given more than 50 conference workshops on theory, tools, and applications on achieving performance excellence. He has taught thousands of people at every level from hourly to CEOs and has helped improve processes in the supply chain (marketing, vendors, R&D, engineering, QA/RA, manufacturing, and sales), support functions (logistics, HR, IT, legal, and finance), and management. He is a 3-year Baldrige examiner and 5-year Connecticut Award for Excellence senior examiner and trainer. Castañeda-Méndez is a GE-certified Master Black Belt, MBB/BB/GB trainer, and facilitator. He has two master's degrees (statistics and mathematics), a triple major bachelor's degree (mathematics, philosophy, and psychology), and a permanent secondary teaching certificate (mathematics), all from the University of Michigan.